Metamorphosis
The Reality of Existence and Sublimation of Life

(Volume 5)

The Absolute Freedom of Saudi Being and Nothingness

Shan Tung, Chang

美商EHGBooks微出版公司
www.EHGBooks.com

EHGBooks 公司出版
Amazon.com 總經銷
2023 年版權美國登記
未經授權不許翻印全文或部分
及翻譯為其他語言或文字
2023 年 EHGBooks 第一版

Copyright© 2023 by Shan Tung, Chang
Manufactured in the United States
Permission required for reproduction,
Or translation in whole or part.
Contact：info@EHGBooks.com

ISBN-13：978-1-64784-183-6

Table of Contents

Table of Contents I

Preface 1

Part 1: Existence and Nothingness 6

Section 1: Background and Origin of Scharthe's Thought ... 7
Section 2: Overview 14
  (1) Criticism of Husserl's Phenomenology 17
  (2) Critique of Hegel's Dialectic 20
  (3) Critique of Heidegger's Theory of Existence 22
Section 3: Exploration of the scope and main idea of the pursuit and research 46
  1. Absolute - the view of uniqueness 49
  2. Relativity - There is no "absolute reference system 52
  3. A form of thinking that reflects the nature of things 52
  4. Relative rhetorical devices of language 57
Section 4, the truth, the purest, most consistent with the actual truth 61
  1. Truth that is unique in nature 61
  2. Truths that are opposed to each other in nature 63
  3. The quality of arbitrariness 67
  4. Relative truth is knowledge 72
Section 5: The position based on which problems and issues are analyzed or criticized 77

Part 2: "Existence and Nothingness" works ideas ... 97

Section 1: Phenomenological Monism Replaces Philosophical Dualism 97
  (1) "Being-for-itself" and "being in itself 98
  (2) Emptiness and self-deception 101
  (3) The Phenomenology of Three-Dimensional Time 105
  (4) Human existence and freedom 107
Section 2: Introduction. 111
  (1) The concept of phenomenon 128
  (2) Apparent ontology. 157
  (3) The Phenomenon of Ontology and the Divide of Being 166
Section 3: The Presence of the Self and Perception before Reflection 171
  (1) Consciousness must be aware of something. 174
  (2) Existence of the perceived object 203
  (3) The collapse of traditional beliefs and the passion that man is useless 206
  (4) I love the anxiety that cannot be satisfied by clinging.. 212
Section 4: Ontological Proof 222
  1. Connotation of Ontology 222
  2. Unitarianism and Pluralism 223
  3. Materialism and idealism 224
  4. 'being in itself' 227
Section 5: Philosophical irrationalism that explores the meaning of human existence 229
  01. Definition 229
  02. Existence before essence 229
  03. absurdity 233

Section 6: What Existentialism Wants to Achieve. (1) Confrontation..................264
  (1) Confrontation with essentialism, holistic philosophy, and static metaphysics. 264
  (2) Kant's a priori turn 274
Section 7: The actual existence of freedom as 297
  01. Existence in oneself..................297
  02. Being for oneself..................316
  03. Consciousness and Nothingness..................334
  04. The way of existence of "in oneself" and "for oneself"....350

Part 3: The structure of Sartre's theory of existence. 375

Volume 1: 'Nothingness', an analysis of 'self-deception'; an argument for nothingness, and an incisive analysis of self-deception. 375
  Section 1: Nihilism in Philosophy..................375
  Section 2: The Concept of Essentialism..................396
  Section 3: What is the nature of consciousness?..................420
  Section 4: Consciousness, subconsciousness or unconsciousness. 432
  Section 5: Consciousness from the Objective Material World 444

Volume 2: "Being-for-itself", "Time"-"Holism..................525
  Section 1: The innovative view of the argument "being-for-itself". 525
  Section 2: The Existential Implications of Schart's Existentialist Philosophy 545
  Section 3: Sartre discusses "being-for-itself" from three levels 557

Section 4: Hegel's Self-possession and Self-existence......... 573
Section 5: Sartre's "being-for-itself" and "being in itself.... 617
Section 6: Being and Time 629
Section 7: Possible causes of errors in logical thinking. 823
Volume 3: the relationship with "others", the idea of "co-presence". 861
Section 1: Relationship with Others................................... 861
Section 2: Objective reality and formal reality................... 870
Section 3: Influence on future generations 895
Section 4: The whole universe is reality............................ 913
Section 5: Obstacles to Egoism 918
Section 6: Shame and Shame.. 930
Section 7: Relationships with "Others.............................. 939
Section 8: The idea of "co-existence"................................. 945
Volume 4: Possession, Acts and Existence, the relationship between 'acts of being' and 'freedom'................................975
Section 1: Possession, Acts and Being................................ 975
Section 2: The relationship between "being-for-itself" and "freedom" 986
Section 3: The Comparison of Sartre and Descartes in Freedom 1068

Conclusion: A metaphysical conclusion is drawn on the relationship between "Self-being" and "Self-activity".. 1087

1. The main philosophical thinking of Saudi Existentialism 1097
2. A metaphysical conclusion to the whole book. 1116

References 1137

Preface

Existential philosophical thinking is a study of universal and fundamental problems, including the doctrines of existence, knowledge, value, reason, mind, and language, but it is a doctrine that is closest to the actual state of human existence.

It starts from the logical conclusion of human beings' knowledge, experience, facts, laws, cognition, and traditional hypotheses about natural and social phenomena.

It starts from the most abstract concept, but lands on the perception, emotion, and rationality hardened by human's daily life experience. It is irrational logical thinking, yet it has a system and an argument. It is summarized from actual verification or deduced from concepts.

According to Heidegger, the greatest mistake in the development of philosophy is that people forget the question of the existence of life itself. Why is there such a doubt?

When we go to a hypermarket and park our car, we see a whole lot of cars parked in the parking lot. Don't we say every day that there is a long list of shopping carts, that there is something here on shelf A, that there is something on shelf B, that there is something hanging on the shelf, that there is something piled up in the open area of miscellaneous goods?

However, we often use this "is", this "in", this "there", that is, the English BE, the French ETRE, the German SEIN, but it is used by us as a The German SEIN is used as a preposition to indicate what something is.

It is often overlooked that something must exist before it can be something. In Heidegger's view, "being" and "is" are abstract, universal ideas, concepts that serve to specify the domain or class of entities, events, or relationships that are self-evident. Therefore, people do not want to pay attention to it anymore.

The concern for the form of thought that directly reflects the nature of things is what something is. That is, the concern for the nature of things. When we say this is a piece of paper, we are concerned with what the paper is, not what "is" is. This is where the distinction between ontology and existentialism comes into play.

In the case of ontology, any object of knowledge must have its own definite essence, and once these essences are lost, the object ceases to be itself. Therefore, all theories that fail to give an essential account of the object of knowledge are invalid.

Existentialism, on the other hand, says that there is no such thing as a fixed, unchanging, innate essence, but only existence. It is only existence that gives rise to the possibility of the nature of things.

The nature is a possibility, and it is existence that makes the possibility of the nature a reality. So, in the most explicit words of the Saudis, "existence précède l'essence" (existence precedes essence).

But here the word existence has changed. The original French text of Schartre's famous book "Being and Nothingness" is l'Etre et le Néant, and this être is the supreme universality that we have described above.

But it is impossible to talk about being without being, without having any being. If we talk about existence, we must talk about it together with something that exists, that is, with the existent.

There are many different kinds of beings in the world, but there is only one kind of being that can comprehend existence and inquire into existence, and that is man. Only man can do something about his own existence. Animals and plants exist, but it is impossible to do something about this existence.

It is only because man does something with his being that man is man. Thus être, the most empty universal, becomes at once the most concrete being, the human reality. être becomes existence.

They make an entity or substance what it is, and it necessarily exists; without it, it loses its identity.

So existentialism existentialisme is also known as actualism. This existence, which Sartre calls "existence prior to essence," is existence, which directly defines human reality.

Sartre's main philosophical work is etre e neant (Being and Nothingness). What Schatt calls neant is not unreal, inscrutable, or completely absent, but is somewhat like what Buddhism calls 'emptiness', which is non-absence and non-existence, not complete absence.

In Buddhism, it is called "naughty emptiness". In fact, 'emptiness' is the ideas, culture, morals, customs, arts, institutions and behaviors that have been passed down from generation to generation, from history to history.

It has an invisible influence and control on people's social behavior. The 'ontology', as religious philosophy is customarily called, is the "ontology of things" (Ontology) without knowledge.

This difficult and paradoxical concept is the result of generations of leading thinkers who have revealed the meaning, role, and influence of things on human beings by reflecting the characteristics and connections of objective things in the human mind and by appreciating the hardening of daily life experiences.

In a broad sense, cognition includes all cognitive activities of human beings, that is, the collective name of mental phenomena such as perception, memory, thought, imagination, and the understanding and production of language, which becomes "phenomenon" (phenomenon), that is, "color world", or what Schatt calls "existence" (etre).

Part 1: Existence and Nothingness

Most of the ideas of Saudi philosophy come from Husserl and Heidegger. The "emptiness" of Heidegger's Sein und Zeit (Being and Time) is very similar to Buddhist thinking, and is analyzed in comparison with Taoism in Book 6 of The True Meaning of the Metamorphosis of Life.

It is not surprising that many Saudi ideas are analogous to Buddhist ideas, given the influence of Heidegger on Saudi Arabia.

In fact, it is clear from this that Heidegger and Scharthe had quietly incorporated Zen philosophy into existentialism after a turbulent period, when the Second World War was in full swing in 1943.

In the midst of the massive tragedy of the war, millions of innocent women, children, and elderly were subjected to sudden and unfortunate disasters and changes, and from time to time they were confronted with the sadness and joy of boundaries.

The inevitable choice between life and death, the long experience of fear, anxiety, loneliness, and absurdity, and the unexpected experience of the meaning of life's existence, have led to a different and deeper understanding.

Although the people of this period had the high technology of the times and the baptism of the industrial revolution, they were confronted with their own lives. But while they felt speechless in the face of the existence of their own lives, they also discovered with amazement that they had a spiritual dimension.

With the loss of the spiritual territory of religion, which provides solace to the soul, man has almost become a fragmented 'being'.

It is in this paradoxical context that Schart's Being and Nothingness takes shape.

Section 1: Background and Origin of Scharthe's Thought

Jean-Paul Sartre, as a contemporary French existentialist, is not only a philosopher, but also a literary and political activist. Sartre was a philosopher, a literary scholar, and a political activist, all of whom were of great importance in the fields of philosophy, literature, and politics.

In terms of his spiritual legacy, the best known are his literary works and certain important philosophical writings.

Among his literary works, the most famous are La Nausee; Nausea (1938) and Le Mur; Wall (1939), both published in 1938, and his philosophical work, Being and Nothingness, published in 1943.

His literary and philosophical writings are not the product of a deep contemplation disconnected from society; rather, they are a body of thought that emerged from the documentation of significant events and facts of the past, in particular, from the contextual developments of the time.

From 1918 to 1978, when Existentialism was born and died with the waves, it has been closely related to the changes in the social situation of the country during the whole sixty-year period.

Simply put, existentialism emerged out of the

unprecedented catastrophe of World War I, the scourge of war that brought great disaster to mankind, the global economic panic between 1929 and 1939, the shadow of the Great Recession, and the formation of a financial system that produced a crisis.

The crisis in the formation of the financial system was spread in the aftermath of the unprecedented massacre of mankind caused by the Second World War and its horrific atmosphere.

In World War I, more than 70 million people fought in the war, almost 9 million were killed, more than 20 million were wounded, almost 4 million were disabled for life, and more than 10 million civilians died of starvation and disease as a result of the war.

World War II had a profound impact on human beings, and the bloody killings and huge damage caused by the war have been reflected in all aspects of human social life for a long time after the war. World War II caused the death of 50 million people worldwide, more than any other war in history.

Among the worst civilian casualties in history were the Nazi German massacres of Jews and other Eastern

European races, the Japanese massacres of countless Chinese and Korean civilians, and the Allied bombings of civilian targets in Germany and Japan at the end of the war.

This war was also the first war in which the number of civilian deaths greatly exceeded the number of combatant deaths. This was the price of war, leaving countless people displaced and suffering from its consequences.

Families were destroyed, relatives were displaced, and countless people died. In the face of the suffering caused by such a catastrophe, the social diagnosis and treatment itself was completely unable to provide a cure for itself.

Moreover, the alienation left behind after the war has subsided has exacerbated the social conflicts. While suffering from mental and physical pain and material deprivation, people also saw the moral degradation of society, the distrust of people and the lack of encouragement and sponsorship from each other.

Under such historical conditions, some people are bound to lose trust in their own lives, in the life that exists in itself, and feel a complete loss of confidence in the value of life.

Some people do not trust others, and even feel that they hate and exclude others, and only believe that they are reliable; some people feel a complete loss of confidence in group human activities and living together, and have to use a certain way or a certain way to solve their own problems.

Some people feel that their life experience is unrealistic and unpredictable, that there are no rules and regulations, that they may be destroyed or extinguished at any time; and so on, with unexplained worries and anxieties.

All these negative human emotions are the raw material of existential thought, and they are also the nutrients for the creation and development of existentialism.

Of course, in Schart's literary works, one can feel the absurdity, uncertainty, anxiety, and helplessness of Schart's writing about the existence of human life itself.

The Wall, a short story published in 1937, is a collection of five short stories: Le Mur (The Wall), La Chambre (The Room), L'Intimé (The Intimacy), and L'Intimacy (The Wall). L'Intimité, Erostrate, and L'Enfance d'un chef (The childhood of a leader).

Among them, "The Wall" was first published in August 1937 in the Nouvelle Revue française; the French writer and Nobel Prize winner Gide considered it a masterpiece.

The theme of the novel is a bit like Victor Hugo's The Last Day of a Condemned Man, but The Wall is the last night of three condemned men. It has also been suggested that The Wall is similar in structure to the second part of Camus' The Stranger.

Both of them are about death row inmates facing death, remembering the past, and thinking about the illusion and untrustworthiness of the world.

Camus also rated "The Wall" very highly, even more highly than Schartre's better-known work "Vomit".

(1936-1939) The Wall is set during the Spanish Civil War, when three men who were to be executed at dawn by Franco's Spanish Phalangists were waiting to die.

This setting shows that the story is based on the real political situation. In the story, the protagonist (narrator), Pablo, is the only one who has a chance of survival among three people trapped in the basement of a hospital waiting to die.

Before he is shot, a Phalangist officer tells him that he can trade the death of another man for his own life if he gives up a revolutionary, the location of his friend.

Pablo did not give up his friend, not because he thought that the morality among friends, the revolutionary cause, needed his friend more, i.e., noble ideals such as being more important than one's own life.

Rather, he thought that no one's life was more important than another person's life, so he did not give up his friends. However, in the last moments before the shooting, he wanted to play a joke with fate.

He told the officer that his friend was hiding in a cemetery. When the other two men were taken out to be shot at dawn, death did not come to Pablo as expected.

When his friend learned of Pablo's arrest, he moved to a new location, which happened to be the cemetery. Pablo never intended to give up his friend, but by chance, his joke became a reality.

When Pablo learned the truth, he cried with laughter in the garden where the survivors had gathered. Set in the Second World War, this is the story of Ibbieta, who is forced by the fascists to give up his comrade Ramon Gris.

Pablo chooses death and refuses to confess.

But after making a false confession in an attempt to trick the enemy, it turns out to be true by chance, and Gris is sacrificed. Finally, he laughs at the absurdity of the world.

Obviously, in this work, Sartre points out the uncertainty and absurdity of human existence, because there is no rational reason for a person's survival or death.

All order, rules, and principles have disappeared for people in the midst of war; a wave of uncertainty of life ensues.

Therefore, it is clear and easy to see or feel that Saudi literature is always written with the goal of depicting the human condition in its situation.

Section 2: Overview

Schart's concern for the situations (mostly unfavorable) in which people find themselves in their lives or in their work is also present in his works that analyze and reflect on fundamental questions about life, knowledge, and values.

In particular, his famous work, Being and Nothingness, published in 1943, is a complete philosophical system of existentialism.

In Being and Nothingness, Scharthe launched a pro-existentialist study of the universe, principles and principles of life, covering such topics as consciousness, cognition, social philosophy, self-deception, the existence of "nothingness," Freudian psychoanalysis, and free will.

As a prisoner of war from 1940 to 1941, Schart read Heidegger's Being and Time, which uses Husserl's phenomenological approach to examine ontology.

Schat returned his philosophical exploration to his understanding of the book, not only understanding Being and Time, but also being able to see things from Heidegger's point of view, being able to understand the important content of Being and Time, and agreeing with him in some parts.

Thus, he follows the theme of his novel, that the existence of man is only an absurd existence, because he feels that he is in a chaotic order that is completely devoid of a causal world, very different from the ordinary natural world.

Man's existence is only an absurd existence because he does not know what his fate will be in the next moment. Man's real existence is only the ability to live freely every minute and every second, whether he is in good times or bad.

For the future, there is no hope to support me now; for the past, there is no sign to tell me that my past is continuing.

The real existence of man is only in the present, in the moment. Nothing (culture, education, religion, etc.) can guarantee my existence, only I can determine my own existence.

Although the existentialist philosophy of Scharthe was influenced by the objective historical conditions of the time. However, the formation of a philosophical system was often influenced by some European and American philosophers who preceded Scharthe.

In the contents of Schart's Being and Nothingness, as read in prison, Heidegger begins Being and Time by saying that Western philosophy has forgotten "being" for too long. In fact, what Western philosophy has forgotten even longer and more deeply is 'nothingness'.

It is no wonder that this question, which has been asked or answered for almost a year, has not received a relative response. That is, what was the positive assessment of 'nothingness' in Western philosophy before existentialism? No specialized studies in the West seem to have seen the opposite in the Eastern tradition.

The most important significance of Schart's "Being and Nothingness" for Western philosophy is that it reverses the absence of positive evaluation of "nothingness" in Western philosophy before existentialism, and considers the experience and lessons from the quenching of past life experiences, from which the value and understanding of the existence of the self and the meaning of true life existence are summarized.

(1) Criticism of Husserl's Phenomenology

"Existence" is not some kind of metaphysical discussion of things that exist. Philosophers usually divide existence into two categories: the external things in the human mind, such as all things in nature and the human body; and the inner world of human beings, such as joy, anger, sorrow, desire, emotion, belief, memory, and so on.

In other words, I can go beyond the cup to its existence and ask questions such as "does the cup exist", but I cannot jump from "cup-phenomenon" to "existence-phenomenon" and ask questions such as "does existence exist".

It is because the widely existing "existence of phenomenon" can reveal that there are entities that exist, or "phenomena of existence" that are explicit and not abstract and general, and can reveal that there are entities that exist, or "phenomena of existence" that are explicit and not abstract and general.

It can point out or elucidate the nature of things that are not easy to see, "the cup that exists"; but it cannot restore things to their original condition or shape as "the cup that exists", which lacks another widely pre-existing "phenomenal existence" to reveal it.

As long as it is a fact that can be observed and observed, it must be supported by an existence that goes beyond the fact that can be observed and observed: whether it is consciousness, language, or the concrete reality and condition that exists in front of us, there is a richer existence than the "cup exists".

Therefore, it is possible to explain, speak and prove the existence of the Cup by using one's own words, texts or other symbols to explain known facts and principles and principles. Unfortunately, Saud only speaks of two kinds of existence, and does not go any further to concretize them.

This is not to say that the present is a special manifestation; nor is it to say that existence is the essence hidden behind the phenomenon. Rather, it means that existence is beyond human awareness and is the basis of awareness.

This is what Schart is calling self-existence. This existence is neither passive nor dynamic, for it is all about the act or the instrument that allows man to interpret known facts and principles and principles with his own words, texts, or other symbols.

If this existence is dynamic, the end, the means, based on the subject, the subject must exist first; if this existence is passive, as an object, it should exist first. Then, this existence is not affirmative or negative, it is just it, it only exists.

(2) Critique of Hegel's Dialectic

The Lesser Logic is the work that best represents Hegel's mature system of logic and marks the pinnacle of classical German philosophy.

In the Lesser Logic, pure 'being' and pure 'nothing' are regarded as the same point of view, both of which are purely general and unimaginable, and have no content at all.

Schart's Being and Nothingness unifies this contradiction, the issue of positive and negative, in a higher sphere. However, Hegel repeats Spinoza's words.

All stipulations are negations. Nothingness, then, still follows existence in the logic of rational thought: it can be negated only if it exists first.

Hegel considers it unimportant and intentionally omits or ignores it. "Nothingness" is the impracticality and inscrutability of an object, and to say that "this cup does not exist" is simply to say that the cup does not exist in the present moment, in the immediate fact and condition that "the cup" exists.

Hegel also relies on negation to realize the contradiction between one form of energy and another, but he believes that the original and unique nature of existence is to transcend to the essence. The ability to think, to judge, to reason, etc., is based on regulations, and remains beyond the metamorphosis of these regulations.

Then, in a normal state of thinking, in order to obtain the desired result, a person has the confidence and courage to face the present situation calmly, and quickly and comprehensively understand the reality to analyze a variety of feasible solutions, and then judge the best solution, and implement it effectively.

Why can't "existence" be defined as revealing the inner feelings of thoughts, feelings, life experiences, etc.? And how can the ability to think, judge, and reason metamorphose beyond the self, so that it can continue to reflect the dialectical development of objective things through concepts, judgment, and reasoning, i.e., the reflection of objective dialectics?

Thus, Schartre changed the order of Hegel's original statement: all negation is stipulation. Nothingness can only function by contrast with existence, and nothingness is only the surface of existence. One can add to Schart's explanation of each other.

Negation is the pointing out or elucidation of the nature of things that are not easily seen, of other possibilities, not the cutting off of connections before and after, in order to point out or elucidate new possibilities of the nature of things that are not easily seen.

If we say "nothing exists", it is still a state of existence. Whether it is a pure, bright or unreal thing, or a holographic photograph, illusion and deception can only be established based on reality.

In the process of cognition, people reflect the process of reality through concepts, judgments, and reasoning in a form of expression that is closer and more accurate than the ancient metaphysics and Adorno's understanding of negation in modern social-critical theory.

(3) Critique of Heidegger's Theory of Existence

In Being and Time, Heidegger uses various human beings with physical existence, or explicit, rather than abstract, generalized emotions, to establish an intellectual mastery of nothingness, indirectly or invisibly, to correct Hegel's error that

Nothingness is its own impractical, inscrutable vagueness, supported and constrained by human transcendence. For example, it describes the relationship between two things, such as God and the world, animals and plants, the cognizer and the cognized, etc., where one exceeds the other or is external to the other.

It also implies that there is a discontinuity, or a gap, between these two things. Nevertheless, there is a way to get from the one to the other, and this transition is either in the physical or in the cognitive sense.

The transcendence of the so-called "nothingness" is the opposite of immanence (the latter emphasizes staying within, or residing above), but in fact the two are also complementary. For example, God is transcendent, above the world, the highest being and the ultimate cause.

Yet He is also immanent, for He is also within the world through participation and causality. Thus, the transcendence of 'nothingness' is a fundamental concept in theological and religious discussions of God, as well as in philosophical discussions of knowledge and existence.

Schart pointed out that for Heidegger: any stipulation is transcendent. Heidegger says that man is "here" and "in the world", but precisely "other" and "out of the world": "He reveals himself to himself from the other side of the world, and he looks back at himself from this horizon to restore his inner being".

It is only Heidegger who says that man is "a temporal being of long standing". Next, Scharthe argues his own view of things or problems from the perspective of Being and Nothingness.

The infinite nature of existence leads to the fact that man is always separated from the other parts or the whole, the subject, "as he sees himself". It is not the movement that precedes the world, nor the world that precedes the movement, but it is the transcendence of the self over the whole of what exists, what could exist, that constitutes the world, in a broad sense, the whole, all, everything.

This transcendence is the very negation of existence-virtualization. As a philosophical meaning, it is the ultimate form of skepticism. The idea that the world, life (especially human beings), exists without objective meaning, purpose, or comprehensible truth.

Rather than being one's publicly stated position, it is a contradictory opinion. For example, Dadaists claim that the Dada movement is not an art movement but an anti-art movement.

Sometimes they took parts of other works and pieced them together, much like found poetry, and in this way they undercut the meaning and definition of art.

At other times, Dadaists were concerned with aesthetic trends in order to avoid them, trying to make their work unintentionally and aesthetically valuable.

There is also deconstructionism, which is dissatisfied with thousands of years of Western philosophical thought, challenging the traditional unquestionable philosophical beliefs and denouncing the Western metaphysical tradition since Plato.

The punk movement, which began in England as a musical rebellion, was a resistance to existing forms of popular music, including progressive rock and heavy metal.

But the message is the same; it is subversive, rebellious, and anarchistic. It draws on themes such as confronting social problems, the oppression of the lower classes, and so on.

Punk culture is a message to society, which means that not everyone is well off, not everyone is the same.

These movements are nihilistic in nature, and nihilism has been defined as a characteristic of the social dynamics of certain times.

Just as Busia called postmodernity a nihilistic era, some Christian theologians and authorities assert that modern and postmodern times are considered nihilistic by many critics because of their rejection of God.

Although postmodernism has been ridiculed by some as nihilism, it does not fit the nihilist formula in so far as nihilists tend toward defeatism.

The postmodernist philosopher seeks to find and celebrate the forces and causes of the diverse and unique human relationships he explores.

In ethics, the term "nihilist" or "nihilism" is used to refer to a person who completely rejects all authority, morality, and social convention, or who claims to do so.

Either by rejecting all established beliefs, or by extreme relativism or skepticism, the nihilist believes that those in control of power are ineffective and should be confronted. For the nihilist, the ultimate source of moral values is not culture or reason, but the individual himself.

Because skeptics do not have to draw any conclusions

about the reality of moral concepts, they do not have to discuss the question of existential meaning and nothingness in the absence of knowable facts.

Although, in principle, postmodernism is considered to be a nihilistic philosophy, it is worth noting that nihilism accepts postmodernism's criticism. Nihilism is a claim to the truth of the universe, which postmodernism rejects.

It considers the nihilization of man and the world as something that arises under certain conditions or as a result of something that is beyond ordinary laws and common sense; as phenomenology calls 'suspension', putting in brackets concepts that have yet to be proved.

If Sartre had known that Skinner had used white rats as his subjects. The response of the lever became a tool for the white rat to operate in order to obtain food. A lever was installed at one end of the box, and there was a trough and a hose under the lever. When the lever is pressed, a food pill or a drop of water can appear in the trough.

When the white rat first entered the box, the activity rate is very high, occasionally pressed the lever, the food will automatically appear, the white rat can get and eat.

After repeatedly increasing the number of times, the white rat can learn the behavior of pressing the lever to get food. Later, when the rat is hungry, once it enters the box, it will take the initiative to press the lever to obtain food. This process is what Skinner calls operant constraint learning. He would say that what this proves is precisely that habit, learning, comes from accidental discovery.

But Hegel was correct when he said that spirit is a negative, not Hedger. Schart found that in the context of the nihilistic era described above, the spirit was formed as Heidegger's.

What characterized Hedegger was the use of an affirmative term that implied negation to describe his "here and now". His transcendence, "the conspiracy of the self beyond ......", instead of laying the foundation of nothingness, is constrained by nothingness within transcendence.

By treating nothingness as the counterpart of transcendence, he has in fact placed nothingness as transcendence into the original structure of transcendence.

Although separated by a century of time and space,

Scharthe interrogates Hegel with a specific analysis through the exchange of ideas: "It is not enough to treat spirit as something indirect and negative; one should point out that negativity is the structure of spiritual existence. What should spirit be in order to make itself negative?

He also interrogates Heidegger through the exchange of ideas in time and space: "If the denial of the existence of things or the truth of things, of their rationality, is the original structure of transcendence, then the ancient structure of "human reality" is not enough.

Then what should the ancient, unexplored structure of "human reality" be if it is to be able to transcend the world? Heidegger's 'nothingness' is a return to existence beyond the world.

It can be pointed out for Sartre that this is actually like Hegel's dialectic, where the spirit completes the whole circle before returning to itself. However, Hedger's approach is neither necessary nor possible in a life where the phenomenon of "this cup does not exist" is always encountered.

According to Schart, negation originates from a

particular transcendence, a transcendence towards another existence. This is not Hedger's indirectness, but the possible negation (non-substantiation) contained in the real, such as "this cup does not exist".

Existence cannot be passive with respect to nothingness, because nothingness cannot act on existence, and it is impossible to pick up a cup with a hand of nothingness, thus claiming that Berkeley's "existence is perceived" is not valid.

"Existence is active in relation to nothingness, but not, as the Stoics advocate, "producing results without changing itself", because existence can transcend itself and create nothingness precisely because it is in a state of flux, of incomplete certainty.

This explains the incompatibility of the monism advocated by Schart, because, whether it is God, the Tao, the Absolute Spirit, or immortal matter, all of them, because of their complete sameness, cannot explain how they can generate consciousnesses that contradict their life existence, such as human beings or animals.

Nor can it be said that they are omnipotent and perfect. Nihilism is a kind of reverse monism with the same problem. For monism is the philosophical doctrine that there is only one origin of the world, a branch of ontology.

Materialistic monism affirms that the origin of the world is matter, while idealistic monism affirms that the origin of the world is spirit. Neutral monism holds that either matter or spirit is the only reality, the nature of what we perceive or know.

Therefore, in this "existence" of self-life, of being in itself, "nothingness" becomes a question that demands an answer or a solution.

The seriousness of the problematic state of affairs, enough to be studied and discussed, or yet to be resolved, implies not only a double nihilization, as Schartrecht says, but a threefold one: first, asking questions that demand answers beyond the real.

Secondly, by asking a question for an answer, the possibility of an answer to a question that has been predetermined to require an answer or an answer becomes available; moreover, the possibility of an answer is diversified. Moreover, the possibilities of answers are diverse enough to give rise to room for further study or discussion.

Thus, the world is not a mass of sameness, but an open diversity. This nothingness is what Descartes, after the Stoics, called freedom.

Schartre corrects Husserl: "Intentionality is not an object presenting a fullness to consciousness, but a nothingness. It is only when you ignore the context, the distance, and even the observer that you can see 'this is the cup'.

However, consciousness cannot assume a state without consciousness, because it requires a consciousness to be present as a witness to the existence of another consciousness.

This is based on the fact that consciousness exists, not nothingness. Therefore, consciousness cannot understand the problem of the origin and emergence of consciousness and even existence. Consciousness cannot put itself in a position of priority over existence and cannot say that "existence is perceived".

This is what Chartres calls "self-referential existence". Edison's achievement is not a matter of luck, but the result of his indefatigable, courageous and innovative self-existence.

Due to his family's poverty, Edison only received three months of elementary school education and began working on trains at the age of 12. Despite this, he still worked hard, hard self-learning.

As a teenager, he had a keen interest in natural science, not only mastering a wealth of knowledge in electricity and chemistry, but also enjoying conducting some "small experiments" in the car and at home. He Light.gif (2355 bytes) "Problem" child

Once, Nancy told him that the friction of fur could generate electricity, Edison was excited to catch two big cats and tied their tails together with copper wire to make their fur rub against each other, resulting in scratches all over.

What's worse, one day Edison finally asked how the gunpowder was made. The first thing that I did was to ask Edison how the gunpowder was made. Yes! The kind mother, Nancy, still could not bear to tell him, and as a result ...... Edison burned down most of the granary, and it was also Christmas Eve, it was really too bad.

As mentioned earlier, Edison really only had three

months of schooling, if not because he thought he had a habit of asking questions, so that the school teacher was very angry, Nancy had to take him home and teach himself. She knew that Edison's love of the brain, good thinking approach to learning, is different from the traditional teaching.

Therefore, she let Edison use her own method to learn knowledge, while teaching him to read Shakespeare, the Bible, history books, etc. She also bought a book called "natural science experiments", Edison was very fascinated by this book, the book of experiments all done. From that moment on, the boy's life was completely changed.

Therefore, in order to gain a sense of existence and to adapt to others, people have to operate and even believe in their own character settings (including the basic settings of the character: name, age, height, etc., as well as the background of birth, background settings of growth, etc.).

To put it simply, it is to create a complete character. In the second volume of Being and Nothingness, Schart refers to this as a kind of self-deception, a belief that already implies disbelief in the outcome.

Schatt calls this self-deception a feeling of optimism in

the process and pessimism in the outcome, a man like Tantalus and Samson's donkey.

The punished Tantalus in ancient Greek mythology, when he was thirsty for water, the water would disappear; when he was hungry for fruit, the branches of the fruit tree would rise.

In the Bible, it is mentioned that Samson hung a carrot on the cart in order to make the donkey push the mill according to his will, and then put the cart with the carrot on it on the donkey's body, and the donkey, in order to eat the carrot, kept moving forward.

So the donkey kept moving forward, and the cart and the carrot moved forward at the same time as the donkey moved forward, and the donkey could never eat the carrot. The verse shows that there is always a shortage and a distance between desire and satisfaction, between object and subject.

What makes a subject a subject and what is related to it is man's knowledge of natural or social things, the meaning he gives to the object, and the spiritual content he transmits and communicates in the form of symbols.

All the spiritual contents that human beings

communicate in the communication activities, including intention, meaning, intention, understanding, knowledge, value, concept, etc., are in the field of meaning. Therefore, from the society of consciousness, people and people or people and things, the reflection of a certain nature of connection, is the basic factor in human social life.

1. The meaning of the existence of nothingness

Meaning is the pursuit, hope, and understanding of human beings in the state of interaction and interconnection between social things, which is revealed through language, and under certain conditions, it also refers to the direction of judgment under the human prograde thinking.

Emotion is given to the meaning. The presence or absence of meaning is related to the degree of expression of human intentionality. The giving, giving, entrusting of emotions (major tasks, missions, etc.)

It is the meaning of the purpose, value, or importance of a thing or action, and the degree of relationship between the meaningful thing in the emotion is the meaning itself. The thing of meaning is the thing that has or has not the purpose, value, or importance of the thing or action.

The whole meaning of things is everything that has or has not the purpose, value, or importance of things or actions, and everything that has or has not the relational expression is the meaning of things themselves or the purpose, value, or importance of actions.

Everything that has or does not have a purpose, value or importance of things or actions has its own internal meaning to be manifested.

The basic meaning of the purpose, value or importance of a thing or action is that the relationship between "something and nothing" is achieved in the consciousness. The main meaning of "there is and there is not" is that "there is and there is not" refers to all things, meaning "all, all, completely". As the consciousness changes, the meaning of "there is" and "there is not" also changes.

As long as the purpose, value, or importance of a thing or action is present in the consciousness, the concept of the relationship between existence and non-existence emerges, and there is meaning.

The purpose, value, or importance of a thing or action has a value of existence in itself, but no meaning exists.

The manifestation of this given emotion in the purpose, value or importance of a thing or behavior is the source of meaning, and the meaning of the purpose, value or importance of any thing or behavior is determined by human emotion.

The existence of the meaning of all things is related to the existence of human feelings, and the basic meaning of this existence is the meaning of existence.

Therefore, the existence of existence is itself the existence of meaning, and when there is no existence, meaning will not appear. Meaning itself is also a kind of emotional trust and empowerment.

It is something in the consciousness that does not really exist. When something exists in consciousness in the form of "something and nothing", that is when the purpose, value or importance of the thing or action is achieved.

When a person expresses the purpose, value or importance of a thing or action in the form of having or not having, it is an expression of meaning.

The purpose, value, or importance of a thing or action varies with the subject's emotion. The purpose, value, or importance of a thing or action given by any subject only indicates the extent of the subject's emotion on the purpose, value, or importance of a thing or action.

The emptiness of the intentionality of the subject is the emptiness of the "existence" view of life, which exists in the life of the self.

It refers to the lack of desire to understand society, the lack of interest in anything, the lack of purpose, and the lack of knowledge or understanding of what one wants for the future.

In short, like to enclose themselves in a space, have no interest in anything, do nothing, only know to consume time, become a Buddhist nerd. It is precisely in this process that the meaning of Sisyphus rolling the boulder emerges, just as it does in the repetition of repetition.

2. The meaning of life existence

Thus, it is clear here that we can easily see or feel the influence of Husserl's phenomenology, Descartes' dualism, and Heidegger's existentialist perspective in Scharthe's book Being and Nothingness.

Of course, Scharthe eventually went far beyond them

and formed a new system of his own. For Scharthe and Heidegger, the thinking on the foundational issues of philosophy shifted from the abstract realm of traditional philosophy, which focuses on transcending phenomena, to emphasizing the "existence" of human life itself and exploring the meaning of concrete life experiences and life existence.

Like Heidegger, Schartre does not think that a universal, fixed, eternal thing (God, entity) can explain human existence; nor does he think that a logical system such as that employed by Hegel can fully explain human thinking, historical events, and natural phenomena.

What can really explain human existence is not through an abstract system, but must be explored in the context of human existence in reality, and the state of human existence must be explained from the concrete experience of human beings.

All the abstract time, space, and scope of concrete experience without human consciousness itself are meaningless and undesirable.

The traditional philosophers' exploration of human existence does not return to the various problems of human existence, but stays in the abstract realm of detachment from concrete experience.

For Schart and Heidegger, the study of the universe and the principles and principles of life established by those traditional philosophers is like building a mansion that can only be seen but not lived in.

It has nothing to do with the real life of people. If one wants to explore the existence of human beings, one must start from the experience of concrete life experiences. In addition, Schart's analytical thinking, which explores and reflects on fundamental questions of life, knowledge, and values, was also influenced to a considerable extent by Husserl.

In his 1936 L'Imagination (Imagination; Imagination) and The Transcendence of the Self, his 1939 Esquisse d' une theorie des emotion (Outline of the Theory of Emotion; The Emotions) in 1939, L'Imagination, psychologie phenomenologique de l'imagination (1940), and Being and Nothingness (1943).

In Being and Nothingness, we can clearly see Husserl's imprint. In particular, his 700-page Being and Nothingness is based on Husserl's view of things or problems from a certain standpoint or perspective of consciousness when discussing human freedom.

There is a significant difference between Schart's existentialism and Husserl's phenomenology. However, in Being and Nothingness, Scharthe attempts to bring the two philosophies together.

The combination of these two holistic, fundamental and critical, inquisitive mindsets about the real world and man is not an accidental event in history, but rather because they are both deeply imprinted by the influence of Cartesian philosophy.

I think, I am" is the cornerstone or the starting point of all existentialist philosophical discourse, and this starting point is not only existential but also epistemological.

Furthermore, the study of universal, fundamental questions includes the fields of being, knowledge, value, reason, mind, and language.

Philosophy differs from other disciplines in that it has a unique way of thinking, such as a critical, often systematic approach, based on rational argumentation.

In everyday language, philosophy can be taken to

mean the most fundamental beliefs, concepts, or attitudes of an individual or group, although this is not the definition here.

The philosophical focus of both phenomenology and existentialism is on the relationship between human beings and the world, and the existential-structure and foundations that support this relationship, as their research directions.

Thus, the study of consciousness and its absence in phenomenology has had a significant impact on the Saudi study of human existence in the world of life experience and experience.

For Saud, Husserl's theory of "intentionality" of consciousness proves that "man is free to act", a person's knowledge of natural or social things, is the meaning that man gives to the object, the spiritual content that man transmits and communicates in the form of symbols.

All the spiritual contents that human beings communicate in their communication activities, including intentions, meanings, intentions, perceptions, knowledge, values, concepts, etc., are necessary for the thesis of meaningfulness.

Although in Saudi philosophy of existence, some people's knowledge of natural or social things is the meaning that people give to the things that are the target of their actions or thoughts.

These include intentions, meanings, intentions, perceptions, knowledge, values, concepts, and so on, all of which are inherited from ideas, cultures, morals, customs, arts, institutions, and ways of behavior that have been handed down from generation to generation and from history to history, in terms of basic issues of meaning, such as types of existence and issues of man and the world.

In the use of language, we can also see the historical inheritance, such as the Saudi borrowing of Hegel's self-denial of consciousness (although the interpretation is not exactly the same). However, the treatment of the problem is quite different from that of traditional philosophy.

In his book Being and Nothingness, Scharthe places the elusive 'nothingness' in the context of existence and encounter, turning the whole 'nothingness' into a kind of 'existence', while the impractical and elusive 'nothingness' is something different from the same kind of thing or from the same kind of encounter.

It is a reality that is different from the same kind of things or ordinary situations, and is constantly entering into other people's perceptions of natural or social things, and is the meaning that people give to things that are the target of their actions or thoughts.

It includes intention, meaning, intent, recognition, knowledge, value, concept, etc., all of which are included in the field of meaning, acting or changing other realities in an indirect or invisible way.

When someone is "absent" (i.e., nothing) and another person has an expectation of his or her "presence" (i.e., a mood of "being"), then the elusive "nothingness" arises, acting or changing in an indirect or invisible way.

The possibility of "nothingness", which is unrealistic and inscrutable, to act or think about "existence" as a target, to act or change in an indirect or invisible way, lies in an individual's own anticipation.

Section 3: Exploration of the scope and main idea of the pursuit and research

Schart's philosophy of freedom was once popular in Western societies. This is because, for people who were in the cruel situation of many disasters and hardships at that time, the desire and need to transcend the barriers of the objective world and to realize the subject was the focus of their ardent expectation.

Therefore, the Saudi philosophy of freedom was born in such an environment of urgent need for autonomy. In other words, Saudi Arabia's philosophy of freedom carries the mission of the times to find an outlet for the existence of life, which means, in all sincerity, how to face the future and live with dignity.

However, it seems to be impossible to talk about freedom in an environment full of social constraints and to break through the bundle of restrictions wrapped in layers.

It is difficult for people who have lived through two world wars to say, with one eye open and one eye closed, that people are free to live and to decide their own future.

It is hard for them not to admit that they are living helplessly in the real world all the time, being moved and determined by the whole environment.

Surviving under the brutal war, staying alive, living safely, and having food and clothing is the greatest fortune, how can we admit that freedom of autonomy is possible?

Even in a peaceful and happy social environment, there is no denying that individual behavior is restricted to some degree by social matrix, let alone in the paradoxical time and space in which the Saudis live!

So, how does Saudi Arabia talk about human freedom? How does man free himself from objective limitations and determine his own future? In his La Transcendance de l'Ego; The Transcendence of the Ego (1936), L' Etre et le N' eant; Being and Nothingness, 1943) and L'Existentialism est un humanisme; Existentialism, 1946, all address the question of freedom demanding answers or replies.

However, Scharthe's philosophy of freedom is not a superficial meaning of known facts and principles and principles that can be explained by ordinary people in their own words, texts or other symbols. In his book "Being and Nothingness", Scharthe refers to human freedom as freedom without any conditions and without any restrictions. Therefore, Scharthe's philosophy of freedom refers to the philosophy of absolute freedom. We cannot help but ask, what is absolute freedom?

Since the relationship between various things in the world is just one of the three relationships, it is very important to understand and grasp the absolute, relative, and arbitrary nature of the "existence" of the life of the self, and to observe the awareness of the cognition. In the cognitive process of human life experience and refinement, there are often insurmountable obstacles to the choice of boundaries in the purposeful activity of the object of action or thought, usually due to the conceptually deviant cognition that confuses the absolute, relative, and arbitrary.

Therefore, to clarify the relationship between these three levels, and to generalize the unique properties of similar things from their many properties on the basis of perceptual awareness, and to form the concept of linguistic words or intentional expressions, it is important to clarify and summarize the cognition of Schart's Being and Nothingness, including the cognition of being and existent, the cognition of existence and temporality, and..., to enhance the existence of self-life in the future. It is very important to clarify the cognitive efficiency of the cognitive awareness of the "existence" of the existence itself.

1. Absolute - the view of uniqueness

Absoluteness means that there is an absolute relationship between different things, which is pure and unconditional, and can exist independently of other things. In the process of human cognition, especially in the long cultural history of the Greek-Roman world (centered on the Mediterranean Sea, including a series of civilizations such as ancient Greece and ancient Rome), many absolute concepts and notions have been proposed, which is the inevitable result of human cognition under the way of thinking of metaphysics.

For example, ontology is an absolute concept that exists independently of all things, and nothing can affect or change ontology, while ontology can eternally and continuously replicate and extend itself, and thus derive the whole world. According to the ontological point of view, there is only the essence in the world, and everything else is only a symbol of the shape and phenomenon of the essence or a concrete manifestation of the essence's function in various aspects. The absolute nature of the ontology will further require a deeper cognition.

The cognition of the "existence" of one's own life existence is only the reflection and description of the functional role of the essence, i.e., the pursuit of cognition without any condition or restriction, which inevitably leads this cognition theory into the dilemma of cognitive enlightenment. Because the observation of absolute cognition requires that there be a distinction between correct and incorrect theories, human cognition should consist of a series of correct theories, or at least the development of theories should be a continuous process of constantly approaching a certain limit of correctness (truth).

Unfortunately, however, the process of human cognition in the pursuit of truth has shown that the various theories that have been generated in the past are neither "correct" nor convergent in their deviations. On the one hand, the existing specious theories cannot stand the test of time and space and will sooner or later be withdrawn from the stage of human cognition; on the other hand, the new theories cannot logically derive the correct thinking from the old ones because their logical foundations are different.

Another manifestation of the absolutization of perception is the requirement that no artificial elements such as hypotheses and axioms be used in the process of human perception. This is the point that positivism, in its opposition to and critique of metaphysics, emphasizes, going from the absolute structure to the other extreme, demanding the abolition of the structure of objective cognition. This is actually a computer, a modern electronic computing machine used for high-speed calculations, which can perform numerical calculations, logical calculations, and memory storage functions.

It is a modern, intelligent electronic device cognitive method that can run according to a program and automatically and at high speed process large amounts of data, i.e., all data are input into the computer without "bias" and the computer finds the interconnections (laws) completely "objectively". The fatal weakness of this objective approach to cognition is that it must assume that the world is continuous and that the parts can represent the whole. This is because the information we can input and process is limited, while the information that exists in reality is infinite. Of course, this assumption is not allowed by the absolute nature of knowledge. Thus, the finiteness of the collected information and the infinity of the existence of reality make the absolute nature of objective cognition a dilemma that cannot be solved.

2. Relativity - There is no "absolute reference system

Relativity refers to the existence of a comparative relationship between two opposing things of different nature at the same time. The breadth and depth of objective knowledge is always limited by history and has an approximate and relative nature. Knowing the relative meaning of the development of things can prevent people from becoming rigid in their thinking, stagnant and unable to start and only understand the local, not the whole area.

3. A form of thinking that reflects the nature of things

Relativity means that there is a relative relationship between different things, and this relationship is both interactive and conditional, and the mutual relationship between various things will change with the change of environment and conditions. The rhetorical technique is called contrast, which is also called relative.

The use of a thing has two opposing nature at the same time relative, can be good and bad, good and evil, beautiful language ugly, this kind of opposition to highlight reveal, give people a deep impression and enlightenment. It is a form of free composition, which is not limited by the structure of the composition of the body and the composition of multiple parts, but by the comparison and contrast of various aspects of the form itself, such as size, density, emptiness, conspicuousness, shape, color and texture. Harmony is to seek similarity, while comparison and contrast are to seek difference.

White and black, cold and hot, dry and wet, and day and night in nature are all unified by comparison and contrast. The premise of comparison and contrast is that a thing must have two opposing properties at the same time, and that there is a comparative contrast in the conceptual parameters. For example, in our daily life, if someone is said to be tall, then the people around him must be shorter than him, or he is obviously taller compared to what you imagine to be his normal height.

This illustrates that comparative contrast requires a

reference to the subject of the comparison, or a specific group, without reference to a single body can not constitute a comparative contrast. A comparative contrast is in fact a comparison, which can be significant and strong, or vague and slight; it can be simple or complex.

Relativity. Just give an example.

A professor at a prestigious university wanted to demonstrate his knowledge and deliberately made things difficult for his students by asking a question that required an answer: "Did God create everything that exists? The students had the courage to answer without flinching, "Yes, he created everything. The professor deliberately repeated "Did God create everything?"

"Yes, sir, he certainly created everything." The students replied in a firm voice. The professor joked in a humorous way, "If, according to you, God created everything, then because sin exists, it means that God also created sin.

And because we can define a person by what he or she does, then we can use logic to reason that God is passive. The students were silent, and no one responded to the professor's hypothetical inference. At that time, the professor was very proud of himself and bragged to the students, proudly saying that he once again proved that religious beliefs are a mystery.

At that moment, one student raised his hand and said, "Professor, may I ask you a question?

The professor said, "Of course. The professor said, "Of course.

The student stood up and asked, "Professor, does cold exist?

The professor said, "What kind of question is that? Of course cold exists. Have you never felt cold? As soon as the words left his mouth, there was a burst of laughter... the other students were laughing at the fact that this student was asking a stupid question.

But the young man replied, "Sir, cold doesn't actually exist. According to the principles of physics, what we call cold is just a state without heat. "Every human body or object we study can store and transmit energy, and it is the heat that causes the body or object to have and transmit energy. Absolute zero (-460f) describes a state where there is no heat at all, where everything stops moving completely. Cold does not exist. The atmosphere of the air freezes at a point where space and time meet.

"We coined this word to describe how we feel when there is no heat. The student went on to ask, "Professor, does darkness exist?

The professor said, "Of course it does." The student replied, "Sir! You are wrong again, darkness doesn't exist either. "Darkness is actually the state of absence of light. We can study light, but we can't study darkness. In fact, we use Newton's theorem to divide light into different colors to study the wavelengths of different light waves.

"But you can't measure darkness. A beam of light can break the darkness and light up the world. How do you study how dark a place is? By studying how much light there is. Isn't it? So darkness is a state that humans use to describe the absence of light.

Finally, the young student interrogated the professor: "Sir, does sin exist?

Now the professor, no longer sure of his answer, said calmly, "Of course, I have just said it. We see sin everywhere, like people abusing people every day, and the world is full of violence and crime all the time. These are all sins.

The student replied, "Sin does not exist, or at least it does not exist by itself. Sin only exists because of the absence of God, like coldness and darkness, which is a word used to describe the state of being without God. God did not create sin, just as cold comes from the absence of heat and darkness from the absence of light, but sin is the result of people not knowing God and lacking God's love in their hearts.

The professor sat down in silence. The class applauded with a call. The student's name was Albert Einstein. This is a true story.

4. Relative rhetorical devices of language

In the discourse of linguistic description, comparison and contrast is one of the basic combinations of lyrical discourse. It is the combination of two opposing natures of a thing at the same time, combining words and phrases with opposite sensory characteristics or moral meanings to form a contrast and strengthen the expressive power of lyrical words.

There are two kinds of contrasts in terms of the way they are constructed: 1. Contrary contrasts; 2. Anti-object contrasts. The contrast also has the meaning of contrast, so that the characteristics or nature of the opposite or opposing things emerge, more distinct and prominent. The difference between comparative contrast and foil is as follows.

In order to highlight the main thing, use similar things or the opposite, different things as a companion, "bake the clouds to the moon" expression. The main purpose is to highlight the main role of the thing being accompanied.

The two opposing things are placed side-by-side, without distinction between the main and the guest, and without primary and secondary relationships.

2. The contrast depicts two things; the contrast makes the subject obvious, using another thing to accompany or contrast to highlight the original thing, which needs to present the characteristics of two things; the comparison contrast refers to a substance, or a whole and another substance, or another whole compared with each other, or another whole.

or another whole is compared with each other, or refers to the existence or change depending on certain conditions. Therefore, the two opposing natures of one thing at the same time can be two things, or two different aspects of one thing.

The rhetorical effect of setting off is mainly in highlighting the positive or negative things, expressing strong thoughts and feelings, and deepening the central idea of the article. The most important thing is that it is not just a matter of the nature of the person, but also a matter of the nature of the person.

The rhetorical effect of comparison and contrast is to suggest the nature of things in a comparative way, making the good appear better and the bad appear worse. The rhetorical effect of comparison and contrast is to suggest the nature of things in a comparative way, to make the good appear better and the bad appear worse.

Thus relativity and absoluteness are the products of logical thinking, as much by necessity as by chance. It is precisely because logical thinking produces the concept of relativity that there are so many positions, and so many differences of opinion and disputes between people.

Scientifically speaking, any movement and evolution of anything is the inevitable result of the joint action of various factors. Therefore, in science, there are only unknown factors, and there are no accidental phenomena.

The so-called chance is just an unexpected necessity.

If we insist on discriminating between the relative and absolute concepts, we will never get a unified understanding. It is also impossible to thoroughly solve the problems of things that are the target of action or thought.

To completely solve social problems, we must thoroughly clarify all aspects of the things that are the target of the actions or thoughts that cause the problems, and eliminate them one by one, or change the negative factors into positive ones. This requires the establishment of a scientific way of thinking.

The so-called scientific way of thinking refers to the way of thinking of scientists, which is based on the relationship between the quality and quantity of things themselves.

This way of thinking is fundamentally different from the philosopher's logical thinking, which is based on the logical relationship between things, which is comparative thinking.

The simplest example is the ratio of size, for example, the size of a small elephant to that of a wolf-dog, the small elephant is bigger than the wolf-dog that ordinary people imagine, but compared to the elephant, it is obviously smaller than the elephant.

Section 4, the truth, the purest, most consistent with the actual truth

Truth, the purest, most consistent with the actual truth, that is, the objective things and their laws in the correct reflection of the human mind. It refers to the completely correct reflection of the infinite development of the objective world. Truth, i.e., the only true truth that is eternal and unchanging.

Truth can be divided into relative truth and absolute truth: relative truth is a limited truth that is established under certain conditions; absolute truth is the absolute nature of truth, i.e. absolute truth is a broad truth that is not limited in any way.

1. Truth that is unique in nature

Absolute explanation: unconditional; absolute advantage without any restriction; certainty, certainty. It is the same as "relative". It means without any condition, without any restriction.

The explanation of the truth is the correct reflection of objective things and their laws in the human mind, and you say a truth in its purest form. Absolute truth means a completely correct knowledge of the infinitely developing objective world.

This knowledge is constantly approached through the process of infinite development of human understanding. Absolute truth has two other meanings:

(1) Any truth is an objective truth, a true truth, that is, a correct reflection of objective things and their laws in human consciousness, containing objective content that does not depend on human will, and is separated from absurd errors in principle.

This is absolute and unconditional, the correct reflection of objective things and their laws in people's consciousness, and science is a method that is tested by practice and infinitely close to the truth.

(2) the process of acquiring knowledge through the formation of concepts, perception, judgment or imagination and other mental activities of people's consciousness.

(3) The process of acquiring knowledge through mental activities such as forming concepts, perception, judgment or imagination, i.e., the ability of the mind to perform the mental function of information processing is infinite, and it can be infinitely close to the objective world.

In a broader sense, knowing includes all cognitive activities of human beings, i.e. the nature of perception, memory, thinking, imagination, understanding and production of language, etc. It is also absolute and unconditional, in the sense that acknowledging the knowability of the world is the same as acknowledging the absolute truth.

2. Truths that are opposed to each other in nature

Relative truth is the relativity of truth, which refers to people's relatively correct knowledge of limited objective things at a certain historical stage and under certain conditions:

Relativity of truth, also called relative truth, refers to people's knowledge of the objective world and its laws with approximation and conditionality. It includes: (1) the approximation of truth content; (2) the finiteness of specific truths.

(1) the approximation of truth content; from the perspective of specific things and their processes, the infinite diversity of specific forms, structures and properties of the material world determines that any truthful knowledge is only a correct reflection of certain aspects and degrees, and is only approximate in nature and needs to be further deepened and enriched.

(2) Finiteness of concrete truth: In terms of the knowledge of the whole universe, the infinite nature of the universe in space and time determines that any truthful knowledge is a correct knowledge of a part or an aspect of it, and the truthful knowledge in a certain space and time can never exhaust the material world, but can only be developed and perfected continuously.

Absolute truth and relative truth reveal two different aspects of thinking in the process of development of things, but there is no fixed boundary between them, any truth has two attributes, absolute and relative, and is a dialectical unity of absolute and relative.

(3) Absolute truth and relative truth are interdependent and mutually inclusive. Any truth is a correct reflection of objective things within certain conditions, degrees and scopes, and therefore has

relativity.

But it is a correct reflection of objective things, and cannot be overturned within its limits, so it is absolute, and absolute truth is contained in relative truth.

(2) Relative truth to absolute truth dialectical transformation, truth is a development of things through, both the order and context and the way, method, always in the transformation and development from relative to absolute, any objectivity, is a central concept of philosophy.

It refers to the reasonableness of thinking or judging something from different viewpoints or perspectives, the nature of a thing that exists independently of subjective thought or consciousness, corresponding to "subjectivity.

Objective truths are not affected by subjective means such as human thoughts, feelings, tools, or calculations, but can maintain their truthfulness. Therefore, objective truth is one of the interrelated things in the process of transforming from relative truth to absolute truth.

The metaphysical view of truth, the logic that separates the absolute and relative nature of truth, refers to rational or abstract thinking, which reflects the regularity of objective things in the form of a theory.

The past events and actions of human society, as well as the systematic recording, interpretation and study of these events and actions, include two levels of meaning: the historical development process of objective reality, and the historical development process of human understanding. The role or influence of the things in question can produce two types of one-sided interpretations of the truth.

First, the unilateral opinion or statement of absolutism, which exaggerates the absolute nature of truth and denies the relative nature, thus denying the development of truth; dogmatism, also known as "fundamentalism". It is a form of subjectivism.

The main characteristic is to treat books and theories as dogma, rigid thinking, all from the definition, formula, not from the actual, opposed to the specific situation specific analysis, denying that practice is the standard for testing the truth. Absolutism is a form of expression.

Secondly, relativism is a unilateral opinion or statement that exaggerates the relativity of truth and denies the absolute nature, thus denying the existence of objective truth. Pragmatism, on the other hand, is based on the fundamental principle of determining beliefs as the starting point, taking action as the main means, and obtaining practical results as the highest goal.

Pragmatism is derived from the Greek word for action. Therefore, pragmatism refers to the interpretation of behavior and action, and is concerned with whether the action can bring about a certain practical effect, that is, with the direct utility, benefit, usefulness is the truth, and uselessness is absurdity and error. This is the typical representative of relativism.

3. The quality of arbitrariness

Among the various relationships of human cognition, apart from the absolute and relative relationships discussed earlier, there exists another kind of relationship, namely, arbitrary relationship. The so-called arbitrary relationship is commonly referred to as arbitrariness, which, as the name suggests, refers to the relationship between something and something else.

As the name implies, it means that something is not connected to anything else, so its state of existence is free and not influenced and bound by any other external things.

For example, when a person walks in space, he or she can walk freely without any effort because there is no gravitational attraction of the earth, so the walking of a person in space is arbitrary (in fact, it is a more difficult thing for a person to walk in space because of the absence of friction and the lack of fulcrum for action).

In general, the study of arbitrariness, is no human knowledge of natural or social things, is the meaning given by man to the object things, is a symbolic form of human communication.

It is the spiritual content that human beings transmit and communicate in the form of symbols. Here, the reason why we have to study and comment on arbitrariness is because of the following two discretionary issues. In order to make a decision.

First, to understand better and more deeply: in order for us to understand better the absolute and relative nature, because in the real world, there are various kinds of actions or influences on things, which are no more than absolute, relative or arbitrary relations.

Only when these three kinds of relations are taken

into account, it is possible to make our examination of the state of interaction and mutual influence between various things - the connection of a certain nature between people and people or people and things - logically complete, so that our understanding is complete and comprehensive.

Second, to obtain higher cognitive efficiency: in the process of human cognition, even if it is arbitrary, it still has a certain observation of the process of acquiring knowledge through the formation of concepts, perception, judgment or imagination and other mental activities, that is, the mental function of the mind to process information.

That is, the value of the mental function of information processing by the mind, which allows human beings to observe the cognitive process in the specific, get rid of the ideas, culture, morality, customs, art, institutions and behavior passed down from generation to generation and from history.

The social behavior of people have invisible influence and control the role of the concept of bondage, can be maximized, to give full play to the subjective initiative of human cognition, so that humans get a higher cognitive efficiency.

Arbitrariness in its daily application to reality (Latin: Realitas) means "something that exists objectively" or "a condition that fits the objective situation". Reality is the sum of all actual or existing things, as opposed to things that are entirely imaginary.

The term reality is also often used to denote the ontological state of things, including whether they exist or not. Philosophical questions about natural reality, existence, or being can be discussed in the context of ontology (one of the major branches of Western philosophy, metaphysics).

Ontological questions are also found in many fields of philosophy, such as philosophy of science, philosophy of religion, philosophy of mathematics, and philosophy of logic.

Related questions include whether only actual objects are real (physicalism), whether reality is intrinsically immaterial (idealism), whether unobservable entities exist that are assumed to be consistent with scientific theory, whether there is a God, whether there are other abstract objects, and whether possible worlds exist.

In a broad sense, "reality" includes all things that can

be observed or explained in their own words, texts, or other symbols as known facts, principles, and principles, so it includes both existence and nothingness.

Reality in the narrow sense has different conceptual levels in philosophy, including phenomena, facts, reality, and axioms. Philosophy considers that there are two kinds of reality, one is the reality of nature and the other is the reality of thought (including linguistic and cultural aspects)

On the one hand, ontology is the study of what exists, but there are various ways of expressing it, such as being, existing, "what is", and reality, etc. The task of ontology is to describe what the most universal realities consist of and how they relate to each other.

Some philosophers distinguish between reality and existence, but many analytic philosophers avoid using the words "real" or "reality" in their ontological discussions and use "existence" instead.

An important question in analytic philosophy is whether existence is one of the essences of things, and most philosophers believe that it is not, although this idea has changed somewhat in the last decade.

On the other hand, when discussing objectivity in metaphysics and epistemology, the question of whether "reality" depends on ideological and cultural factors is involved. Such as feelings, beliefs, religious and political movements, and worldviews.

4. Relative truth is knowledge

Realism (French: Réalisme), also known as actualism, holds that cognition among human beings, our understanding and perception of objects, is consistent with the actual existence of objects independent of our minds.

It is generally defined as being about reality and actuality, while rejecting idealism and giving special importance to concrete facts, which cannot be impractical, inscrutable, or abstract. Truth is always repeatedly relevant at any time and place.

However, the search for truth, as well as the pursuit of the ideal, will always provoke the curiosity of philosophers and scientists to explore the truth. Human beings must strive for knowledge and improvement. Absolute and Relative Truths reveal two distinct and interrelated epistemological domains in the development of truth.

Thus, for us, in a restricted situation, freedom is mostly a relative freedom. That is, we can sometimes make choices and sometimes not.

This is a philosophical concept that gives rise to different perspectives when it is understood. Relative means conditional, temporary, finite, and particular; absolute means unconditional, eternal, infinite, and universal. Relative and absolute are philosophical aspects that reflect two different aspects of the nature of things.

Realism, on the other hand, holds that reality exists independently of any feelings or beliefs. As long as the object, existence, and fundamental characteristics can be distinguished, it proves to be independent of any feelings, language, beliefs, or other human factors, and can be called the reality of this object.

Therefore, it is universal in the world because the world is so large that it is an arbitrary relationship between things that are far apart in distance or level, spatial distance or temporal difference.

It is just that, in general, we treat such situations as mere backgrounds for cognition and are indifferent to what is in front of us, not including these unrelated events as objects of awareness.

For example, when we watch a game, our attention to whether or not one of the participants is a national athlete or a relative of our own is very different from how we feel when we watch it. There are many similar situations, such as watching the news, treating patients, adjudicating cases, and managing people.

In addition, according to the third law of natural philosophy, there is interdependence and interpenetration among individuals, and arbitrariness is not exclusive or absolute.

An object can be unrelated to some objects, but it cannot be unrelated to all objects. There is no such thing as absolute arbitrariness, because the world lives in a way that does not destroy nature's ecology, and allows nature to flourish in a sustainable way.

Another manifestation of arbitrariness is arbitrariness, such as subjective idealism, which is obviously anti-scientific when it comes to understanding and explaining an event without the constraints of the real world, and which is based on mere subjective speculation.

However, under certain conditions, arbitrariness is

useful for human observation and cognition. For example, a certain object as the scope of study, based on experimental and logical reasoning, to seek a unified, accurate objective laws and truth is constructed. This requires a certain degree of arbitrariness in the establishment of human theories.

Among the three relationships mentioned above, only relativity maintains a certain degree of cognitive flexibility while appropriately maintaining the tension required for cognition, thus greatly enhancing the efficiency of human observation and cognition, and thus gaining a wide scope of application.

In fact, absoluteness and arbitrariness are only descriptions of perceptual cognition, reflecting the opposites and randomness of nature on its surface.

Only relativity truly reflects the organic and systematic nature of nature, and thus forms a unified observation of cognition. Since unified cognition is the necessary condition for maximizing the efficiency of observation and cognition, the essence of human observation and cognition is the ability to observe the nature.

Therefore, relativity and the resulting two-dimensional cognition lay a solid philosophical foundation for human beings to transcend opposition and deepen the cognition of observational awareness.

Relativity and absolute reciprocity are the dual attributes of the same thing that are both interrelated and distinct from each other, as one substance or a whole and another substance or another whole compare with each other, or exist or change depending on certain conditions.

All things contain both relative and absolute aspects. Therefore, we can perceive that the specific things in the universe and their change process are conditional, limited and relative, while the existence and development of the whole universe is unconditional, infinite and absolute.

The absolute exists in the relative, and is expressed through countless relative. There are various manifestations of relativity and absoluteness: motion is absolute, stillness is relative; the universality of contradiction is absolute, the particularity of contradiction is relative; the infinity of space-time is absolute, the finiteness of space-time is relative.

Barmenides of Elea (ancient Greek: Παρμενίδης ὁ

Ἐλεάτης, c. 515 - 445 BC)[1] was an ancient Greek philosopher of the 5th century BC, one of the most important "pre-Socratic" philosophers, and a member of the Eleatic school.

He was born in the Elia region of southern Italy, and his main work is the "On Nature" written in rhyme, of which only remnants remain today.

He called "existence" as the spiritual essence as absolute, and believed that there is only "existence" and no "non-existence", and that "existence" is complete, infinite, and immovable.

Section 5: The position based on which problems and issues are analyzed or criticized

Everything is One

Parmenides opposed Heraclitus' view of universal change. He believed that the diversity and variation of all things is only an illusion, and that there is only one thing in the whole universe, eternal and indivisible, which he called "one".

He argued that either the existent exists or the

existent does not exist, but that the existent does not exist is wrong. If we accept Heraclitus' view, we will think that things will change and grow old and die.

But, according to Parmenides, if we think that things will disappear, we make the mistake of thinking that "the existent does not exist".

Truth and Opinion

Parmenides' "being" is achieved through the path of logic. Since "being" as a coefficient ("is") is the definitive embodiment of any linguistic expression, it gives thought its own definite object.

Perception is in "non-being" due to flux and cannot be determined by thought. Therefore, Barmenides emphasizes that "existents" can only exist in thought and speech.

The concrete things that are objects of sensation and are in the flux of birth and death are "non-existent" because they cannot be expressed and fixed exactly in words.

In his view, both the Miletus school, which takes the changeless "non-existence" as the original of everything and advocates the existence of "non-existence", and

Heraclitus' fire-principle theory are false and absurd "opinions". The only "theory of truth" is to insist that "existent" things exist and "non-existent" things do not exist.

Protagoras, as a wise teacher, devoted himself to the connection between morality and politics. He opposed Socrates' theory of innate morality and advocated that "virtue is teachable".

Unlike other masters of sophistry and rhetoric who offered a clear practical approach, Protagoras tried to conceive of universal and broad human phenomena (such as linguistics and education); he was also interested in orthography.

His famous quote: "Man is the measure of all things: the measure of the existence of the existent and the measure of the non-existence of the non-existent. Like many of the remnants of pre-Socratic philosophers that have survived to this day, this sentence has no context and its exact meaning is open to interpretation. Plato categorized Protagoras as a relativist philosopher.

Protagoras was a proponent of agnosticism. In his last essay, "On Gods," he wrote: "As for the gods, I am not sure that they exist or that they do not exist, nor do I dare to say what they are like.

For there are many things that hinder our sure knowledge, such as the obscurity of questions and the shortness of life. There are craters on the Moon named after Protagoras. In the case of the moon, there are craters named after Protagoras, and in the case of the moon, everything is seen as merely relative, with no absolutes, and "everything is only relative truth".

The 17th century Dutch materialist philosopher Spinoza was a monist or pantheist. He believed that there was only one supreme entity in the universe (later called the Spinoza entity), the universe itself as a whole, and that God and the universe were one and the same.

His conclusion is based on a set of definitions and axioms, reasoned through a science of norms and criteria for the validity of thought and argument, which traditionally includes principles of definition, classification, and proper use of the laws of thought, as well as objective principles of regularity and argument.

Spinoza believed that God includes not only the material world (extensiveness) but also the spiritual world (thinking). He considered human intelligence as a component of the highest physical intelligence.

Spinoza also believed that this entity is the "inner cause" of everything, that it rules the world through natural laws, so that everything that happens in the material world has its own necessity.

God is the only one in the world who has complete freedom, while man can never obtain free will, although he can try to remove the fetters of the external.

If we can see things as inevitable, then the easier it is for us to become one with God. Thus, Spinoza suggests that we should look at things "sub specie aeternitatis" (under the eternal and unchanging phase).

According to Spinoza, everything perceived by the senses can exist only and only through the existence of a supreme entity, and everything in the world in the pure sense (without reflecting on the supreme entity) is only an illusion.

In his book "Ethics" he explains that "existence belongs to the nature of the entity" Spinoza gives an example with the circle.

If a circle is likened to a supreme entity (or God), and a

cross is arbitrarily drawn within the circle to form a right angle (signifying the formation and persistence of all things in the world), then there are an infinite number of right angles within the circle. But if the circle (the highest entity) did not exist, there could be no angle, and there could be no so-called "right angle" (of everything).

Although the right angle did not exist before it was drawn, the idea of the right angle itself was implicit in the circle, and the idea of everything, according to Spinoza, was originally implicit in the Supreme Entity. Hegel points out that his concept of God can be called "worldlessness".

Spinoza believed that all things (not only humans) "strive to preserve their own existence". This effort he calls conatus. This Latin word means both effort and inclination.

This double meaning is what Spinoza wants, because he believes that this effort of self-preservation actually comes from the tendency of the nature of all things. Everything by its nature strives to preserve itself. Therefore, the effort to preserve oneself is the nature of all things.

In ethics, Spinoza argues that as long as one is

subject to external influences, one is in a state of slavery, and as long as one is in agreement with God/Nature, one is no longer subject to such influences, but can be relatively free, and thus free from fear.

Spinoza also asserts that ignorance is the root of all sin. On the subject of death, Spinoza famously said, "The free man thinks least of death, and his wisdom is not a meditation on death, but a contemplation of life. He lived this motto thoroughly throughout his life, and always faced death with equanimity.

Spinoza was a thoroughgoing determinist, believing that the occurrence of all things that have happened is absolutely inevitable.

He even believed that human behavior is already completely determined, and that freedom is our ability to know that we have been determined and to know why we do what we do.

So freedom is not the possibility of saying "no" to what happens to us, but the possibility of saying "yes" and understanding that something will have to happen that way.

Spinoza's philosophy is very similar to the Stoics, but

he diverges sharply from the Stoics on one important point: he completely rejects their view that motivation can triumph over emotion.

Instead, he argued that the emotion could only be replaced or defeated by another, stronger emotion. He argues that there is a difference between active and passive emotions, between factors that play an important role in something, or that can form an important part of something.

The former are relatively capable of explaining known facts and principles and principles in their own words, texts or other symbols, while the latter are not.

He also believed that knowledge with real motives of passive emotions can be transformed into active emotions, thus foreseeing a key idea of Sigmund Freud's psychoanalysis to call the 'entity' as matter the absolute, considering this entity to be constant and immutable.

German classical philosophy generally refers to the philosophy of Kant, Fichte, Schelling, Hegel, and Feuerbach, and represents the highest stage of modern Western philosophy.

It inherited the rationalist tendency represented by

the German philosopher Leibniz, and at the same time was influenced by the empiricism and skepticism of the famous philosopher Hume in the Scottish Enlightenment.

In addition, the Enlightenment literature represented by Lessing and Goethe also exerted a considerable influence on German classical philosophy. (Spinoza's fatalistic ideas are sometimes considered to be one of the important sources of classical German philosophy.)

Under the combined influence of these ideas, classical German philosophers summarized and explored a range of philosophically significant issues, although most of them were often considered to be idealistic in general terms, but their claims were not unified.

Thus, in classical German philosophy, the term "absolute" is used more often. Kant speaks of the "absolute imperative" to refer to an a priori, supreme principle of the norms and standards of behavior that should govern the common life of human beings.

In epistemology, he also refers to the elusive "object-self" as an absolute "something on the other side". Schelling calls a supra-rational force the Absolute, which he regards as the original source of all things.

Hegel's entire philosophical system is a self-movement of the Absolute Spirit, which he describes as the original cause and inner essence of everything in the universe, which existed independently long before the emergence of nature and man, and from which the whole objective world was derived or transformed.

According to Hegel, every aspect of philosophy and every field of logic is a stage, an aspect, an expression of the development of the Absolute Spirit. As for the "absolute spirit" itself, Hegel cannot say anything about it.

But Hegel's idea that the absolute universal is not only the same in itself, but also different, and that the absolute and the relative are opposing unities, has the rational idea of dialectics.

Metaphysics denies the dialectical unity of the absolute and the relative, and separates the absolute from the relative, or thinks that the absolute is the absolute, and is detached from the relative, and thus goes to absolutism.

The absolute and the relative are separated, or the absolute is the relative, and it is detached from the relative, which leads to absolutism. Absolutism exaggerates the absolute aspect of things and uses it to deny relativity, while relativism exaggerates the relative aspect of things and uses it to deny the absolute.

Both are one-sided and one-sided, a distorted reflection of the objective and a castration of the absolute and relative dialectics of things and knowledge.

In particular, the Saudi Arabia is in an era of terror, where life and death are often not decided by oneself, and under such conditions, the Saudi Arabia is still singing the rhetoric of "absolute freedom"?

The focus and purpose of this essay is to explore the "absolute freedom" in Saudi Arabia. In discussing the absolute freedom of human beings, Saud takes the existence of human beings (existence for oneself) as the starting point.

The basic concepts he uses, such as freedom, choice, and responsibility, do not have the moral and value meanings that people usually understand, but are in the same existential domain as self-existence.

In other words, in order to value a certain aspect and try to make it happen, to satisfy the demand for absolute human freedom, one must first explore human existence (the real existence of human beings), which is more in line with facts, reason or some kind of accepted standard.

Generally speaking, Scharthe's existentialism is deeply influenced by Heidegger and has a deep imprint in the process of re-invention and metamorphosis. However, Scharthe's philosophy retains characteristics that are quite different from those of Heidegger. The most obvious is that his emphasis on "absolute freedom" is the real existence of human beings.

In fact, Scharthe is the only one among the existentialists who has been labeled as a more thorough researcher of "absolute freedom. His predecessors, including Heidegger, also did not have a deeper understanding of absolute freedom.

For Schart's philosophical view of Being and Nothingness, "absolute freedom" is "the reality of man". However, the so-called human reality is not simply what it is in everyday life.

For example, when someone says that you are real, it means that you are a real person in people's minds, not a hypocrite! It is easy to get in touch with people, and people like to deal with people like you, this is a way of dealing with people, and a positive attitude in life.

Human beings do not exist like ordinary objects; rather, the real nature of human beings is explained by the way they exist in real situations.

He emphasizes that "being" is for the sake of "being", and that the process of "being" is in itself, rather than "being" as "what is". For example, 'reality' is a definition of human expression.

Mostly it means that someone's words and thoughts are consistent, that is, someone's words and actions are consistent, words and thoughts are consistent. It means that the 'being' is the 'being' of 'what is'.

Schatt uses "being for itself" to express Being, not as what it is (e.g., a table, a chair), but as "how to be" in each particular and physical existence, or explicitly rather than abstractly and generally, in the process of "the existence of self-life itself," which reveals the real actual existence of Being itself, the real.

In short, according to the philosophical viewpoint of Schart's Being and Nothingness, it is not whether a "being" is merely surviving, but whether he lives his life meaningfully.

It is whether he lives his life meaningfully, whether he can create his own characteristics, whether he constructs the value of his own existence in the real situation.

According to Schart, man's existence is not a natural thing, but rather a creative force that creates his own "being" and determines his own nature and characteristics.

It means that "the process of being" itself, or the process through which things develop, is a possible "act" of life that is chosen towards the existence of self-existence itself.

It is self-choice, self-creation. According to Schart, the existence of one's own being is in the domination of one's own being, in the mastery of one's own being, in the determination of how one's future "should be? This is his "Being and Nothingness".

This is the core of the concept of absolute freedom in his book "Being and Nothingness". Freedom is the ability of man to act, to be the absolute creator of his own actions.

Then, how can man, who exists in his own being, not exist as a natural thing, but create his own act of existence "absolutely"?

In order to answer this question, the philosophical thinking of Schart's "Being and Nothingness" returns to consciousness itself. Since the freedom of man is the absolute creation of the self, all limitations and fetters cannot exist.

Neither can a perfect God nor nature exist, and consciousness is the pure nothingness of the existence of the ego life itself.

It can be said that this nothingness of consciousness means that "absolute freedom" is a prerequisite for discussing the things that are the goal of action or thought.

Consciousness is nothingness without any content, and it is because consciousness has no content that it cannot find any physical existence, or any definite, rather than abstract, generalized, person or thing to rely on, without any reason to support its "existence".

Therefore, the consciousness of existence that exists in the self-life of man is a free existence. All acts of consciousness thus arise instinctively and naturally, without external influence.

As far as the instinctive, spontaneous behavior of the thing that is the goal when acting or thinking is concerned, Chartres is clearly influenced by Descartes' philosophy. Descartes, starting from the suspicion of everything, found that consciousness has a power of choice

Beyond any determinism. Whether there is a deceiver or not, the existence of the human ego cannot be denied, and the ability to doubt freely is absolute and instinctive.

Thus, the concept of absolute freedom proposed by Scharthe's philosophical thinking in Being and Nothingness is basically an inheritance of the model first created by Descartes, and this prototype has often been repeatedly imitated and recreated by subsequent authors.

Scharthe goes further to explain and justify this absolute freedom. In Being and Nothingness, we can clearly see the shadow of Descartes.

Scharthe makes a series of theoretical statements, called presuppositions, in an attempt to prove the truth of the conclusion that existence for itself (i.e., consciousness without content) is an absolute freedom.

Both the analysis of "human reality" and "absolute

freedom" are analyzed in terms of the structure of consciousness; the "intentional" structure: all consciousness is consciousness of something.

In Schartre's eyes, the description of the existence of man's self-existence itself must be analyzed in terms of the structure of consciousness, even as a being that can determine its own nature by the existence of its self-existence itself.

For this intentional structure indicates that consciousness is a kind of activity of making nothingness, in which one makes the world of things, which is the target of one's action or thought, nothing, and at the same time one makes the existence of one's own being itself nothing.

Nothingness establishes negation, and negation establishes a negator, and man, as a negator, is thus distinguished from all things that are the object of action or thought, and becomes a contentless, non-substantiated, deprived being of absolute freedom, without any support or basis.

Saudi Arabia thus equates absolute freedom with the nihilization of consciousness. At the same time, we can see that Husserl's theory of intentionality plays an important role in the Saudis' esoteric account of absolute human freedom.

Thus, the main focus of this paper is on the Saudi "absolute freedom" and its close relationship with consciousness.

How does man define "being" in the self-life existence itself, as an absolutely free being? What is the real existence of man?

In what way does man exist in this real-life experiential situation? How does man attain freedom in the real situation? All these questions about the human being, about freedom, are closely related to consciousness. Why is the intentional structure of consciousness an absolute freedom in the view of Chartres?

Why does the fact that consciousness is nothingness indicate the absolute freedom of man? And what is the relationship between this consciousness of nothingness and freedom? What is the role or effect of freedom on things between freedom and intentional action, etc.?

The point of this thesis, which is a study of the

principles and principles of the universe and life, is to point out the correlation between consciousness and absolute freedom, and further to show the correlation between absolute freedom and action. This is to show the strength and direction of the linear relationship between the two or several random variables.

For Schart's philosophical view of Being and Nothingness, to state that the existence of human ego life itself, the existence of being is absolute freedom, must be based on an exploration of the structure of consciousness. This study of the principles and principles of the universe and life will be discussed in detail in this scholarly article.

In the process or linguistic form of applying certain reasons to support or refute a point of view, usually consisting of a thesis, a thesis, an argument, and a way of arguing, the relationship between Husserl's "theory of absolute freedom" and that of Descartes and Schartes will also be compared.

The "absolute freedom" of Scharthe's philosophical view of Being and Nothingness has, to a considerable extent, influenced contemporary art.

This is because Schart's emphasis is on a freedom

that is free from all fetters, including its own restrictive nature. This freedom is not based on reason.

In fact, this absolute freedom means that reason is excluded from consciousness and that all ideologies, forms, and thought processes, such as ideas, cultures, morals, customs, arts, institutions, and behaviors, which have been passed down through the generations and through history, are abandoned.

The philosophical view of "absolute freedom" advocated by Scharthe, the true nature of human beings, only refers to this freedom without boundaries, and this freedom without boundaries was soon appreciated by artists in France and other European countries at that time.

In particular, one of the pioneers of post-modern art, the "New Fiction School" and the "Absurdist Drama", in fact, inherited and developed the Schattian concept of absolute freedom, which was repeatedly imitated and recreated by subsequent authors.

Part 2: "Existence and Nothingness" works ideas

The second section, the idea of the work "Being and Nothingness

Section 1: Phenomenological Monism Replaces Philosophical Dualism

In the course of its development, philosophers often fall into the dilemma of dualism. Husserl, by reducing being to a series of manifestations of being, tries to get rid of the philosophical dualism of internal and external oppositions, of potential and activity, of finite and infinite, which Sartre considers a great progress.

First, the phenomenological monism gets rid of the dualism between the internal and external oppositions of being. Sartre points out that if the exterior of being is understood as a skin that conceals the true nature of the object, then there is no skin at all.

If, on the other hand, this true nature is really the secret of things, and since it is within things, one can only foresee and assume it, but never reach it, then this nature does not exist either.

The same dualism of appearance and essence is denied in phenomenology. The nature as a series of principles is only the connection of the manifestations, that is, the nature itself is a manifestation. The existence of the appearance reveals itself as much as its existence reveals its essence. This is Husserl's intuition of essence.

The opposition between the finite and the infinite is not valid either. Sartre replaces the binary opposition between finite and infinite with "infinite in finite". The present exists as the manifest, that is, the phenomenon expresses itself on the basis of existence. It is in refuting the philosophical dualism with Husserl's phenomenological theory that Sartre establishes his phenomenological monistic philosophy of existence.

(1) "Being-for-itself" and "being in itself"

In Being and Nothingness, Sartre analyzes and critiques the philosophical propositions of Beckley's "To exist is to be perceived" and Descartes' "I think, therefore I am" to find a kind of I-thought that is not reflection, and seeks to obtain an ontological proof of existence from the perceiver's pre-reflection. He seeks to obtain an ontological proof from the existence of the perceiver before his reflection. He defines consciousness as transcendental from Husserl, and considers this to be Husserl's most important discovery. Being is an indispensable ground for being, being is everywhere for being and yet nowhere.

There is no being that is not a certain way of being, no being that is not grasped by both revealing and concealing this way of being. Consciousness can always transcend being, but not towards its being, but towards the meaning of that being. To deepen the philosophy of being, Sartre proposes the theory of "being-for-itself" and "being in itself". This is the essence of Sartre's philosophy of existence.

"Being in itself' has three fundamental characteristics: "Being exists, being is self-existent, and being is what it is."

The first characteristic, 'being in itself', means that existence can neither be derived from possibility nor subsumed into necessity. Necessity concerns the relationship between ideal propositions, not the relationship between existents. A phenomenon of being can never be derived from another being, because it is a being. This is the contingency of being in itself. "Nor can 'being in itself' be derived from a possibility, which may belong to another realm of being-for-itself.

The second characteristic, that being in itself is self-existent, cannot be explained by creationism. Even if existence is said to be created, existence cannot be described and explained in terms of creation, because it regains its existence outside of creation. That is to say, existence is non-created. Neither can existence be said to be self-caused in the way of consciousness; existence is neither passive nor active, but both passive and active belong to man.

Man is active, and the means used by man are passive. Existence is itself, and the oneness of existence means that it is united with itself without any distance. Here there is no relationship between existence and itself, it is an interiority that cannot be realized by itself, it is a certainty that cannot affirm itself, it cannot be active, because existence itself is full, existence is self-existent, and that is all.

The third characteristic of existence is that it is what it is. Existence itself is opaque, the Self has no mystery, it is solid. In this way, being-itself is a rigid, absolutely unified metaphysical kingdom without any connection, change or development, but without any distinction or variation.

"Being-for-itself' and 'being in itself'

The opposite of 'being-for-itself' is very different. Self-being is not existence, it is non-existence, it is the negation of existence, it is nothingness. Being-for-itself is a "non-autonomous" absolute, also known as a non-physical absolute. Its reality is purely interrogative. It raises questions because it is always in question itself, its existence is never given, but is questioned.

Because the nothingness of difference always separates it from itself, "for oneself" is always left undecided, because its existence is an eternal postponement. oneself" itself.

(2) Emptiness and self-deception

Nothingness comes from negative judgment. Consciousness cannot produce negation except by taking the form of negation, by which a being or a way of being is presented and then thrown into nothingness. Sartre then argues for the dialectical concept of nothingness.

He points out that there is a certain parallelism between the actions one takes towards existence and those one takes towards nothingness, which makes one wish to see existence and non-existence immediately as two complementary components of the real, like darkness and light.

In short, these are two concepts of complete simultaneity, which are somehow united in the production of being, and it is therefore futile to examine them in isolation.

Sartre analyzes the Hegelian philosophy in which pure being is pure abstraction, absolute negation, non-being, so that pure being and pure nothingness are one and the same thing.

Hegel makes the transition from being to nothingness, that is, his definition of being itself contains negation, a repetition of Spinoza's formula: all stipulations are negations.

From a phenomenological conception, Sartre argues that being and nothingness can be conceived as complementary; they are two equally necessary components of the real thing. It is not necessary to make the transition from being to nothingness, as Hedger does.

In Hedger, being and non-being are no longer empty abstractions; for this, even when existing face to face with nothingness, it is man who is in constant anxiety.

This is the manifestation of human reality in nothingness, and it is only in nothingness that existence can be transcended. Anxiety is the discovery of this double and continuous nothingness.

Being and nothingness cannot be separated, and man is the being that brings nothingness into the world. Man's existence is lack. Nothingness allows man to transcend and create, to be free.

The theory of self-deception is a unique theory constructed by Sartre from the standpoint of phenomenology, using the psychoanalytic method of Froude.

He states, "Man exists not only as a being who makes negation manifest in the world, but he is also a being who can adopt a negative attitude toward the self." "We are happy to admit that self-deception is a lie to ourselves.

Lying is not an attitude of denial; the essence of lying

lies in the fact that the liar is fully aware of the truth he is hiding." Self-deception appears to have the structure of a lie. The difference between self-deception and lying is that in self-deception the person is only hiding the truth from himself.

Thus, there is no duality between the deceiver and the deceived in self-deception. Rather, self-deception inherently involves a singularity of consciousness.

Using Froude's psychoanalytic theory, Sartre points out that in order to escape difficulties, people naturally turn to the subconscious.

Psychoanalysis replaces the notion of self-deception with the notion of a liar without a liar, replacing the duality of the deceiver and the deceived with the duality of "I" and "this," creating an autonomous consciousness of self-deception between the subconscious and the conscious.

Sartre points out that self-deception is a structure that originates in human reality, a structure that is what it is not and is not what it is. The purpose of self-deception "is to make me be what I am in the way that 'I am not what I am' or not what I am in the way that 'I am what I am'."

The greatest problem facing self-deception is belief.

And the fundamental problem with self-deception is believing. Truly speaking, self-deception cannot believe what it wants to believe. But it is precisely because it admits that it does not believe what it believes that self-deception is called self-deception.

(3) The Phenomenology of Three-Dimensional Time

According to Schart, time is an organized structure, and the past, present, and future as the three elements of time are not a collection of "materials" that must be put together. It is the self-referential transcendence to its own possibilities that leads to the question of temporality.

In order to clarify the nature of time, time is divided into three dimensions so that it can be discussed separately. On the question of the "past", there are two views.

One is Bergson's and Husserl's theory of the existence of the "past", that is, although it has turned to the past and merely ceased to be active, it still exists.

The second is Descartes' view that the "past" ceases to exist. Sartre disagrees with both views, arguing that the study of the "past" is inseparable from the study of the "present," that is, that the present exists as the basis of its own past, and that the past is meaningful only in its connection with the present.

Sartre also points out that the word "was" is a way of being. As an intermediary between the present and the past, "was" is neither entirely present nor entirely past in itself, in fact it cannot be either present or past, but it is an indication of the temporal connotation of existence.

The present is "for oneself". "For oneself" is the face of the whole "being in itself" in the present. The same "for oneself" faces all existence at the same time. "for oneself" is detached from oneself, tends to existence, and is in existence but is not this existence. "For oneself" is to appear to existence in the way of escape. It is impossible to grasp the present in an instantaneous way. The present must be grasped in the connection of the past, the present and the future.

The present itself cannot be grasped; once it can be grasped, it becomes the past, and when it is not grasped, it remains the future. This is "what is not and what is not".

According to Sartre, what is called temporality is such

a continuous self-denying way of being that is self-referential.

Therefore, the real starting point of time is the future, not the past. The reason why the past cannot be the starting point of time is that time is not a self-existent being; it can only be a self-referential holistic, indivisible, and constantly self-denying way of being. Sartre also explains the problems of static temporality, the dynamics of temporality, primordial and psychological temporality, and the time of the world.

(4) Human existence and freedom

Freedom is a fundamental concept in Sartre's existential philosophy. Sartre gives a new meaning to freedom as an activity of pure consciousness characterized by human subjectivity and transcendence, which is the same as "for oneself".

That is, freedom is not a certain nature of existence, but human existence itself, and human beings are free. Freedom is not something that one pursues or chooses, but is something that one has in oneself, and from which one cannot escape.

The first condition of action is freedom. For all activity

should be intentional, that is, purposeful and motivated. Several important propositions of Sartre follow.

First, existence precedes essence. Freedom is without essence, and freedom is the basis of all essences. Therefore, man is destined to be free.

Second, freedom is the freedom of choice. This choice has no support point, and it prescribes its own motives, so it may appear as absurd. This absurdity is not because it is an irrational existence, but because it does not have the possibility of not choosing.

Again, freedom and responsibility. Since freedom is the freedom to choose, some people think that freedom means freedom to do whatever one wants.

Sartre disagrees with this view and repeatedly points out that in making any kind of choice, one has to be responsible for oneself, for others, and for society. This responsibility is of a particularly special kind, and it is childish to say that "we did not ask to be born", which emphasizes people's dispassionate simplicity.

The book "Being and Nothingness" consists of an introduction and four volumes. An exploration of 'being' using a phenomenological approach.

The first volume argues for the elusive problem of 'nothingness' and provides an incisive analysis of 'self-deception'.

The second volume argues for 'being-for-itself' and presents an innovative view of 'time' with a 'holistic' understanding.

The third volume argues for the relationship with 'others' and presents the idea of 'co-presence'.

In the fourth volume, the relationship between "being-as" and "freedom" is demonstrated, and the relationship between self-possession and self-activity is used to draw a metaphysical conclusion to the whole book.

The classical philosophy Aristotle defines existence in its various forms as follows: an existence is any thing that can be described as "being" or "having", and which, using various rhetorical devices, can be visualized, acted upon, or thought about as an object.

In order to study the domain of being, we must first determine under what circumstances we can describe things as 'being' or 'having', and what the nature of this describability is.

By 'domain' we mean one of the largest classifications of things - 'things' in this case means any object that can be called but cannot be reduced to other categories. Thus the first three volumes of Schart's Being and Nothingness address the fundamental point of view of phenomenological ontology through Aristotle's categories of the various forms of 'being'.

For in existentialist philosophy, which follows the classical school, the concept of scopology (Greek: κατηγορια) is used to classify all existence in the broadest sense. For example, time, space, quantity, quality, relations, etc. are all scopes.

In taxonomy, scope is a generic term for the highest level of categories. It is different from the academic classification of learning according to disciplines, and from the encyclopedic classification of knowledge centered on nature and human beings, because it is a philosophical classification system that focuses on the distinction of the essence of being.

Aristotle's scopology, which defines the domain of 'being', proposes the first philosophical classification system and helps to motivate philosophers to consider what is the object of study of universal, fundamental problems in the disciplines of being, knowledge, value, reason, mind, language, etc. It also helps to determine the structure, character, and laws of self-referential 'being'.

Thus, in Schart's book "Being and Nothingness", the phenomenological "psychoanalysis of being" is used to describe the rationale and the meaning of the criteria that should be followed when dealing with the interrelationship between human beings and society.

"The most important part of the book is about the "freedom of will" of human beings, and the eternal possibility of autonomous action, which should be regarded as the essential characteristic of "self-being".

Section 2: Introduction.

Translated with www.DeepL.com/Translator (free version)

Before reading Schart's "Being and Nothingness" in depth, it is important to understand that existentialism emphasizes the existence, dignity and value of human beings, and fully represents the spirit of humanistic education. It is concerned with the "subject person", the "concrete and living person".

Existentialism does not explain the world, but the meaning of "existence" in the existence of the self, pointing to the transformation of a position of consciousness from one of subordination to subordination to one of autonomous hostility.

In the introduction, Schart begins his study of the existence of 'being' in the egoic being itself, which refers to the spirit of the unknown, or the act of searching for things, or the search for answers.

He begins with a contextual or methodological analysis of dualistic ideas of right and wrong, of reaction, and of words and actions, and then rejects them in order to induce a reflection on phenomena and existence.

Dualism is a branch of ontology, which holds that the origin of the world is two entities, consciousness and matter, and that the meaning of an entity is something that is distinguishable and exists independently within itself. It does not require the existence of an inner law or reason of things.

It is opposed to monism. It is an attempt to reconcile the philosophical views of materialism and idealism. Since dualism is a philosophical doctrine that the world has two separate origins, consciousness and matter, it emphasizes that matter and spirit exist equally and fairly.

In essence, it insists that consciousness exists independently of matter, and that matter and consciousness are not related to each other (or are mutually presupposed), which makes it easy for some materialists to consider it idealism.

But in fact, dualism reflects the nature of things in the form of thinking. On the basis of perceptual understanding, people can generalize the unique properties of the same thing from the many properties of the same thing.

In the form of a letter or a short and trivial chapter, the concept of thought in the mind is conveyed to others, and it is decided that it cannot be biased to the position of any one of them.

Although the cognitive ability of human beings towards the environment and the self, as well as the clarity of cognition, we have been able to understand, identify someone or something with certainty, not only understand the matter, the phenomenon is not only a phenomenon, existence is not only existence, but also be able to see from the other side's perspective.

And to be able to see things from the other person's point of view, to be able to understand the cause and effect of this matter, but also to understand all mental activities as well. Such as perception, memory, imagination, etc., as well as projecting beyond the present situation into the metamorphosis.

However, to restore the being to its original external condition or shape, the various manifestations it exhibits are still not sufficient to explain the existence of the being.

In order to better explain known facts and principles and principles in our own words, texts or other symbols, we use analogies with similar characteristics, with the help of the familiar act of "drawing lots".

The existence of things compared to (such as for the purpose of explaining) the circular "barrel of the label", and its various individuals themselves, with an expected psychological.

It is clearly shown that the "sticker" (for example, bamboo, wood, paper, etc.) with a specific mark is drawn by a person and placed in it.

Theoretically, the result is random, and it can be said that any one of the drawn sticks can exist as a whole "poem", and all the "poems" placed together constitute the existence of a complete "barrel".

However, the difference lies in the fact that the different "signature poems" are interpreted by different subjects, and they show a tendency to change over time and space, just as the primitive tribal peoples lacked sufficient knowledge and therefore used the signs of nature to indicate their actions. They are also used as targets for action or reflection.

However, natural signs are not common and must be tested in a human way, so the method of divination was born. Divination is the movement and change of things from the outside world, to non-human spirits, to find out what you want to know.

It differs from prophecy in that there are usually ambiguous answers, allowing the diviner to find a reasonable explanation.

Thus, existence is manifested through its existence, but beyond these external manifestations, there is an infinite whole that can be explained, revealed, and perceived.

To see "phenomenon" is to understand it deeply.

Therefore, how to look at the externally revealed "phenomena" is an inescapable problem in the deeper understanding of the original existence of the "existence" of the life of the self. "A 'phenomenon' is not a fact, and a fact is not a 'being'.

"The transcendence of "phenomenon" requires the subject of "phenomenon" to look beyond the momentary "phenomenon" and see the infinite tendency of the whole condensed in it, so as to grasp the sequence (connotation) under the "phenomenon".

However, the inherent contradiction between the 'phenomenon' and the whole is that a certain object defined by the 'phenomenon', even though it is considered to have a transcendental nature towards the whole, cannot be distinguishable and independent.

But it need not be a physical being. But it does not need to be a physical being to manifest a theory in a whole, or to elaborate its essence, but only to reveal it in a specific 'phenomenon', which tends to lead the understanding of existence to a dualism of existence and manifestation.

The instability of individual manifestation has already explained the transient nature of "phenomenon", while thinking about an inner law or sequence of reason that transcends individual manifestation and infinitely reveals things has already made the continuity, stability, and eternity of things, firstly revealed.

In this context, the discussion of all things existing in heaven and earth is a step away from the framework of the social matrix of dualism (the philosophical doctrine that there are two independent, parallel existences and developments in a diverse world) and a step away from the existentialist path.

Husserl's phenomenology has shown how the restoration of the essence of being is possible, and in Scharthe's case, the "phenomenon of being" is not an infinite extension of the "being of being" in its essence.

Being is neither the nature behind all human activities that touch people, things, and objects, nor the meaning of the things that are the object of action or thought.

"Existence is not only revealed by the "phenomena"

and objects that are "in the field" of consciousness, but also by the extension of the meaning that is "not in the field".

Thus, what is revealed is a kind of existence, a concrete manifestation of a theory or an elaboration of its essence, but existence is not only the present existence, existence is an infinite serial extension of manifestation.

The object of consciousness has only the characteristics and essence of "all things that exist in heaven and earth", and the moment it is realized, it inherently implies the tendency to extend infinitely beyond the phenomenon.

The proposition of "existence and phenomena" itself requires the perspective of "beyond existence" to be perceived.

The transcendental nature of existence is the basis for the awareness of the "phenomenon" of existence. Philosophically, Berkeley built his entire philosophical system with God at its center.

He is known for his opposition to the independent existence of the material world. Finally, on this basis, we can see the narrowness of the view that "to exist is to be perceived".

Berkeley argues that the stars in the sky, the mountains and rivers on earth, and all the objects contained in the universe have no independent existence outside of the human mind, but exist only in the sense that they are known and recognized by the human mind.

Therefore, if they are not really known to man, if they do not really exist in the spirit of all other created things, then they do not exist at all, as Berkeley states his proposition in A Treatise Concerning the Principles of Human Knowledge.

He says that when people say that a thing exists, it is only because that thing is perceived by us. To say that a thing exists is to say that we have seen it, heard it, smelled it, been aware of it, etc., and that is the meaning of existence.

To say that something can exist apart from mental perception is simply incomprehensible. Berkeley admits that his theory is flawed if one can conceive of something that can exist even without being perceived.

In fact, Berkeley wanted to point out that even if one could conceive of something that would exist even if it were not perceived, one would actually still be perceiving that thing.

Of course, Berkeley's argument that "to exist is to be perceived" has been refuted by many later philosophers, including the English philosopher G. E. Moore, who argued that how can things exist only when they are perceived?

He gave a famous refutation, he said here exists a person (refers to Moore himself), this person raised his hand, has proved to understand the existence of the external world.

Another philosopher is Jean-Paul Sartre, who thinks that Berkeley's sophistry has confused the existence of perception and perception, and that perception and perception are two different things, both distinct and interrelated.

Thus, in the extension of being, the unperceived, absent being can also be a being, transcending the inner sequence of phenomena, and thus be condensed into the being, contributing to the constitution of being.

If one continues along the path of Berkeley's theory, one can easily move towards "epistemology", and therefore, Scharthe introduces "consciousness" as an analytical medium.

"Consciousness is the reflection of the objective material world by the human mind, the sum of various mental processes such as sensation and thought, as well as the ability to recognize the environment and the self, and the degree of clarity of cognition.

As a kind of existence, it cannot achieve knowledge of something by itself. Before realizing the function of consciousness, it must "exist" as an existence that can be known (consciousness can know things).

In short, consciousness is like a microscope without a target. It can clearly see what is presented under the microscope, which is not visible to the naked eye.

It is because such a tiny being can be seen through the microscope that tiny things can be seen. This act of awareness cannot exist a priori, before the thing to be known.

To think of consciousness only as an 'individual or structure known through the microscope' would simplify the process of 'knowing an individual or structure'. Here is an example. Let's imagine that consciousness is a light source.

When the light is projected and dispersed, it points to the whole world, in the broadest sense of the word, the whole, all, everything. It is the sum of all existence, not just a specific object.

At this point, consciousness cannot move from the unconscious state to the unconscious state with the being it is conscious of, otherwise, [object-being] will easily fall into the dualism of cognition.

Therefore, before "consciousness" is realized, we do not know the existence of the object, and therefore, first of all, there is an awareness of consciousness, which is itself a state of "existence".

When the light source of consciousness falls on a specific object, the light source beam or particle stream is focused on the being, and consciousness temporarily captures the knowledge of the being and comes to a concrete conclusion: the being!

However, it is important to clarify that light illuminates only parts of the existence of people or the environment other than oneself, and that these illuminated parts correspond to the "presence" of everything that exists in the world.

The existence that is captured by the consciousness is known through the external manifestation of the being, while the part that is not reached by the light corresponds to the "absence" of people or the environment other than oneself.

Thus, the existence of being is not only manifested by being. The existence of everything that exists in heaven and earth is perceived by consciousness, just as the existence of the illuminated object is focused by the light source, and has the characteristic of dynamic reflexivity. (Reflexivity means that the actor is consciously confronted with the world; the actor tries to understand the meaning of the situation he is confronted with, and takes an appropriate response according to his understanding of this meaning, and also pays attention to the effect of this response on the situation, adjusting his action accordingly or further determining the meaning of the situation.) The light source cannot dynamically illuminate an object before light occurs, nor can consciousness dynamically illuminate existence before consciousness is

(Reflexivity means that the actor is consciously facing the world; the actor tries to understand the meaning of the situation he is facing, and according to his understanding of this meaning, he takes an appropriate response, and also pays attention to the effect of this response on the situation, adjusting his actions accordingly or further determining the meaning of the situation)

The source of light cannot dynamically make an object light up before light occurs, nor can consciousness dynamically make existence conscious and perceive before consciousness is known.

Consciousness" is the boundary [limit] that can be reached by the cognitive ability and clarity of cognition of the environment and the self, not the awareness of "existence", but the awareness that "consciousness can know".

The moment the light beam is focused on an object, the existence of the object is projected back to the light source through the particle flow.

Consciousness captures the moment of existence of "being" and the manifestation of "being" is fed back to consciousness through the consciousness of existence, which leads to the conclusion that "the existence of "being" is conscious".

In the social experience of daily life, everyone has a pre-determined routine of daily life, and we can grasp certain ways of thinking based on the existence of "being" in our own being.

We can grasp the meaning of the quenching situation of our daily life experience according to our own existence. A wise person, when acting or thinking about the target, will certainly understand the boundary choices he or she faces.

They try to capture the meaning of the moment, and then take appropriate action in response to that meaning. The "existential" response that exists in the self-life itself is also involved in the formation of this meaning.

At the same time, the meaning that emerges from the whole situation of boundary selection serves as a quenching chain of daily life experience, and sets up the real meaning of life existence that exists in the "existence" response of the self-life existence itself.

In other words, the social action, in participating in the formation of the meaning of the situation, also makes the response itself be put into the context of the situation, which in turn sets up the meaning of the "existential" action that exists in the self-life existence itself.

There is a second layer of reflexivity here, which is the process of social action to clarify the true meaning of self-life existence.

Let us try to understand the relationship between "being and consciousness" and "being and perceived" in a simple way by using a popular folk tale of a blind man feeling an elephant.

Just as an elephant reveals itself to be an elephant through the synthesis of its parts, so "being" synthesizes itself into a perceivable whole through all the parts of its existence that can be conscious, and which infinitely tends to the limit.

The elephant is partly the elephant that the blind man perceives, but the whole that the elephant presents as a "being" is not the whole that the blind man subjectively perceives as an elephant.

In this regard, the elephant cannot be equated with the synthesis of subjective impressions that the elephant is aware of, so that "being" is not the same as a simple collection of subjective impressions that can be perceived.

The perceived "being" is the "being" that the perceiver

is aware of, and the perceiver only accepts the feedback of his own consciousness, only perceives his own perception, and stubbornly takes his own perception as his perception.

Thus, "existence" cannot be simply reduced to the total union of various perceived perceptions. The average human perceiver, ignoring the perceived "being," is a passive being, just as an elephant is passively perceived by a blind person.

However, perception is an overall view of the external world when external stimuli are applied to the senses, and a detailed analysis is made by the human brain along the lineage, from a certain cognitive understanding.

It organizes and interprets the sensory information of the outside world for us. In fact, it is the psychological study of behavior and mental processes of cognition (Cognition) which refers to the formation of concepts, perceptions, judgments or imagination and other mental activities through the experience of previous daily life experiences.

The process of acquiring knowledge, experience, and lessons, in the moment when perception occurs, "existence" contains both active and passive, those perceived "existence" are actively detached from its perceiver, independent of "existence", and "exist" in the life experience of "existence" that exists in the self-life existence itself.

(1) The concept of phenomenon.

In the introductory part of the exploration of "existence," the main focus here is on Husserl's commentary, which replaces the phenomenon-substance dualism with phenomenal monism and reduces the essence to the recognition of phenomena as "back to the thing itself" on the basis that it is well done.

He also argues against the dualism of manifestation and "existence" (where "existence" is solidity, that is, the essence is seen through the phenomenon) by using the example of electric current and force. The essence as manifestation is no longer opposed to existence, but becomes a standard by which "existence" sees things.

Then the present is all that is manifested and perceived by a thing, that is, it reveals itself in such a way that it can be seen, heard, smelled, and touched by the person it shows, and it is the absolute expression of the fact that it can be observed and observed by itself.

Phenomenology (English: phenomenology, from the Greek phainómenon, meaning "that which appears", and lógos, meaning "study") is one of the most important philosophical schools of the 20th century, formally founded by the German philosopher Georg Husserl.

He was influenced by Franz Brentano and Bernard Bolzano, who believed that every perceptual image formed through perception is a past impression of something reappearing in the realm of consciousness, which refers to the perceptual image formed in the mind based on perception. This includes memory images and imaginary images.

The former refers to the image of a perceived thing that is not in front of us, but reappears in the mind. The latter refers to the perceptual image or memory image, the processing of certain transformation, and the formation of a new image.

Representation is intuitive, but not as clear, complete and stable as perceptual image. It has a certain generality and is the intermediate link from perception to thinking.

According to the different sensory channels that are

dominant in the formation of an image, it can be divided into visual, auditory, motor, olfactory, taste, and tactile images.

The individual uses sensory, perceptual, thinking, memory and other mental activities to perceive and understand his or her physical and mental state and changes in people, events and objects in the environment.

At the same time, it also advocates the existence of "truth itself," that is, the idea of an objective being that transcends time and space and the absolute and universal existence of the individual, and proposes the study of the nature of consciousness, or describes the fundamentals and laws of a priori and absolute knowledge.

"Phenomenology is the philosophical study of the structure of experience and the structure of consciousness. As a philosophical movement, phenomenology was founded in the early twentieth century by Edmond Husserl, and was later adopted by him in the German city of Gothenburg.

It was later developed by a group of his followers at the Universities of Göttingen and Munich in Germany. Phenomenology has since spread to France, the United

States, and elsewhere, and has expanded far beyond the context of Husserl's earlier writings.

"Phenomenology, according to the vast majority of philosophers, should not be seen as a unified movement, but rather as a collection of authors who share a common family of tacitly recognized similar ideas, but who, while seeking differences within the same, also have their own significant differences.

Thus: "A single, final definition of "phenomenology" is dangerous and may even be as contradictory as the lack of thematic focus, which is often logically impossible to judge right or wrong. In fact, it is neither a doctrine nor a philosophical school.

Rather, it is a style of thought, a method, an open, ever-new experience that leads to different conclusions and leaves those who want to define the meaning of "phenomenology" at a loss.

According to Husserl's conception, "phenomenology" is first and foremost a systematic reflection and study of the structures of consciousness, of the phenomena that appear in various acts of consciousness. "Phenomenology" can be clearly distinguished by a Cartesian analysis, which views the world as a collection of objects, objects, and objects in continuous action and reaction.

In the "phenomenological" monism that replaces the philosophical dualism, Scharthe uses his "phenomenological" ideology to conduct an ontological inquiry.

In this exploration, one discovers the basis of 'negation' and upholds the characteristics of the basis of the negation of all nihilization, the ability to recognize the environment and the self, and the clarity of cognition. Scharthe begins by questioning 'existence'.

The question of 'existence', which exists in the life of the self itself, offers the possibility of a negative answer. The question becomes a bridge between two non-Beings that can communicate.

The "existence" of the existence of the self-life itself is the non-existence of knowledge; the existence of the self-life itself is the possibility of the non-existence of the self-life itself.

It is because the person who asks the question does not know whether the answer to the question will be affirmative or negative. Such a question shows that we are enveloped by the unrealistic and inscrutable "nothingness".

In fact, it is the non-existence of our own being that uses external stimuli to establish specific responses or patterns of behavior that limit us. For example, water can change into: solid ice, liquid water and gaseous water vapor.

1. The water in the freezer of the refrigerator is frozen into a solid substance in the form of ice.

2. When you open the freezer of the refrigerator, a white smoke descends, that is also the water vapor in the air condensed into small droplets of water when it is cold.

3. Boiling water generates water vapor, which comes outside, thus reducing the original amount of water, and this phenomenon of liquid heating into gas is called "vaporization".

Of course, in "being in itself", there is no negation, but as long as there is a relationship with consciousness, for example, consciousness asks questions about it, then the "existence" of the self-existence itself is established as a basis for negation of non-existence.

In order to ask a question, there must be the

possibility of negation, and the necessary condition to be able to say "no" is

The necessary condition for being able to say 'no' is that the non-being that exists in the life of the self is always present in us and outside of us, and the 'being' that exists in the life of the self is the elusive 'nothingness' that entangles the 'being'.

After examining and criticizing Hegel's and Heidegger's idea of the elusive 'nothingness', Schartre elaborates his own view of the origin of 'nothingness': In order to be able to question the existence of 'being' in the self-living being itself, it should be stipulated in some way that the elusive 'nothingness' must show the possibility of negation.

"The nature of "nothingness" is such that the unrealistic and inscrutable "nothingness" can be nullified, the inscrutable "nothingness" can be assumed by the existence of the self-life itself, and the inscrutable "nothingness" can be continuously supported by the existence of the self-life itself.

Through the nature of the elusive 'nothingness' itself, which is the existence of the self-life itself, it is determined that nothingness itself does not exist, it is 'being in existence', and it cannot be the self-vanishment of the existence of the self-life itself, but there must be a kind of existence (which cannot be 'being in itself') into things.

That is to say, the elusive "nothingness" from which existence comes into the world should make the elusive "nothingness" into nothingness in the elusive "existence" that exists in the existence of the self-life itself.

And this "existence" that exists in the self-life itself is also its own inscrutable "nothingness". The process of detachment from existence, which requires the pursuit of existence "being in itself" through nihilization (of oneself and the world), is the process of making the elusive "existence" of the self existent itself. That is, the existence that makes the elusive "nothingness" appear in the world.

This means that the actual existence of the being in itself is not to eliminate the being in itself, but to be free and to do everything without any obstacle, but to change the relationship between the being and the existence.

For the actual existence of being in itself, to place an object outside of being is to place oneself outside of the circle of being. At this point, the actual existence of "existence" that exists in the existence of the human ego itself escapes from the constraint of this existence.

It is outside of what the autonomous intention can do within the limits of its own power and ability. He cannot move above himself, but escapes through the inscrutable "nothingness".

Here is the original starting point of Schart's intentional freedom: "The actual existence of the 'being' of man's self-existence itself secretes an elusive 'nothingness' that makes itself independent, a possibility that Descartes, following the Stoics, called freedom. Freedom is in fact the necessary condition for the nihilization of the inscrutable 'nothingness'.

Translated with www.DeepL.com/Translator (free version)

Here is the original starting point of Schart's intentional freedom: "The actual existence of the 'being' that exists in man's self-existence itself secretes a 'nothingness' that makes itself independent and inscrutable, and for which the possibility of

Descartes, following the Stoics, called it freedom. Freedom is, in fact, the necessary condition for the nihilization of the elusive 'nothingness'.

Dualism is a word with multiple meanings. Ontological dualism is the counterpart of monism, which holds that the world consists of two indispensable and mutually independent elements, while monism holds that the origin of the world is unique.

The ontological monist believes that both materialism and idealism are ontologically dualistic as long as they consider matter and consciousness to be independent of each other.

Dualists, on the other hand, advocate the coexistence of duality. The idea of dualism was proposed by the Greek philosopher Plato, who believed that there are two worlds, one is the rational world of the soul and the other is the real world of the body.

Only the world of the soul is the "real world", and therefore the world of the senses is only a shadow of the world of the soul. Therefore, in the process of philosophical inquiry into the development of thought, philosophers are often caught in the dilemma of dualism.

Therefore, Husserl attempted to reduce "being" to a series of manifestations of being, trying to get rid of the philosophical dichotomy between "internal" and "external", the dichotomy between the potentially inscrutable "nothingness" and the actual existence of existence, and the dichotomy between finite and infinite.

Schart believes that this is a great advancement in superconscious thinking. This is because "Superconsciousness" is about how to use the superconsciousness of the self to foresee the future, to change the self, and to control destiny.

First of all, the "phenomenological" monism gets rid of the philosophical doctrine that the world has two separate original beings, "internal" consciousness and "external" matter, and emphasizes that matter and spirit exist in equal and fair opposition.

Schart also points out that if the "appearance of being" is understood through analytical study as a skin that covers the "true nature" of the thing that is the object of action or thought, forming a holistic concept, then there is no distinction between "internal" and "external" epidermal beings.

On the other hand, if the "true nature" of the actual existence of the "existence" of this self-existence itself is really the secret of the nature of things, because it is hidden in the "nothingness" of things, which is impractical and inscrutable, within things, then one can only see the "true nature" of things that are not in sight.

People can only use their past memories or similar experiences to conceive specific images of things that are not in front of their eyes, to anticipate and assume their existence, but they can never achieve it.

"Phenomenology also denies the dualism of "appearance" and "essence". The essence, as a principle of the series of domains, is interpreted only as the association of appearances, that is, the essence itself is an appearance.

The existence of the present reveals itself as much as it reveals its existence, its essence. This is what Husserl meant by the intuition of nature as opposed to each other, as dependent on certain conditions, and as changing with certain conditions, such as the opposition of big and small, beauty and ugliness.

Duality refers to the two contradictory properties inherent in the thing itself, i.e., a thing has two opposing properties at the same time. Therefore, the opposition between finite and infinite is not valid either.

Schartre replaces the duality of finite and infinite with the mutuality of finite and infinite. The act of existence of the present is the existence of the apparent. This means that the present is something that is distinguishable and exists independently within itself.

But it does not need to be the basis of physical existence to express the essence itself. It is by using Husserl's phenomenological theory to criticize and refute the wrong statements or actions of philosophical dualism that Scharthe establishes his phenomenological monistic philosophy of existence.

The main point here is Schart's commentary on Husserl, in which Schart replaces dualism with phenomenological monism and advocates the philosophical doctrine that the world has two separate and distinct essences (i.e., matter and spirit).

That is, any religious doctrine or philosophical doctrine that divides the universe into two separate parts, restoring the essence to its original state or shape, that is, the phenomenon of "returning to the thing itself".

By the example of electric current and force, the

dualism between phenomena and existence (where existence is solidity, that is, essence seen through phenomena) is refuted.

The essence as a manifestation is no longer in opposition to existence, but becomes a self-variable that changes as existence changes, so that the phenomenon is itself revealed as it appears, and is an absolute expression of itself.

The phenomenon here is a phenomenon in the purely phenomenological sense, that is, the "thing in itself" in Husserl's "back to the thing in itself", that is, "the existence of consciousness directly from the living experience of existence.

The romantic works of Prussia are a profound reflection on the sentimental time and memory based on art, and a profound observation on love and jealousy.

Under a gray perspective characteristic of Prussia, in which homosexuality plays an important role, he also mixes numerous examples of failure and sadness for a life of emptiness, refuting the dichotomy between the potentially inscrutable "nothingness" and the existence of real activities.

Here the present is a phenomenon in the purely phenomenological sense; that is, the "thing itself" in Husserl's "return to the thing itself," that is, "to the conscious presenter in direct experience.

Schart draws on Proust's example to argue against the potential-activity dichotomy, but in many small ways Proust is dealing with Freudian territory: the "instinctual depth" and "primitive state" that Proust refers to when people are asleep.

It is reminiscent of Freud's "subconscious", a state of mind that lurks beneath the consciousness without being conscious of it and cannot be directly observed by others, that is, something that cannot become conscious at all under normal circumstances, for example, a desire that is repressed deep inside and not realized.

It is the so-called "iceberg theory": the composition of human consciousness is like an iceberg, only a small part of which is exposed to the surface (consciousness), but the vast majority of which is hidden underwater and influences the rest (unconsciousness).

The subconscious, according to Froude, has an active role in exerting pressure and influence on human [personality] and [behavior]. In his exploration of the human mental realm, Froude used the occurrence of any event

The principle of determinism, which holds that things happen for a reason, includes the principle that there are external conditions that determine the occurrence of an event, including decisions made by humans of their own free will. The principle of determinism is that things happen for a reason, not because other events happen.

Various theories of determinism are woven throughout the history of philosophy. The direct opposite of determinism is non-determinism. Determinism is also often contrasted with free will.

Determinists believe that free will is an illusion and that the human will is not free. Seemingly trivial things, such as dreams, slip of the tongue, and penmanship, are determined by underlying causes in the brain, but are merely manifested in a disguised form.

Thus, Froude proposed the hypothesis of unconscious mental states, dividing consciousness into three levels: conscious, subconscious, and unconscious.

He also noted that the memory of everything that

exists in the mind is related to the process of analysis, synthesis, judgment, and reasoning on the basis of representations and concepts, and many other factors in the process of cognitive activity.

Finally, Marcel Proust did write, very unconsciously, about the typical psychoanalytic figure: maternal and paternalistic, sexually deviant, obsessive, depressed, and neurotic.

All this seems to be a complex psychological process of seeking or establishing rules and evidence to support or determine a belief, decision, and action, which strongly corroborates Freud's doctrine.

Marcel Proust, on the other hand, wanted to write about a human experience, not a psychoanalytic novel.

Fortunately, Freud is more understandable to the general public than Einstein, and it seems that this kind of approach to everyone's daily life is more convincing.

Freud was talking about subconsciousness, unconsciousness and consciousness, while Marcel Proust was generally talking about "subconsciousness" which is submerged under the consciousness without consciousness and cannot be directly observed by others.

But there is no Freudian imprint in Marcel Proust's viewpoint and terminology, nor is there any subtle sign that he was aware of Freud's work at the time of writing.

His novel becomes a "series of unconscious novels" because it is, in itself, a "thing in itself" within a "return to the thing in itself," that is, a phenomenon of forgetting and remembering "in direct experience of the conscious presenter," as he later explained in an interview with Burroughs.

He argues that there is a practical function of reality or the future in our logical thinking, which is easily lost from our minds, but can be called back by memories controlled by subjective consciousness.

Sometimes we can "return to the "thing itself" in the "thing itself", i.e., "to the conscious presenter in direct experience", by chance in memory. However, we do inadvertently discover that there are other things that we cannot call "forgotten" because we never remember them.

Because we never remember them, they are also powerfully complete through the tendency to see things in a way that is not based on the subject's own needs, it is the memory of the viewpoint, experience, consciousness, spirit, feelings, desires or beliefs that the individual can have.

And up to the surface, there is no repression in them, nor is there a comprehensive awareness and recognition of the individual's own physical and mental state and changes in people, events, and things in the environment, using sensory, perceptual, thinking, and memory mental activities, subconscious repressive forces, moral norms, or analytic methods of curing them, and sexual impulses do not play any role.

1. Stream of consciousness

Marcel Proust attempts to capture the infinite richness of emotions, and to capture the infinite richness of actual feelings generated by the individual's mental activities such as sensation, perception, thought, and memory, the integrated awareness and recognition of changes in one's physical and mental state and in people, events, and objects in the environment, and the recognition of various feelings, memories, and ideas.

In order to record all the feelings of a person's life together. The idea of the novel must have a certain span of time and space. Marcel Proust does not record the feelings of "the presenters of consciousness in direct experience" at one time and one place, but "the gathering of feelings at different moments". That is to say, the mental activity of many different moments makes up the whole life of a person.

It is not possible to transform the actual time into mental time "for the conscious being in direct experience" through rational thought activity, or even through conscious recollection under the constraints of reason.

Marcel Proust himself once said: "Our efforts to recall the past are always in vain, and we rack our brains to no avail. It is hidden beyond the mind, beyond the power of the mind, in something we do not expect (in the feeling it gives us).

The thing that is "present to the consciousness in direct experience" is something that we encounter by chance before we die, or we don't encounter it even after we die, and that was many years ago.

In addition to some scenes and episodes related to my going to bed, the first part of Marcel Plutus's long novel "Remembering the Years". Gombray: a sleepless night, a never-ending memory.

A time when carriages and automobiles alternated generations, a time before the World War, a time when I longed most for my mother's visits before bedtime, and a world of inaccessible adults in the attic, when the rest of my life was long gone, gone.

"If someone were to ask me, I would probably say that there are other things, other lights, in Gombray. But what I remember is only the result of intentional recollection, aided by wisdom, and what I recall intentionally is not as audible as the past".

Marcel Proulx recovers the "sound" of the past through the unconscious association of the "Gaelic house - St. Mark's Church in Venice" from the senses of taste, smell, hearing, and touch.

This kind of unconscious association fully revives the past in the inner world, and the famous "Little Madeleine's Dim Sum" is a typical example of this kind of unconscious association.

"The spoonful of tea with its fine crumbs touched my palate, and immediately I was shaken, and I noticed that something extraordinary had happened to me". "The smell of tea awakened the truth in my heart. It was this cup of tea and this piece of refreshment that opened the gate of his association with the life of "Gombray".

As a result, the streets, alleys, gardens and past events of Gombray emerged from the touch of tea cups and lips, giving rise to a scene of infinite memories, reviving the long-lost memories of time and space.

Madeleine's treats become a "catalyst", an elixir, which unexpectedly induces and fractures countless living but long-buried contents of life, reminding the protagonist of the beauty of the hawthorn hedge in his hometown;

The protagonist walks into the courtyard of the Gellhunt House, and as he steps on the uneven stones, he recalls a past event in Venice; then he walks into the small living room and hears the sound of a spoon hitting a basin, which reminds him of the brakes of a train; soon, he wipes his mouth with a napkin, which reminds him of a peacock's tail, with its blue and green feathers.

This is how the whole work is composed of unconscious associations, one thing inducing another, one ring leading to another, forming a very specific stream of the author's memories.

This is the reason why Reminiscence of Years Like Water is classified as a stream-of-consciousness novel. In this unconscious association, in the stream of memories, the author resurrects the actual time he has experienced by forming a mental time.

The actual time sequence is linear, while the mental time is formed by unconscious association, where one thing induces another thing.

Presented as a montage believe that time and space relative to the long shot film expression method, it through a series of different locations, different distances, different angles, different methods of shooting, a number of short shots down the combination, use to edit into a film with a plot.

From there, the pacing and narrative style that influenced Proust's "Memories of a Time Like a River" uses this reversal of chronology.

For example, Swan's love was originally something the protagonist heard about and happened before he was born, but the story was told before, while the protagonist grew up and fell in love with Swan's family and with Swan's daughter, and only later learned about Swan, but put it in the back.

This kind of cross-talk reflects the illogical memory, that is, the subconscious memory that Sartre wants to express. In Marcel Proulx's view, remembering a person is a matter of memory. For Marcel Plutus, remembering a person or an event is not something that can be done all at once.

It is a crossover, a reversal, a confusion, and in the articulation between practices, no necessary connection can be seen, but as a whole, this does not prevent the novel from being an organic whole.

Thus, Schart's turn of phrase asks, "Is the restoration of all objects or phenomena that exist objectively in nature to their original condition or shape, to their various manifestations, the elimination of all dualism? In fact, it is the dualism of replacing abstract or inanimate things with concrete examples as finite and infinite.

This statement means that we can see the phenomenon of an infinite sequence of gradual appearances. It is equivalent to looking at a finite cube itself from the perspective of an infinite phenomenon.

The objectivity of the theory of reality replaces the

reality of replacing abstract or inanimate things with concrete examples. To this extent we should grasp the phenomenon of "red" with the impression of red in order to get beyond things. That is, to grasp the transcendent finite phenomenon with an infinite perspective. It is the infinity of the finite.

Then the infinitude of Marcel Plutus' potential is in fact transcendence (finitude) and infinity. But this finitude appears as existence (internal) and infinity appears as manifestation (external).

It is back to the dualism of internal and external opposition. To resolve this dualism, the first conclusion of the theory of phenomena must be that the manifestation does not return to being in the same way that the Kantian phenomenon returns to the essence.

If the manifest is not opposed to being, the manifest being/(the phenomenal being) becomes the "thing itself" in the "return to the thing itself," i.e., "the one who presents itself to consciousness in direct experience.

It is equivalent to the angle of the light projected from the two ends of the object (top, bottom, or left, right) at the center of the human eye, looking at a finite cube itself, and getting an infinite phenomenon. The objectivity of the phenomenal theory replaces the actuality of things.

To this extent, we want to obtain the transcendental thing-philosophy. Philosophical theories other than phenomenology may also deal with the 'transcendent' in a second sense, though.

For example, the empirical philosophy represented by Hume tries to exclude all transcendence, but Hume's theory uses transcendental concepts that are not found in phenomena, such as "habit", "human nature", "sensation", "stimulus", etc.

Thus, a fundamental starting point of Kantian philosophical theory is the idea that the ability to transform sensual intuitions (habitual experiences) into knowledge - purely intellectual concepts (i.e., "domains")

and the ability to put into practice ideas outside of knowledge (e.g., God, mind, freedom) - purely rational concepts - are functions of reason that are inherent in human beings and without which we cannot understand the world.

And Husserl still calls it psychology, which belongs to the transcendental thing to be excluded as the goal when acting or thinking. That is, the mastery of phenomena beyond the finite with an infinitely macroscopic field of thought or knowledge.

It is the infinity of the finite. Then the inexhaustibility in the Prussian potential is in fact transcendence (finitude) and infinity.

But this finitude appears as existence (internal) and infinity appears as manifestation (external). The dichotomy of interior and exterior returns.

Therefore, in order to understand the relativity of this dualism, it is necessary to understand Kant's epistemology and ethics, which argue for knowledge and morality respectively, and which critique and assimilate English empiricism (Hume, Berkeley) and European rationalism (mainly the rational tradition of Wolf-Leibniz).

He had a profound influence on German idealism (Fichte and Hegel) and Romanticism. Epistemology and ethics constitute the two major parts of Kant's philosophy, the former on the "phenomenal world" and the latter on "freedom of will" being diametrically opposed and dichotomous.

The intermediation of the two becomes the concluding

idea of Kant's "critical philosophy," the communication and unification of nature and freedom, in the Critique of Critical Force. Thus, the first conclusion of the theory of phenomena must be that "the manifestation does not return to being in the same way that Kant's phenomena return to the essence.

If the manifestation is not opposed to existence, the existence of the manifestation [the existence of the phenomenon] becomes the only person or thing to which the study of action or thought, as all human activity, can be directed.

2. The phenomena of existence and the existence of phenomena

In this section, Scharthe distinguishes between the phenomena of existence and the existence of phenomena at the ontological level. Schatte first asks "Is the phenomenon of existence the same as the existence of the phenomenon", that is, the revealed existence, and the existence of the manifest being.

Is the existence of the revealed being the same as the existence of the manifested being? Sartre draws on Husserl's "intrinsic intuition" and Heidegger's "human reality is existential/ontological". But here Sartre asks, "Is the present of being to the present of being?

Sartre delineates ontology, the study of philosophical concepts such as being, existence, becoming, and reality. It includes questions such as how to classify entities into basic categories and which entities exist at the most basic level.

Sometimes ontology is called existentialism, and it belongs to the main branch of philosophy known as metaphysics. At the basic level, there is a distinction between "phenomena of being" and "phenomenal being. Scharthes first asked whether "the phenomenon of existence" and "the existence of phenomena" are one and the same.

In other words, is the "existence" that reveals the existence of the self and the "existence" of the manifested being the same? Schart is drawing on Husserl's "Intuition of essence" and Heidegger's "The actual existence of the 'existence' of the human ego-being itself is the theory of existence (ontology).

However, Schart is asking paradoxical questions verbally and explaining in detail why he verbally asks the question, "Is the essence of "being" the same as that of "the phenomenon that can exist".

(2) Apparent ontology.

From this perspective, Sartre makes three inferences about the phenomenon itself The phenomenon of the essence, the manifestation, is not in the object, but in the meaning of the object.

The essence is not hidden behind the object; it is the intentional reference of intentionality. Existence is a phenomenon of existence only when it is revealed in a direct way.

Existence is not equal to nature, to character. Being = existence itself. The phenomenon of being needs to be revealed in a direct way. At this point the phenomenon of existence does not cover up or reveal existence.

Existence is some kind of lack, it is some kind of condition to be revealed, waiting to be revealed. For example, you have a sticky piece of clay that can be made into anything. The prerequisite for making anything is that you have a piece of clay that is sticky.

So the sticky clay is waiting to be pointed out or elucidated as the nature of all objects or phenomena that exist objectively in nature.

From this it follows that [the existence of phenomena] and [the existence of phenomena] are two different things.

First, [the existence of phenomena] requires the supernatural nature of existence, which does not mean that existence is opposed to phenomena or manifestation is a return [to the phenomena of existence].

Secondly, [the phenomenon of being] can only manifest itself in an immediate way, like a flash of enlightenment.

Scharthe also points out here that if [the existence of phenomena] is defined as an ontological structure, then it is only a changed form of Beckley's "to exist is to be perceived", while the content and essence remain the same.

To exist is to be perceived

Berkeley's most familiar philosophical proposition is that to exist is to be perceived (esse est percipi, to be is to be perceived). Berkeley believed that the stars in the sky, the mountains and rivers on earth, and all the objects in the universe have no independent existence outside the human mind.

Their existence lies in the individual's use of sensation, perception, thinking, memory, and other mental activities, and the comprehensive awareness and knowledge of his own physical and mental state and the changes of people, things, and objects in the environment.

Therefore, if they are not really known to man and do not really exist in the spirit of all other created beings, then they do not exist at all.

In A Treatise Concerning the Principles of Human Knowledge, Berkeley explains his own understanding of natural and social phenomena in accordance with existing empirical knowledge, experience, facts, laws, laws, laws of knowledge, and tested hypotheses.

The proposition of logical inferential summation of natural and social phenomena in accordance with existing empirical knowledge, experience, facts, laws, perceptions, and tested hypotheses, by means of generalization and deductive reasoning.

He says that when people say that an objective object or phenomenon exists in nature, it is only that thing that we perceive, and to say that a thing exists is to say that we have seen it, heard it, smelled it, and used our senses and perceptions.

The meaning of existence lies in the fact that we have seen it, heard it, smelled it, and used our senses, perception, thinking, memory and other mental activities to perceive and recognize the changes in our state of mind and body and in the environment of people, things and objects.

If something can exist without mental perception, it cannot be analyzed in detail and understood from a certain cognitive point of view.

Berkeley, in his work "Principles of Human Knowledge," admits that his theory is flawed if one can conceive of something that can exist even without being perceived.

In fact, Berkeley wanted to point out that even if one can imagine the brain's overall view and understanding of the external world without external stimuli acting on the senses, organizing and interpreting the external sensory information for us, something will still exist.

There is still something that exists, in fact, there is still a direct reflection of that something in the brain through the sense organs in the surface characteristics of objective things.

"Existence is perceived" has aroused the refutation of many later philosophers, including the English philosopher G. E. Moore, who argued that how can things exist only when they are perceived?

He said that there exists a person here (referring to Moore himself), and this person raised his hand to prove the existence of the external world. Another philosopher, Jean-Paul Sartre, thinks that Berkeley confuses perception with the existence of perceptions, and that perceptions and perceptions are two different things, but they are related to each other.

2. The existence of God

Some people questioned how the existence of the external world and all the objects or phenomena that exist objectively in nature could be the result of the brain's view and understanding of the external world as a whole when stimulated by the external senses, organizing and interpreting the sensory information of the external world for us.

If I talk to my father and my father is there, but once I turn around and leave, does it mean that my father does not exist? If I look back at my father's back and my father is there again, does it mean that my father has disappeared from the world for a short time?

Of course Berkeley would think that Father still exists, because even though Father has left the sight of my consciousness in the series of processes of perception, sensation, attention, and awareness of internal and external information, I can still perceive their existence with my hands and body;

There are things that we cannot touch with our hands and bodies, such as distant objects or landscapes, but we can perceive their existence with our eyes; there are things that we cannot see with our eyes or touch with our hands and bodies, such as songs, music, words, etc., but we can perceive their existence with our ears.

Likewise, there are other people who can see my father, so what my father perceives when he leaves me is done by the action of his own mind. The human mind interprets and deciphers the stimulus signals and produces various sensations in the mind.

This change of sensation still exists, which is a subjective reflection of the human mind to external things.

Of course, other creatures also have similar perception processes, but humans have not yet understood and deciphered them.

And even if no one in the world perceives the Father, the Father still exists, but what is that all about?

Berkeley says that the external world exists, that the stars in the sky, the mountains on earth and Father exist, even when humans and other created beings do not perceive them, and no doubt when we open our eyes in the morning, we cannot prevent things from coming to us.

But that does not mean that Berkeley has changed his position and admitted that his theory is wrong. He argued that things exist outside of human perception because they are perceived by a higher mind, and that higher mind is, in Berkeley's view, God.

Berkeley further used his own proposition to prove the existence of God. He argued that we can clearly perceive that the external world is so regular and orderly that we can infer the existence of a higher spirit, something that is distinguishable from and within itself, and exists independently.

But it need not be a physical being. In particular,

abstraction is usually regarded as the entity that guarantees the existence of all objects or phenomena that exist objectively in nature, and that spiritual entity is nothing other than God.

3. Matter does not exist at all

The philosophical interest of Berkeley's holistic, fundamental and critical inquiry into the real world and man is to prove the existence of God and to oppose the existence of matter, which is clearly reflected in his philosophical proposition "to exist is to be perceived".

As is well known, Locke introduced the distinction between first and second nature. The first nature refers to the inherent properties of objects that do not depend on human perception, and Locke considers that extension, shape, and motion belong to the first nature of things. Berkeley points out that first nature is only an idea that exists in the mind.

According to common sense, when I say that the trash can has a certain social significance or influence, the commonness and inevitability of the occurrence, that is, the affirmation of the objective existence of the trash can.

And it has the commonness and inevitability of all the

objects or phenomena that exist objectively in nature, and then I see it, so I have the concept of the garbage can in my mind, and then I can say that my extended concept is similar to the extension of the garbage can.

But Berkeley thinks that a concept can only be similar to a concept, and it is impossible to compare a concept with something of a different quality. Moreover, when I say that the garbage can is an objective existence, the garbage can is already a product of the mind and no longer exists independently.

Therefore, it is impossible for the archetypes of observation, memory, and concepts to exist without the mind, whether it is the observation of a particular object or a righteousness. In this way, Berkeley argues that Locke's second nature, like the first nature, is the mind's concept of observing and remembering particular objects or doctrines.

One believes in the reality of material entities, which are philosophically defined as the necessary essence of things, without which things cannot be the basis of things, and which support the many different properties of things, especially the first nature of matter.

Berkeley considers the word "support" as a figurative term, a literary expression of the metaphor of the house. In fact, one has no idea how material entities "support" the nature of things. Even if one can conceive of the concrete existence of material facts and conditions in front of one's eyes, this is a process of human cognition.

It is an expression of the self-cognitive consciousness to abstract and generalize the common intrinsic characteristics of the perceived things from perceptual awareness to rational awareness, and to be able to explain, in this way, something that is distinguishable and exists independently within itself.

But it does not need to be a physical being. In particular, abstraction is often seen as a "support relationship" between entities and first nature, but Berkeley sees these material entities as ultimately just products of the mind's observational thought and memory of particular objects or meanings.

(3) The Phenomenon of Ontology and the Divide of Being

This reminds me of my previous experience with the weak meditation practice. It is said that a Buddhist Zen master did not use words to teach the Way, but had a different method, namely, enlightenment through observation of awareness and personal realization.

This Master Yuan believed that if one does not thoroughly understand the pure truth of goodness and the essence of emptiness, one cannot interpret the word "Buddha" (the enlightened one of truth) even if there are many words. This Zen master is the famous Dharma Master.

The story takes place at the time when he was about to enter into extinction and pass away, he gave a difficult question to his disciples.

It is not that Dharma Master deliberately made things difficult for them, but sometimes Dharma Master needed to verify whether his disciples had understood the flesh and blood or the essence of the teachings he had passed on, and whether he could really leave with peace of mind.

One day, he called all four of his disciples, who were usually close to him, to his meditation room and said, "I can feel that my days are numbered and that I will soon pass away. I have nothing to regret in this world, and now that I am going to the Western Pure Land, I have no fear.

However, I have called you here today because I am

worried whether you will get the essence of my righteousness. First, tell me what you have learned from your recent practice, and what you have achieved or demonstrated about the pure spiritual essence of the highest good.

The four disciples lined up in front of Master Dharma's bed and waited in silence for the teacher's lecture, none of them speaking first. Looking at their faces, the four disciples lowered their heads one by one, and no one made a sound.

When they saw this, Dharma Master smiled slightly and said, "It's nothing, I'm also full of righteousness, no need to worry and worry. First disciple, tell us what you have gained from your recent practice.

Hearing the teacher's name, the eldest disciple, Dao Dai, did not shrink from the back of the room, but came forward and said, "In the face of the text, we should give and take, we should not give up everything, nor should we listen to everything. We should use the text as a tool to facilitate our righteousness.

After hearing this, Dharma Master did not say anything, but moved his gaze to his second disciple,

Master Nyi, and said, "Second disciple, tell us about it.

The second disciple said, "I have a different opinion from that of the Master. I think that words are not desirable, and it is up to us to decide whether words are desirable or undesirable.

Just as an enlightened person can no longer see the troubles of the world, so too can words be invisible. After hearing this, the face of Dharma changed a little, but the change was not so great that it could not be noticed unless one looked closely.

Dharma Master turned his head and looked at his third disciple, Dao Yu, and said, "Do you think what they said before was right?

Third disciple Dao Yu shook his head and said, "No, I think the world is like mountains, rivers, wind, forest, water and fire, all are visible but inaudible. If you want to see something, you can't just explain it through the surface or words, you have to rely on your inner feelings.

After listening to the third disciple, Dao Yu, the Dharma Master smiled in a rare way and went on to ask the fourth disciple. He said, "What do you think about the degree of realization of the pure spiritual essence of the supreme goodness, as described by the three brothers before you, or the state of manifestation?

The fourth disciple, Hui Ke, did not say anything, but just performed the most respectful salute on the spot, got up, and remained standing there.

When he saw how the fourth disciple behaved, Dharma finally laughed and said, "Yes, you have got the essence of me in terms of how far you want to go or how you want to behave in terms of realizing the pure spiritual essence of goodness.

The other three brothers have only obtained the skin, flesh and bones. From here we can feel that Dharma's meditation was difficult and difficult to hear.

The last disciple of Dharma was the second ancestor of Zen, Master Hui Ke.

For a person who understands the true meaning of life, the "existence" of one's own life and the enlightenment of one's mind about the things that are the goal of one's actions or thoughts is the root of everything.

Section 3: The Presence of the Self and Perception before Reflection

We introduce Husserl's Intentionality, which is the ability of the mind to represent or present things, attributes or states. Simply put, much mental activity is about the external world, and intentionality is this "about".

Brentano defines it as one of the characteristics of a "psychical phenomena", thus distinguishing it from a "physical phenomenon". He uses expressions such as "connectedness to content," "pointing to objects," or "intrinsic objectivity.

All mental activities such as sensation, perception, thinking, memory, etc., the individual's comprehensive awareness and knowledge of his own physical and mental state and the changes of people, things and objects in the environment are the special functions and activities of the human brain of something, which is the reflection of the objective world unique to human beings.

In philosophy, when both consciousness and thought are used to refer to the reflection of the human brain on objective things, they can be used in common, but the scope of consciousness is broader, including perceptual and rational cognition, and also includes human emotion, will, conscience and so on.

Consciousness itself is transparent, empty and contentless. Then consciousness, as pure consciousness itself, is a positional consciousness of the world, so it must eliminate things in order to restore the position of consciousness to the world.

In order to restore consciousness to its pure state of positional consciousness of the world. But here Schart pointed out that positional consciousness itself is unconscious. That is, it cannot be aware of the present, it is a consciousness that points to a certain location.

Therefore, it is the intention, method, route, plan and order that command the perceptual organization to actively collect knowledge about the properties and regulations of the object, and prepare the thought for the knowledge, which is generated by the analysis of the knowledge acquired by the perceptual organization.

The role and influence of the object on the subject is the direct reason that causes the subject's thinking organization to analyze and process the knowledge acquired by the perceptual organization, thus generating the awareness of knowing

It is also the thinking activity of the human brain to reflect the characteristics and connections of objective things and to reveal the meaning and role of things to people.

In a broad sense, cognition includes all cognitive activities of human beings, i.e. perception, memory, thinking, imagination, understanding of language, and generation of other mental phenomena, and the sufficient necessary conditions for the unification are

It is the human cognitive ability of the environment and self, as well as the degree of cognitive clarity, to their own consciousness of this consciousness, what exactly does this mean, in fact, can use their own words, words or other symbols, the known facts and principles, principles into the interpretation of.

Consciousness is the reflection of the human mind to the objective material world, and all human activities are directed to people or things, all objects or phenomena that exist objectively in the natural world, that is, individuals use sensory, perceptual, thinking, memory and other mental activities, their own physical and mental state and the environment of people, things and changes in the process of comprehensive awareness and understanding.

(1) Consciousness must be aware of something.

Individuals use sensory, perceptual, thinking, memory and other mental activities to become aware of and recognize their own physical and mental states and changes in people, events and things in the environment, and must be aware of the environment and self-cognition.

The cognitive activities, including sensation, perception, memory, imagination, and thinking, are more conscious, rational, and effective due to the existence of self-consciousness. Metacognition is the awareness of cognitive processes.

In order to avoid such a way, the individual uses mental activities such as sensation, perception, thinking, memory, etc., to become aware of and recognize the changes in his physical and mental state and in the environment of people, things and objects, which can only become a direct relationship between the self and the ego, not a cognitive relationship.

Schart is in fact correcting Descartes' mistake of

introducing into the self-consciousness the objective things that are the target of action or thought, the process of the human brain reflecting the activities of objective things in a certain space and time.

It is actually a copy of Descartes' "I think, therefore I am/I suspect, therefore I am". In this step, Sartre introduces the idea that people know themselves and are concerned about themselves, especially about how others perceive their appearance, or their behavior.

When a person is aware that he or she is being watched or observed, there may be an unpleasant sense of self, a sense that "everyone is watching" him or her. Some people may have a stronger sense of self for transformation than others.

In order to solve the dilemma of being right, Schart said that all positional awareness of objects is at the same time non-positional awareness of oneself.

(Non-positional consciousness of oneself is, in fact, the aforementioned reflective consciousness.) Non-reflective consciousness makes it possible to reflect on one's ability to recognize the environment and the self, as well as on the clarity of cognition.

But this reflective consciousness should not be seen as a new kind of consciousness, but as the only way of being that makes consciousness of something possible.

Schart thought that before Descartes' I-thought, there should be a kind of primordial consciousness (non-reflective consciousness), which is unconscious, conscious of the location of objects, transcendent, and a prerequisite for reflection.

This kind of individual's integrated awareness of one's own physical and mental state and the changes of people, events and objects in the environment by using sensation, perception, thinking, memory and other mental activities exists in itself, and is first, similar to Husserl's pure consciousness. The existence of all consciousness is the existence of consciousness as a "being".

This section focuses on the above reflection, and the noun interpretation of the reflected individual's integrated awareness and recognition of his or her state of mind and body and the changes in people, events, and objects in the environment, using mental activities such as sensation, perception, thinking, and memory, and the two selves of Je and Moi in The Transcendence of the Self.

Non-reflection and reflection are mutually possible. Non-reflection provides the material for reflection, and reflection provides the meaning for non-reflection by listing the basic attributes of an event or an object to describe or regulate a word or a concept.

Kant's "I-thought" is a quantitative indicator of the probability of something occurring, contained in something and predicting a trend, which is an objective argument rather than a condition for subjective verification.

Descartes' and Husserl's "I think" is a real situation of things, which can refer to events that have happened in the past, or to statements that are verified and neutral, and in science, to the observation of provable concepts.

People have talked about the need for "I think" facts, and I think this is correct. However, "I think" inevitably has a name. In "I think," there is an "I" (Je) who thinks.

The "I" that we achieve here is extremely pure, and a kind of "self-reflection" starts from the "I-thought". It can refer to events that have occurred in the past, or it can refer to verified and neutral statements, which in science means that the concept of provability is.

Whenever we grasp our thoughts either by direct intuition or through the intuition of memory, we grasp an "I" (Je), which is the "I" of the grasped thought and which manifests itself as an "I" beyond this thought and all other thoughts.

For example, if I were to recall the overall view and understanding of the outside world in my mind when the external stimuli were applied to my senses on the light train the year before, as a landscape for organizing and interpreting the sensory information of the outside world.

Then I could remember the landscape clearly and vividly, but I could also recall the landscape I had seen. This is what Husserl calls the possibility of reflection in memory in "The Internal Consciousness of Time".

In other words, I am always able to carry out any kind of memory activity in a personal way, and the "I" appears immediately. This is the reality of what affirms the right of Conde, which can refer to events that have occurred in the past, or to statements that are verified and neutral, or in science, to the assurance of provable concepts.

In this way, among all my cognitive abilities and cognitive clarity of the environment and the self, it seems that there is no such thing as an integrated awareness and knowledge of one's physical and mental state and changes in people, events, and things in the environment that I do not grasp for the purpose of "I", using mental activities such as sensation, perception, thinking, and memory.

However, it should be noted that all descriptions use various rhetorical devices to visualize and elaborate on things. The rhetorical devices included are metaphor, simile, exaggeration, double entendre, prose, etc., which can describe people and things.

The descriptions can make people or things vivid and concrete, and give people a clear feeling of the author who has "I think" considers "I think" to be a kind of reflective operation.

In other words, it is a second-level activity. This kind of "I-thought" is a kind of mental activity in which the individual uses his or her senses, perceptions, thoughts, and memories.

The ability to recognize the environment and the self, as well as the clarity of cognition, is the result of the individual's awareness of his or her own physical and mental state and of changes in people, events, and objects in the environment.

The cognitive activities of people, such as sensation, perception, memory, imagination, and thinking, are made more conscious, rational, and effective by the existence of self-consciousness.

Metacognition is the cognitive consciousness of the cognitive process as the object of the subject's reflection of various things and phenomena in the objective world, their relationships, processes, nature and laws.

We all agree that the belief in "I think" is absolute because, as Husserl says, there is an indissoluble unity between the reflective and the reflected consciousness (so that the reflected consciousness cannot exist without the reflected consciousness).

Therefore, we have no lack of opportunities to face the synthesis of these two kinds of individual mental activities, such as sensing, perceiving, thinking, and remembering, and the synthesis of awareness and knowledge of changes in one's physical and mental state and in the environment of people, events, and things.

Among them is the cognitive ability and the clarity of

the cognition of the other person, the environment, and the self. In this way, the basic principle of phenomenology that "any consciousness is a consciousness of something" is protected.

However, when I realize that "I think" is my reflection on the cognitive ability of the environment and the self, and the clarity of cognition itself, it is not considered as an action or reflection when the objective exists in all objects or phenomena of nature as a goal. What my reflective consciousness affirms.

It includes the integrated awareness and recognition of the individual's own physical and mental state and the changes of people, things, and objects in the environment by using mental activities such as sensation, perception, thinking, and memory.

It is because the cognitive ability of my reflection on the environment and the self and the clarity of cognition are the comprehensive awareness and recognition of my own personal state of mind and body and the changes of people, things, and objects in the environment by using sensory, perceptual, thinking, and memory mental activities, which is non-locative awareness.

The cognitive ability of my reflection on the environment and the self, as well as the clarity of cognition, is only directed at the individual's integrated awareness and recognition of his or her physical and mental states and changes in people, events, and things in the environment by using sensory, perceptual, thinking, and memory activities.

The reflected consciousness itself was not a positional consciousness before it became the reflected one. The consciousness that is called "I-thought" is not the consciousness of thought. Or rather, consciousness is not its own thought, which is raised by the act of the subject.

We are forced to ask ourselves whether the thinking "I" (Je) has the same cognitive capacity and clarity of awareness of the environment and the self as the two overlapping human beings.

Or should there be a "I" that is reflecting on the individual's use of sensory, perceptual, thinking, and memory mental activities to become aware of and recognize changes in his or her physical and mental state and in people, events, and objects in the environment?

Any reflective individual who uses mental activities

such as sensation, perception, thinking, and memory to become aware of and recognize his or her own state of mind and body and changes in people, events, and things in the environment is actually unreflected in himself or herself, and a new act is necessary to present it.

In addition, there is no infinite regression here, because an individual uses mental activities such as sensation, perception, thinking, and memory to become aware of and recognize his or her own physical and mental state and the changes of people, events, and things in the environment.

There is no need to reflect on one's cognitive ability and clarity of cognition of the environment and the self, because an individual uses sensory, perceptual, thinking, and memory mental activities to perceive and recognize his or her own physical and mental states and changes in people, events, and things in the environment. Only it does not manifest itself as a target when acting or thinking.

But is it not the activity of going back and rethinking the past and learning from it that gives birth to the "Moi" in the reflective person's ability to recognize the environment and the self and the clarity of that recognition?

Thus, it is explained that any thought that is grasped intuitively possesses an "I" (Je) that does not fall into the difficulties we have pointed out earlier. Husserl was the first to admit that an unreflected thought must undergo a radical change in the process of becoming reflected.

But should this change be limited to the failure of a "simple, unpretentious, straightforward character, which at the same time can be interpreted as an uncomplicated, insubstantial way of thinking, often referring to proposals that are not mature enough"?

Isn't the most important change the manifestation of the "I"? Obviously, one should have recourse to the existence of entities, or explicit, rather than abstract, generalized experiences, and this kind of knowledge or skill derived from personal practice may seem impossible.

For this type of experience is reflective, that is, possessing an "I" (Je), according to a precise and brief description of the intrinsic characteristics of a thing, or of the content and extension of a concept.

However, any unreflected individual using mental activities such as sensation, perception, thinking, and memory, the integrated awareness and recognition of his or her own physical and mental state and changes in people, events, and objects in the environment, because it is the person's ability to recognize the environment and the self as well as the clarity of cognition that is not the subject of the self

It is a comprehensive awareness and recognition of the state of one's body and mind and the changes of people, events and things in the environment by using mental activities such as sensation, perception, thinking and memory.

In order to do so, it is only necessary to try to reconstruct the complete moment in which this unreflected individual's integrated awareness and knowledge (by definition, what is always possible) of his or her state of mind and body and of changes in people, events, and things in the environment, using sensory, perceptual, thinking, and memory mental activities, appears.

For example, if I have just closed my eyes and crossed my legs, with my hands in a certain position, and I have broken my delusion, I have to try to remember the environment in which I sat in meditation, my posture, how

I closed my eyes and crossed my legs, and to reflect on whether or not I have reached the realm of my expectations by breaking my delusion.

Therefore, what I want to revive is not only these external details, but some kind of heavy and unreflected mental activity of the individual using sensation, perception, thinking, memory, etc., and a comprehensive awareness and recognition of the state of one's body and mind and the changes of people, things and objects in the environment.

It is only through the cognitive ability and the clarity of cognition of the environment and the self that the target of various actions or thoughts can be recognized by the external stimuli acting on the senses.

Only when the external stimulus acts on the senses, the brain's overall view and understanding of the external world, for us to organize and interpret the external sensory information.

Because these actions or thinking as the target things, always with the individual use of sensory, perceptual, thinking, memory and other mental activities, their own physical and mental state and the environment of people, things and changes in the comprehensive awareness and understanding.

This comprehensive awareness and recognition of one's own physical and mental state and changes in people, events, and things in the environment by using sensory, perceptual, thinking, and memory mental activities should not be set as the object of my reflection, but rather the opposite.

I should focus my attention on the thing that is the object of the re-evoked action or reflection, but in some kind of complicity with it, and in a non-locative way.

In the process of inventorying the cognitive ability of the person to the environment and the self and the clarity of cognitive content, the individual is not allowed to use such mental activities as feeling, perception, thinking, memory, etc.

I do not let this kind of individual use of sensory, perceptual, thinking, memory and other mental activities, comprehensive awareness and recognition of my own physical and mental state and changes in people, things and objects in the environment, disappear from my sight. The result is beyond doubt.

When I read (books, newspapers, files, etc.) and comprehend their contents, there is a comprehensive awareness and understanding of the book, of the personal use of sensation, perception, thinking, memory, and other mental activities of the protagonists of the novel, of their own physical and mental states and of changes in people, events, and things in the environment.

The "I" (Je) does not reside in this kind of consciousness. This kind of individual uses sensory, perceptual, thinking, memory and other mental activities to become aware of and recognize their own physical and mental state and the changes of people, things and objects in the environment, and is the ability to recognize the environment and the self as the target of action or thought.

It is the sum of senses, thoughts and other mental processes, and the degree of clarity of knowledge and understanding of things, which is the non-locality awareness of oneself.

I can now turn these non-positive grasps into things that are the target of action or reflection on a topic and declare that there is no "I" (Je) in the unreflected individual's integrated awareness and knowledge of his or her physical and mental state and the changes in people, events, and things in the environment, using sensory, perceptual, thinking, and memory mental activities.

Then Schatt began to argue that we cannot define happiness in terms of reflective happiness, "i.e., when someone asks you if you are happy now and you think about it, I am happy now.

"Nor can we use the idea of a happy feeling first, and then happiness comes in.

The individual uses mental activities such as sensation, perception, reflection, memory, etc., to perceive and realize that he or she is happy in his or her state of mind and body and in the environment. This is still in the duality of reflecting and being reflected. In fact, it is an inseparable existence, and this whole body exists for the sake of reality.

In Being and Nothingness, Sartre uses the ontological dimension of the individual's mental activities of sensation, perception, thought, memory, etc. to

In Being and Nothingness, Sartre explores in greater depth the existence of the individual who uses mental activities such as sensation, perception, thought, and memory to perceive and recognize changes in his or her own physical and mental states and in the environment of people, things, and objects.

The relationship between the non-reflective individual's integrated awareness and knowledge of his or her physical and mental states and the changes in people, events, and objects in the environment and reflective consciousness is further analyzed, thus providing new explanatory ideas for understanding the problem.

At the beginning of his holistic, fundamental, and critical inquiry into the real world and human beings, Schart insists on the fundamental point that the mere ability of human beings to recognize the environment and the self, as well as the clarity of cognition, is the result of the non-reflective individual's use of mental activities such as sensation, perception, thinking, and memory.

This is because a pure person's cognitive ability and clarity of cognition of the environment and self does not set itself as its object.

The ability of pure human beings to recognize the environment and the self, as well as the clarity of cognition (non-reflective consciousness), is in fact a prerequisite for the possibility of reflective consciousness. However, Scharthe at that time put forward the idea that

The cognitive ability and clarity of cognition of the environment and the self is the individual's comprehensive awareness and recognition of his or her physical and mental state and the changes of people, things, and objects in the environment by using mental activities such as sensation, perception, thinking, and memory.

He pointed out that consciousness (consciousness is the awareness of oneself) is the integrated awareness and knowledge of one's own state of mind and body and the changes of people, things, and objects in the environment by using mental activities such as sensation, perception, thinking, and memory.

The degree of cognitive ability and clarity of cognition of the environment and the self is completely different from that of the reflective person because the individual uses mental activities such as sensation, perception, thinking, and memory to

The cognitive ability and the clarity of cognition of the environment and the self as a person who uses mental activities such as sensing, perceiving, thinking, and remembering to become aware of and recognize the changes in his or her physical and mental state and people, things, and objects in the environment.

But the problem is that if we insist on the definition of intentionality (which is what Sartre always insists on), the individual uses his senses, perceptions, and memory.

If we stick to the definition of intentionality (which is what Sartre always insists on), we say that an individual uses mental activities such as sensation, perception, thinking, memory, etc., to become aware of and recognize his or her own physical and mental state and the changes of people, things and objects in the environment.

In this case, how is the cognitive ability and clarity of cognition of the environment and the self the consciousness of the self?

In other words, when Schart said that the cognitive ability and clarity of cognition of the environment and the self is the ability of the individual to use sensory, perceptual, thinking, memory and other mental activities to

What does it mean when one uses mental activities

such as feeling, perception, thinking, and memory to become aware of one's state of mind and body and the changes in people, events, and objects in the environment? "Consciousness" - John Hiller explains it in layman's terms as.

"the state of perception, sensation, or awareness that continues during the day after awakening from a dreamless sleep, unless one falls asleep again or enters a state of unconsciousness".

The present concept of consciousness is most easily studied scientifically in the area of awareness. For example, someone is aware of something, someone is aware of the self.

The relationship with the self is a direct one, not a cognitive one, which is undoubtedly Schart's new interpretation of Descartes' concept of self-explanation.

As we pointed out at the beginning of the essay, a fundamental point of Descartes' epistemology is the recognition of the self as something that one learns or realizes on one's own, without being taught by others, in the practice of life.

In a word, it is a comprehensive awareness and

"knowledge" of one's physical and mental state and changes in people, events and objects in the environment by using mental activities such as sensation, perception, thinking and memory, which is the view of "wisdom in knowing".

The key point of this view is to consider the cognitive ability and the clarity of cognition of the environment and the self. The essence of this view is that it treats human cognitive ability and cognitive clarity of the environment and self as a mere observer or knower, and correspondingly treats thinking activity and its correlates as an observed or known.

As a result, it not only confuses the individual's ability to use sensory, perceptual, thinking, memory and other mental activities to perceive and recognize the changes in one's physical and mental state, people, events and objects in the environment, and to reflect on one's cognitive ability and cognitive clarity of the environment and self.

It is also difficult to distinguish between the individual's ability to perceive, recognize, and reflect on his or her own physical and mental state, changes in people, events, and objects in the environment, and the clarity of his or her cognitive ability and cognition of the environment and the self using sensory, perceptual, thinking, and memory mental activities. These two concepts.

If we say that the individual uses sensation, perception, thinking, memory and other mental activities, the comprehensive awareness and recognition of his own physical and mental state and the changes of people, things and objects in the environment is a pure negation, that is, to make things exist, and at the same time, it is a false sense of consciousness itself.

Reflection, on the other hand, is a kind of self-referential existence, a personal use of sensation, perception, thought, memory, and other mental activities, a comprehensive awareness and knowledge of one's own state of mind and body, and changes in people, things, and objects in the environment, a kind of re-recognition.

According to Schart, reflection is self-activity, an effort to re-grasp existence, which is the manifestation of the eternal possibility of self-activity.

Self-reflection seeks to internalize one's own existence by going back and rethinking the past and learning from it, for the key to reflection is "being what one is".

The motive of reflection is a dual attempt to [objectify] and [internalize], and in the unity of internalization, it becomes the thing that becomes the goal of one's own self-action or reflection. When the ego looks back and reconsiders the past

When the ego looks back and reconsiders the past and learns from it as a direct thought, feeling, belief, or preference that can quickly emerge without much thought process, for the self that is in its own being.

What it reveals is not the time history of the individual's comprehensive awareness and knowledge of his or her state of mind and body and the changes of people, events and things in the environment by using mental activities such as feeling, perception, thinking and memory, nor is it something that is not distinguishable and exists independently within itself.

But it need not be a physical existence. In particular, abstraction is often regarded as physical historicity, but as some kind of spiritual force that passes, in organized form, beyond those who think about the past and draw lessons from it.

The idea of the universal rationality of concepts, the collection of all universal, positive, and rational ethical forces and concepts, is itself.

If it is said that, as an individual who uses mental activities such as sensation, perception, thinking, memory, etc., to be aware of and recognize the changes in his physical and mental state and people, things and objects in the environment in an integrated way, the ability to recognize the environment and the self and the clarity of recognition, it is the point of the ability of consciousness.

A condition that is different from its own existence (that is, without the ability to recognize the environment and the self, and the clarity of cognition, there will be no one's ability to recognize the environment and the self, and the clarity of cognition beyond the point).

Then it is also the premise that it is possible to go back and rethink the past and learn from it, because the cognitive ability and the clarity of cognition of the environment and the self are the mental activities of the individual using sensation, perception, thinking, memory, etc.

This means that consciousness and self can never coincide. In this sense, awareness is a way of being in the world of self-existence, which embodies the inner negation of [self-existence] and [self-existence].

In other words, awareness is a kind of non-reflective individual's comprehensive awareness and knowledge of one's own physical and mental state and the changes of people, things, and objects in the environment by using mental activities such as feeling, perception, thinking, and memory.

Secondly, reflection is an internalized, object-oriented awareness of the non-reflective individual's use of sensation, perception, thinking, memory, and other mental activities to perceive and recognize changes in one's physical and mental states and in people, events, and objects in the environment.

In other words, it is a kind of re-recognition of awareness (the above-mentioned non-reflective awareness).

It is also a kind of consciousness, but it is a second level of consciousness, which is a testimony to the integrated awareness and knowledge of one's own state of mind and body and the changes of people, things and objects in the environment.

We can see that by distinguishing non-reflective consciousness from reflective consciousness, Schart distinguishes between the integrated awareness and knowledge of one's own state of mind and body and the changes in people, events and things in the environment by using mental activities such as sensation, perception, thinking and memory, and reflective consciousness.

In this way, we can distinguish two kinds of awareness: awareness as non-reflective awareness (including pure awareness and imaginative awareness) and awareness as reflective awareness - a kind of reawareness, a kind of comprehensive awareness and knowledge of the non-reflective individual's mental activities such as sensation, perception, thinking, memory, etc., of his own physical and mental states and changes in people, events and things in the environment. The former is a kind of recognition of a non-reflective individual's use of sensory perception, thinking, memory, and other mental activities.

The former embodies the comprehensive awareness and knowledge of one's own state of mind and body and the changes of people, things, and objects in the environment by using mental activities such as sensation, perception, thinking, and memory as "being-for-itself" and the specific objects in the world.

(Whether it is a comprehensive awareness and knowledge of one's own state of mind and body and changes in people, things, and objects in the environment by using mental activities such as sensation, perception, thinking, and memory, or imaginative knowledge, they all embody "being-for-itself" and "being-for-itself".

They all embody the relationship between "being-for-itself" and the existence that it denies and at the same time brings to the fore. For example, a mere individual uses mental activities such as feeling, perception, thinking, and memory to perceive and recognize his or her own state of mind and body and the changes of people, things, and objects in the environment "I think it is sharp.

(This means that I see the other as an object that is not present, i.e., it means a kind of emptiness or inner negation, and it also means the reference to it and the highlighting of it.)

The latter is an attempt by consciousness to abstract and internalize itself (which is the meaning of Cartesian recognition). Both knowledge as non-reflective consciousness and knowledge as reflective consciousness can be said to be a way of being-for-itself.

Therefore, if for Descartes' philosophical thinking, the reflective individual's integrated awareness and knowledge of his or her own state of mind and body and of changes in people, events, and objects in the environment, using mental activities such as sensation, perception, reflection, and memory, is the most fundamental way of being for the self (Descartes, even when discussing passions, still sees wonder as the first and most fundamental passion), then Scharthe would obviously not agree that this is the most fundamental way of being for the self.

Schart obviously does not agree that the integrated awareness and knowledge of one's own state of mind and body and of changes in people, events, and objects in the environment (both as non-reflective and reflective consciousness), using mental activities such as sensation, perception, thought, and memory, is the most basic way of being-for-itself.

It is because, even as an individual who uses mental activities such as sensation, perception, thinking, and memory to become aware of and recognize changes in one's physical and mental state and in people, events, and things in the environment (non-reflective consciousness), it is only between "being-for-itself" and "being in itself".

We know that, in Schart's holistic, fundamental, and critical view of inquiry into the real world and human beings, emotional and visual consciousness (which, despite their intricate relationship to awareness, are not, after all, awareness)

There is no doubt that both of them are specific manifestations of a certain nature or phenomenon through something, or that the subjective will of human beings manifests the inner [negativity] or [nihilization] relationship between "being-for-itself" and "being in itself" through actions, products, works, and so on.

In other words, if it is possible to determine that a person or thing is a way of being that this person or thing, and not another, is "being for itself," then [the emotion of mental reaction generated by inner feeling] and [the image of the material reproduction of one's visual perception] are also ways of being-for-itself.

(2) Existence of the perceived object

This article focuses on two properties of the existence of the perceiver, the first being relativity. Relativity is not the subjectivity in the Kantian context, but the immanence of the self to the self in subjectivity. This relativity is in contrast to the reflective consciousness of the previous chapter (which in this chapter should be the consciousness of the self).

If all objects or phenomena objectively existing in nature are perceived, there will be perceived things. This is the end of the main empiricist and subjective idealist exposition presented by Beckley in his "Three Dialogues of Hellas and Philonos".

One of the most famous quotes of Bekeley in these idealist principles is "to exist is to be perceived", which means that the world exists solely by the existence of the independent organism that perceives the world, and if there is no perceiving organism, then the world does not exist.

For this theory, Berkeley gave a detailed elaboration and reasoning. Beckley pointed out that there is no first nature at all, because all knowledge is a function of the person who is experiencing or perceiving it.

He pointed out that physical objects are nothing but the accumulation of experienced sensations, which are united in the mind by the force of habit. Otherwise, they cannot exist. A teacup, for example, does not exist until one sees it, touches it, smells it, has its color, shape, and odor.

Without the sensation, or the combination of the responses to the individual characteristics of the objective reality (sound, color, smell, etc.), there is nothing. This is the basic principle of his famous saying "to exist is to be perceived".

His theory is the opposite of materialism. It is a logical deduction of natural and social phenomena according to existing empirical knowledge, experience, facts, laws, cognition, and tested hypotheses, by means of generalization and deductive reasoning.

At the beginning of its birth, it was greatly ridiculed, but with the passage of time and the progress of knowledge, this theory was accepted by the majority of idealists. Scharthe likewise pointed out the error of Becquerel's idealism: "The being that is perceived cannot be restored to the being that perceives.

To be moved by external forces, or to be influenced by others, or to be controlled to act, often means to be unable to create a favorable situation, to make things go according to one's intention, mainly because the existence of anything, or the determination of its properties, attributes, or authenticity, is not a mere, absolute determination.

It must be considered in relation to another thing, providing a perceiver. It is purely the individual's comprehensive awareness and knowledge of his or her own physical and mental state and the changes of people, events, and things in the environment by using mental activities such as sensation, perception, thinking, and memory.

It is decided that the "I" must exist specifically in the immediate reality in order to support a way of being that does not come from my existence. If my existence is not motivated by external forces, or influenced by others, or controlled by actions, often meaning that I cannot create a favorable situation, so that things are carried out according to my own intentions, I become the basis of my various feelings, and everything falls into an impractical, inscrutable nothingness.

(3) The collapse of traditional beliefs and the passion that man is useless

The most important philosophical terms in Schart's book Being and Nothingness are "being-for-itself" and "being in itself", the definitions of self-possession and self-activity.

The definitions of self-possession and self-activity, simply put: "being in itself" is for the moment considered as something objective and external, which cannot be explained in words, and is empty of all humanly given values and regulations, but it is full and complete.

In the context of Schartan philosophy, "for oneself," "negation," and "nihilization" are "a framework for interpreting the resources surrounding the clinging to self-love and providing a proper explanation of the existence of the being in itself.

It is therefore a concept of relative equality, and is only interpreted within the framework of this context, not independently of the framework of the existence of the self-life itself.

Therefore, it is basically a tautology, usually referring

to the desire of the heart that has not yet been pacified. Schartre considers the 'existence' of the human ego life as a 'lacking' existence, which in fact derives from Hegel's ontological prescriptions. Deprivation is not a psychological or physical lack, not hunger, thirst, or insecurity.

It is the ontological stipulation of superconscious metaphysics that the 'being' of one's own life is itself lacking. No matter how much material abundance, physical soundness, and spiritual security you have, your 'being' of your own life is still lacking from birth to death.

The "being-for-itself", that is, the I-love-obsession of man, can never have the fullness of existence as "being in itself". The "being" of one's self-existence is destined to be lacking, which is expressed in the terminology of Saudi philosophy, namely

"Being-for-itself" can never occupy the same place in space as "being in itself", and once it occupies the same place in space, it will erase all mental activities.

Such as perception, memory, imagination, etc. The ability of man to perceive the environment and the self, and the clarity of cognition, the 'being' of cognizing the existence of the life of the self is pure, clear and transparent, without any innate stipulation, which is the phenomenological basis of Schartre's famous proposition - 'existence precedes essence'.

Schatt points out that since consciousness is an empty existence, it is an impractical and inscrutable 'nothingness', but this impractical and inscrutable 'nothingness' is not a lifeless, dead and inscrutable 'nothingness', but a creative and inscrutable 'nothingness'.

In the philosophical thinking of Schart's book "Being and Nothingness", the lack is the root cause of the unsatisfactoriness of the human ego-love obsession.

Fichte is often considered as a transitional figure between the philosophies of Kant and Hegel. In recent years, scholars have re-recognized his position as a result of his profound understanding of self-consciousness.

Like Descartes and Kant before him, questions of subjectivity and consciousness have stimulated much of his philosophical thinking. Fichte also called his philosophy "the science of knowledge". He agreed with Kant's view that "knowledge is the law of nature" and advocated the study of human internal consciousness.

The "principle of absolute unconditionality" is a fundamental proposition or assumption that is the source of all human knowledge and cannot be omitted or removed, nor can it be violated.

Fichte disagreed with Kant on the question of the existence of the Self of all things in heaven and earth. He believed that such a system, which separates appearances from the Self of persons or environments other than oneself, would inevitably lead to a kind of skepticism.

From his philosophical point of view, a rigorous philosophical system should start, as Descartes did, from a supreme and unmistakable self-evidence, a most fundamental proposition or hypothesis that cannot be omitted or deleted, nor violated.

It is a system of rigorous logical reasoning according to its intrinsic necessity. After the question raised by Hume, there exists a logically insurmountable gap between heaven and earth, between all human things and reason.

So he agrees with Kant that only idealism is possible. But he believes that we should abandon the concept of the existence of all human and physical selves in heaven and earth, and replace it with the concept of an absolute self.

This absolute self is not the empirical self, nor the a priori self, but the a priori element of all self-consciousness. This self-consciousness provides the a priori basis for all knowledge, the basis and a priori source of all knowledge and empirical reality, the supreme basis and point of departure in epistemology and the science of knowledge.

Here he integrates theoretical and practical reason and gives the ego a high status, giving it the possibility to act creatively. At the same time, Fichte points out that it is not because the existence of the human ego itself is limited in space and time.

Rather, the proposition should be interpreted in the opposite way: the human self-creates in order to achieve a certain goal, so the human being is destined to have certain limits in space and time.

In other words, the fact that human beings are born with certain limits in space and time constitutes that they are able to actively create themselves in order to achieve a certain goal.

Therefore, in his book "Being and Nothingness", when discussing the lack of human beings, Scharthe used the view of Self-possession and Self-activity to point out that God does not exist.

In his book "Existence and Nothingness," Scharthe uses the ideas of self-possession and self-activity to point out that God does not exist and that human passion is ultimately meaningless and futile.

Schart believes that everything that exists in heaven and earth is either "freedom" or "for oneself". If not "Freedom", it is "for oneself". Suppose that if God exists, God should be a full, solid Self, lacking in nothing; but God should also be a being that can think and act, but to think and act is to be lacking in something.

Shatt Self-possession and Self-activity Philosophical thinking concludes that if God exists, God is just a stone that thinks and acts, but such an absurdity cannot exist, so God is a paradox.

At the same time, Schart pointed out that the passion and effort of "existence" exist in the life of the self, and that what people pursue throughout their lives is nothing more than the desire to become "God", that is, they want to be completely fulfilled, but at the same time they want to be free to choose and think.

To put it more eloquently, people have always wanted to become a God who has the autonomy to stand alone and be at ease.

However, as mentioned above, Schatt believes that consciousness can only be supported by an individual's mental activities such as sensation, perception, thinking, and memory, and can only be an object of comprehensive awareness and knowledge of one's physical and mental states and changes in people, events, and objects in the environment, and cannot occupy the same place in space as it does.

Therefore, in the second half of his book "Being and Nothingness", Schatt writes: "Man is a useless passion after all. Ultimately, it is a meaningless and futile labor

(4) I love the anxiety that cannot be satisfied by clinging

Although Schart has written in his book "Being and Nothingness" that this life existence is almost desperate, but from the overall or macroscopic view of the whole book, Schart believes that the "existence" of man's self-existence itself is a useless and meaningless passion.

However, the process of action or thought, which is formed in the process of practice, the possibility of realization, the desire and pursuit of future society and one's own development, or the goal of plundering, can still find one's strong will to be free.

Because the freedom of will to "exist" in the very existence of human life can lead man to bring all kinds of innovative meanings and values to the inert, rigid nature and human society.

However, Scharthes suddenly and hysterically changes the subject, returning to the quest to reveal the hidden sorrow of the deep inner thought activity, pointing out that freedom of consciousness does not bring happiness and joy at all, but rather a gulf of unsatisfied desires in the heart, which leads to never-ending anxiety and suffering.

In other words, human freedom is the basis for all kinds of meanings and values in the world, but because freedom itself is absolute and ultimate.

Why, according to Schart, does man possess freedom? This is in fact the greatest contingency, because there is no answer at all, and man can only accept this fact.

Freedom is the basis of all values in the world, but it has no basis in itself. In the final analysis, human freedom is an accidental and unintended fact.

Therefore, it has been said: "Man accepts life without choice, then spends it under conditions of helplessness, and finally surrenders it under an irresistible struggle".

Therefore, there is a famous saying in Saudi Arabia: man is condemned to be free. It can even be said categorically that the 'existence' of the self-life itself is condemned to be free from the moment man is thrown into the world.

Moreover, Schart thinks that if man is judged to be free, then he must take the greatest absolute responsibility for his choices and actions as a goal in all his actions or thoughts.

This is the inevitable conclusion of the absolute freedom of the will. In fact, people are often afraid to take responsibility for their own freedom of will, so when faced with the pressure of making choices about boundaries, some people feel anxious and distressed, and then try to avoid making choices.

But Schart's philosophical view also points out that when people feel anxiety because their desires have not yet been smoothed out, they are in fact showing that there is a free "being" in their own being.

As Kierkegaard puts it, anxiety is the dizziness of freedom. Having painfully perceived this meaninglessness, Camus in The Myth of Xerxes claims that "there is only one truly serious philosophical problem, and that is suicide.

Although the "remedies" for such potentially extremely harmful encounters vary, whether it be Kierkegaard's religious "phase" or Camus' persistence, the focus of the vast majority of existential philosophers has been on finding ways to help people.

The focus of the vast majority of existential philosophers is on finding ways to help people avoid negative emotional and unpleasant ways of being. To avoid putting the 'existence' of the self-living being itself in a permanent danger of losing all meaning of 'existence'.

This disintegration of the meaning of 'being' in the existence of the self may pose a serious threat to the balance of the self and the body, and it is contrary to the existentialist belief in analyzing and reflecting on the roots of problems of life, knowledge and values in order to solve the fundamental problems of life.

The belief that it is possible to commit suicide makes all people existentialists. In a meaningless life, for an absurdist, the existence of the basic meaning of a thing must be explained by a higher meaning.

But the meaning of this higher meaning must be explained by a higher meaning than this higher meaning.

This "chain of explanation" cannot reach a final result of definition, so that nothing can have a supreme meaning. Even if this result is found, it may not satisfy us.

It's like saying that a geek or a nerd spends his life in poverty, just lying in bed eating peanuts, looking for water, watching TV, slipping on the phone, playing computer. His life is only in the room that can sustain life.

The daily life is to open the eyes and turn on the TV, looking for peanuts, looking for water, and then the whole day "watch TV, slide the phone, play computer", hungry to find food to fill the hunger, tired to lie down and sleep. We all think that this kind of geek, geek girl repeated life is meaningless.

Why do we have a strong intuition that the life of this kind of geeks and nerds is meaningless: because this kind of geeks and nerds have not tried to do anything in their lives. Even if he loves to watch TV, skate on the phone, and play the computer, how happy he is, and even if he is actively involved in this kind of life, his life is still meaningless.

His life is still meaningless, because this mechanical life is too passive, he did not try to do something.

A person's life is meaningless as long as he does not make any relevant contribution in the world and has an effect or influence on the things in the world, whether he feels happy or voluntary or not.

The second kind of meaningless existence is the kind of life that devotes itself to activities that have no value. Unlike the first type of person who lives and dies in a state of confusion, he not only does not passively eat, drink, and sleep, but also tries to do something amazing in his life.

But he still has doubts about his own life, about the "existence" of his own life itself, and feels that the existence of life is meaningless, because what he does is worthless and not worth doing.

For example, the philosopher David Wiggins cited the example of a farming entrepreneur who spends his life buying land, planting crops, wheat and corn, feeding crops to cattle, sheep and pigs, raising cattle, sheep and pigs, and then selling them for money.

Then they buy new cows, sheep and pigs, buy more land, plant more plants, take more crops to feed more cows, sheep and pigs, then buy more land and make more money...again...; although this mechanical life is not like the first kind of confused life and confused death, the life of looking for peanuts, isolated from everything around them. It has not achieved anything at all.

Farming entrepreneurs seem to have built up some big businesses: they have grown crops, wheat and corn, fed some cattle, sheep and pigs, and made a lot of money, but in this mechanical, repetitive life, the activities they devote themselves to are all meaningless.

Over time, they will not feel satisfied with the business they are engaged in, but will feel that there is "meaning". On the contrary, they will be surprised that the existence of life is meaningless. As a result, they identify with existentialism and nihilism, and life is meaningless and absurd.

The Münchhausen trilemma is a thought experiment in the theory of knowledge, in the meaninglessness of life, for an absurdist, to show that any argument for truth, including logic and mathematics, is impossible.

When a statement is argued, the premises of the argument also need to be argued. The argument for this premise itself has further premises that need to be argued, i.e., the argument is regressive. Thus, there are three possibilities for the finality of any argument.

01. Circular argument: the argument is ultimately supported by itself.

02. Infinite regress: the argument ends up with no end point.

03. Arbitrary termination: the argument terminates on an accepted premise, such as an axiom, a rule, a moral creed, a religious creed, etc.

All three possibilities are less than ideal and are therefore called the trilemma. This dilemma is also called Agrippa's trilemma because the Greek philosopher Sextus Empirico mentioned a similar five-way argument in his book The Outline of Pyrrhonism, and Diogenes Larsio pointed out that it was formulated by Agrippa.

Later, the German philosopher Jacob Friedrich Fries proposed a similar trilemma, which Karl Popper called the "dogmatism-infinite regress-psychism" trilemma.

In 1968, Hans Albert cited Popper in his book Traktat über kritische Vernunft (Critical Rationality) and introduced the term "Münchhausen trilemma".

Moreover, the French existentialist philosopher Camus described this mechanically absurd repetition of life in The Myth of Xerxes (Xerxes): In a universe suddenly deprived of illusions and halos, one feels oneself to be a family member with one's family.

In a universe suddenly deprived of visions and halos, man feels himself to be a person with no family ties, a non-resident, a wanderer in another land, deprived of the memory of his lost home, or of the promises he had made, and of his desire for a land of paradise.

His banishment is irrevocable. In this kind of stage where people and life exist, and where actors and scenes are separated, there is absurdity.

"The notion of 'absurdity' refers to the irrational part

of the world, the conflict and dissociation from our desire to explain everything. According to Camus, a life based on a philosophical proposition that doubts the possibility of objective and reliable knowledge, or even denies it, has no real meaning.

But those who accept the honesty of the "absurd" give meaning to life by their own resistance. This meaninglessness also implies that the world is a place of "right and wrong" and "injustice.

This contradicts the idea that "bad things don't happen to good people" is an abstract, universal idea that acts as a specification of the scope or class of entities, events or relationships.

For the world, it is as if there is no such thing as a good person or a bad person; what happens just happens, and it can happen to any "good" person or "bad" person, as the saying goes: "When the sky is about to send a great task to a man, he must first suffer his heart and mind, toil his bones and muscles, to starve his body and soul, and to do something that will disrupt his work.

The "absurdity" is a common feature of things perceived by human beings in the process of cognition, which is elevated from perceptual knowledge to rational knowledge and extracted from its intrinsic properties.

Many literary works by Kierkegaard, Schart, Heidegger, and Camus use various rhetorical techniques to visualize and elaborate on things, the "existence" of the human ego's life itself, and the "absurdity" of life experience.

Section 4: Ontological Proof

1. Connotation of Ontology

Ontology is the discussion of the essence of the thing under investigation. Ontology investigates the essence of existence, whether it is one thing? Or is it two or more things? This gives rise to the question of the number of one and many. Therefore, as far as the number of ontologies is concerned, they are generally divided into Monism, or Singularism, Bualism, and Pluralism.

These are referred to as Monism and Pluralism. In addition, ontology studies the ontology of existence, whether it is material? Or is it spirit? This gives rise to the question of mind and matter. Therefore, the nature of ontology is generally divided into Materialism and Spiritualism.

2. Unitarianism and Pluralism

Unitarianism is a term used by the British philosopher Walt to avoid confusion between monism and the monism of the mind-matter problem. Monism is the view that the universe is the evolution of a fundamental element, or that the essence of the universe is One Individual Being.

The implication is that it advocates "uniqueness" and "immobility". It is opposed to "many" and "change". The method of thinking about unitarianism is mostly deductive and dialectical. For example, Plato and even the idea of the good are absolute principles, and the idea is a "model" from which phenomena are modeled.

For example, Leibniz believed that the universe is organized by monads, which are infinite in number. The Atomists, Democritus, believed that the universe is composed of atoms, which are infinite, so he was also a pluralist. The pluralist approach to thinking is mostly inductive and empirical.

3. Materialism and idealism

Materialism advocates that "matter" is the only thing that exists, that it is the basic element that constitutes the universe, and that the universe is fundamentally a material world that can exist independently and objectively from the mind.

For example, the natural materialist Feuerbach advocates that the human body is a part of matter, and the brain is a part of the human body, so the brain itself is a part of matter. The mind is attached to the brain, so matter is not the product of the spirit (mind), but the spirit (mind) is the product of matter.

The teleological theory advocates that the spirit or mind is the basic element that constitutes the universe. For example, Leibniz believed that the universe is organized by monads, and that the essence of monads is mind rather than matter, and that all material things can only be regarded as phenomena and not as entities.

The basic entity of these things must be sought in the monads, whose mental capacity is reflected in all things, and the whole universe is the scope of the mind, so it is called teleology.

Therefore, any cognitive ability of the environment and the self, as well as the degree of cognitive clarity, are actions or thinking, as the objective for the target of all objects or phenomena that exist in nature, but also something for the individual to use sensory, perceptual, thinking, memory and other mental activities.

To their own physical and mental state and the environment of people, things and things change in the comprehensive awareness and understanding. Under this definition, there are two ways of interpreting the essence of a thing, or the connotation and extension of a concept.

First, consciousness is a constituent of the existence of the object.

Secondly, consciousness in its deepest nature is related to a transcendent being.

The problem with ontology is that consciousness, as a subjective being, cannot detach itself and in this way set up a transcendental object. The consciousness of an object is present and absent.

The difference between objects and consciousness is that consciousness is absent from the field by its absence, not by its fullness. Therefore, the existence of an object should be pure non-existence, which is defined as a lack. It is something that evades itself, is not given in principle, and is revealed in an image that is constantly passing.

Consciousness is the consciousness of something, which implies transcendence, the constitutive structure of consciousness; that is, consciousness is born supported by a being that is not itself. What Sartre's ontological proof is intended to show is that

Unlike the ontological proofs of theologians or Descartes. Since consciousness is empty, this absolute impersonal subjectivity can only be confirmed in the face of a revealed thing.

"There is no existence for consciousness outside of the explicit obligation to reveal something, to reveal an intuition beyond existence."

Here Sartre applies to consciousness Heidegger's definition of dasein, the possibility of dasein and the attribution of the ego to consciousness, and adds that consciousness is such a being.

It is concerned in its being with its own being, insofar as this being alludes to a being other than itself. This being other than itself essentially refers to the transcendental existence of the phenomena, the Self that exists for consciousness.

4. 'being in itself'

Schart finally brings out here that the object needed by consciousness is essentially "being-for-itself" and "being in itself" waiting to be revealed by consciousness, but there is no antagonistic relationship between existence and consciousness.

The existence of the phenomenon (consciousness) cannot be restored to the phenomenon of existence as inferred earlier. At this point Sartre wants to show that there are two types of Self-possession and Self-activity of existence.

One is that there is no existence that does not exist in a certain way.

Secondly, there is no existence that is not grasped by a way of being that both reveals and conceals existence.

Being and Nothingness consists of an introduction and four volumes. The Introduction, which explores existence in a phenomenological way.

The first volume argues for the problem of

nothingness and provides an incisive analysis of self-deception.

The second volume, which argues for "being-for-itself," offers an innovative perspective on the holistic understanding of time.

The third volume, on the relationship with others, presents the idea of "co-presence".

In the fourth volume, the relationship between being-as and freedom is demonstrated, and the relationship between Self-possession and Self-activity concludes the book metaphysically.

The first three volumes address the fundamental viewpoints of phenomenological ontology. The book identifies the domain of being, the structure, characteristics and laws of "being-for-itself".

The book describes the ethical meaning of freedom by means of a phenomenological "psychoanalysis of being". The most important part of the book is "To have, to be and to exist", which deals with the question of "human freedom", where the author argues that the eternal possibility of action should be considered as an intrinsic feature of self-being

Section 5: Philosophical irrationalism that explores the meaning of human existence

01. Definition

Existentialism is a philosophy that explores the meaning of human existence in this irrationalist trend. Nietzsche and Kierkegaard can be regarded as the precursors of existentialists.

In the 20th century, it spread very widely, and its philosophical ideas on the study of universal, fundamental issues in the disciplines of existence, knowledge, values, reason, mind, language, and so on, continued into the humanism that emerged in the 1960s.

Jaspers and Heidegger, Paul Schart and Camus were its representatives. These propositions have influenced literature, psychoanalysis, and religious philosophy.

02. Existence before essence

The most famous and explicit of these propositions is the maxim of Jean-Paul Sartre: "Being precedes essence". By this he means that there is no innately determined morality or soul other than man himself.

Both morality and soul are created by man in his own existence, in the existence of life itself. Man is not obligated to conform to a moral standard or religious belief, but he is free to act voluntarily after consideration.

Therefore, when we evaluate the merits of people, things, and objects, good, bad, beautiful, or unreasonable conclusions, it is the actions of a person, not his identity, that is evaluated after analyzing whether the object of his thought exists, whether it has a certain attribute, and whether there is a certain relationship between things, either affirmatively or negatively, because the nature of a person is defined by his actions, and "a person is the sum of his actions.

Schart denies the existence of God or any other pre-defined rule. He rejects any "resistance" in life that disturbs and inconveniences, because they reduce man's scope for free choice.

If there were no such obstacles, then the only problem one would have to solve would be to choose which path to take in the existence of the being itself. However, the human being is free to choose in his own existence; even in his self-deception, there is still potential and possibility.

Schart also suggests that "the other is hell". This position on which the study analyzes or criticizes problems and issues seems to be contradictory to the position or perspective of "freedom of choice" for all objects or phenomena or problems that exist objectively in the natural world.

In fact, everyone is free to choose, but everyone has an inescapable responsibility for the outcome of his or her choice. Choice is a voluntary act taken by a person after deliberation.

This action occurs when there are at least two possible choices, and the choice is to achieve a goal. There are several basic conditions that constitute a "choice".

(1) A mental process of deliberation, without which the action would be mechanical or arbitrary.

(2) there is an actual action, if only thinking without specific external behavior, then it does not constitute a choice, so the consideration before the action, only to make a decision, not a choice.

(3) can choose the object can not be only one, must be more than one, choose one of them, only then constitute a choice.

(4) is a purposeful action, choice is a person to achieve a purpose and take action, this action affirms that the actor is in a certain value context.

In a philosophy of analytical reflection on fundamental questions of life, knowledge, and values, the discussion of "choice" focuses on two aspects.

First, the fundamental nature of choice: is it only a mental event or does it include external behavior?

Second, is there freedom of choice?

The latter depends on the object of choice. In daily life, there are many subtle events that affirm man's freedom of choice; but in the larger context, such as man being part of the natural world and bound by the laws of nature, is man's choice only an illusion under the constraints? This is the debate between the Determinist and the free will advocate.

Schart's holistic, foundational, and critical inquiry into the realm of reality and human beings suggests that the greatest problem facing human beings in the process of choice is the choice of others, for each person has the freedom to choose, but the freedom of each person may affect the freedom of others, hence the term "others are hell.

Existentialism does not deny the existence of God. For example, Kierkegaard was a Christian who believed that existentialism was the beginning of the Christian mode of thought.

Nietzsche, in his book The gay science, suggests that "God is dead". Nietzsche does not mean that God is dead on a metaphysical level; rather, Nietzsche realizes that God's death represents a crisis in existing moral standards.

(The question "Why is there anything at all?", or, "Why is there something rather than nothing?" has been raised or commented on by philosophers including Gottfried Wilhelm Leibniz, Ludwig Wittgenstein, and Martin Heidegger - who called it the fundamental question of metaphysics)

03. absurdity

Absurdism (French: absurde; English: absurdism), also known as absurdism, is a philosophical term derived from the Latin word absurdus, meaning musically "out of tune", and is used in existentialism to describe the meaningless, paradoxical, and disordered state of life.

The concept also refers to the irrational part of the world, the conflict with our desire to explain everything, and the departure of each from the other, indicating respectively a judgment of the truth and the continuity of the state.

According to Camus, a life based on skepticism has no real meaning, and therefore, to explore skepticism, we must first define it in a meaningful way.

So what is skepticism? Skepticism and epistemology are two sides of the same coin. Without epistemology (the study of the sources of knowledge, the process of development, the methods of knowing, and the relationship between knowledge and practice), there would be no skepticism.

That is, the objective reason for the existence of skepticism is precisely the tendency of knowledge to see things through the lens of the subject's own needs.

It is the subjective nature of the viewpoint, experience, consciousness, spirit, feelings, desires, or beliefs that an individual can have and the conditionality of the existence and development of things. Perception is, after all, a human perception.

Therefore, the influence of the subject on perception cannot be ignored, and its influence may even be decisive.

The subject's temporal limitations, social position, cognitive perspective, knowledge structure, personal experience, and subjective will and emotions, as above, interfere with the perceptibility of the process of acquiring truth through the formation of concepts, perceptions, judgments, or mental activities of imagination, so that in the end we cannot reach the truth itself at all, but only approach it endlessly.

Because man is not God, the above-mentioned factors that affect the human brain, reflect the characteristics and connections of objective things, and reveal the meaning and role of things for people, are factors that every living person can never eradicate.

They can never be eradicated, but can only be reduced as much as possible. The simplest physical measurement, no matter how precise the measuring instrument is used, the error is bound to exist, we can only get as close to the truth as possible, but never reach it.

This "infinitely close but never attainable" is also the basic idea of limit theory. Skepticism is in fact a rejection of the ancient epistemological paradigm of philosophy (the proposition that thought and existence are naturally unified).

The modern philosophy of holistic, fundamental, and critical inquiry into the real world and human beings emerged because of a break with this paradigm of ancient philosophy.

Modern and contemporary philosophy can be said to be based on the paradigm of the proposition that there is no natural unity of thought and being, as distinct from ancient philosophy.

The creation and formation of this new paradigm is obviously the great work of skepticism. But those who accept the absurdity of honesty will give meaning to their lives by their own rebellion.

This [meaninglessness] also implies that the world is a world of right and wrong and injustice. It has to do with the fact that people, on the basis of perceptual knowledge

The concept that "bad things do not happen to good people" is contrary to the concept that "bad things do not happen to good people", which is unique to the abstraction and universality of many attributes of similar things.

For example, there is no such thing as a good person or a bad person for all things and objects that exist objectively in the orbit of the universe; what happens just happens, and it can happen to any "good" person or "bad" person.

The "absurd", a form of thinking that reflects the nature of things, has always been a level or expression of literary excellence beyond the ordinary. Many literary works by Kierkegaard, Tostoevsky, Scharthe, Heller, and Camus

They all use various rhetorical devices to visualize the absurdity of the world as encountered by man. This is derived from the Latin absurdus, meaning musically "out of tune," and is used in existentialist literature to describe the meaningless, contradictory, and disordered state of life

After one's ability to perceive one's environment and self, as well as the clarity of one's perception, and the painful awareness of the existential meaninglessness of the existence of this life itself, Camus in The Myth of Sisyphus claims that "there is only one truly serious philosophical question, and that is suicide.

Although the "remedies" for such potentially extremely harmful encounters differ in their use, whether it is the religious "stage" of Kierkegaard or the persistence of Camus. Both develop the self in the process of contradicting the external material world.

But the focus of most existential philosophers has been on helping people to avoid bad ways of living and to avoid the long-term danger of losing all existential meaning in the existence of life itself.

This possibility of the breakdown of the existential meaning of life itself poses an important threat to one's own equanimity, which is contrary to the existentialist intention of analytically thinking about and reflecting on the fundamental questions of life, knowledge, and value.

According to one's own beliefs, it is possible to commit suicide for this reason, and this makes all people existentialists. In an existential meaningless life where life exists in itself

An absurdist, who is able to defy the spirit of rape, who, in the face of threats to his own interests and even to his own life, insists on his own values, without compromise, and without being tempted to abandon his principles. Face to face, do not run away, to face "suicide".

1. The Absolute Spirit of Hegelian Philosophy

Absolute Spirit is the core concept of Hegel's philosophy. It means "the uniquely existing, all-embracing, universal essence". From Hegel's point of view, the "absolute spirit", which is not limited by any conditions, is both physical and subjective, and his logical thinking is based on the idea of "absolute spirit".

Hegel believed that his theory could explain the laws of development of all the objects or phenomena that existed in nature.

Therefore, he said, "Everything that is rational is real, and everything that is real is rational.

The meaning of this statement is that all things or phenomena that are objectively present in nature are bound to happen if they are in accordance with the absolute spirit, which is not limited by any conditions or restrictions.

This means that Hegel's view of the absolute spirit is deterministic. Determinism or determinism, also known as Laplace's Creed, is a holistic, fundamental and critical position of inquiry into the real world and human beings.

It is the view that any event, including a human decision made of free will, has external conditions that determine the occurrence of that event, and not some other event.

Theories of determinism are woven throughout the history of disciplines that use analytical thinking to inquire and reflect on fundamental questions about life, knowledge, and values. In direct opposition to determinism is non-determinism. Determinism is also often contrasted with free will.

Determinists believe that free will is an illusion and that the human will is not free; historical dialectic was popular in the first half of the 19th century, when Kierkegaard first questioned the philosophy of the system: "The individual has no place in Hegel's system. The first questioning of the philosophy of the system by Kierkegaard was in the first half of the 19th century: "The individual has no place in Hegel's system.

But the despair that Kierkegaard experienced is different from the despair felt by the general public, and there is a difference between the fundamental and characteristic characteristics of all objects or phenomena that exist objectively in the natural world.

The despair of the worldly people is only the feeling of poverty, frustration, loneliness, pain and other feelings related to the seven emotions and desires, while the despair that Kierkegaard refers to is the soul experience that arises when pursuing the "existence" of the true meaning of life existence in the universe and life itself.

Despair arises from the pursuit of the "existence" of one's own life existence, and it is in the feeling of despair that one can feel one's own "existence".

A person is different from an objective and real table or chair that does not depend on one's subjective consciousness. Does a table or a chair "exist"? To the table or chair itself, the table or chair does not exist.

Because the table or chair itself cannot feel its own existence, only people feel its existence. Therefore, the table and the chair do not have loneliness and pain, while people do.

2. According to Kierkegaard, human "existence" is divided into three levels.

The first level of sensual existence: the sensual

material enjoyment that determines the nature of things solely by one's feelings and perceptions of all the objects or phenomena that exist objectively in nature, taking into account only the personal and concrete facts and conditions that exist in front of one's eyes.

The second level is rational existence: the ability to form concepts, make judgments, analyze, synthesize, compare, reason, and calculate, and after careful consideration of the evidence, to reason out reasonable conclusions, so that one's life is more serious, responsible, and in accordance with social ethics and morality.

The third level is the religious existence: the highest level that human beings should pursue: religion is a kind of doctrine that makes use of human beings' wonder and awe at the mysteries of the universe and life, and is used to teach the world and make people believe. It is a life of prayer and love, a life of consciousness and reverence for God, and a life of spiritual trust.

3. [Empiricism] in the age of Kant; [Rationalism]

In Kant's time, there were two major theories of European philosophical thought that were holistic, fundamental, and critical inquiring into the real world and human beings: first, the [empiricism] developed by John Locke, David Hume, and others; and second, the [rationalism] of Descartes and others.

First, empiricism (English: lang|en|Empiricism), in the related field of philosophy, is a basic theory that knowledge can or should only come from sensory experience.

Empiricism is a view of the theory of knowledge, along with rationalism and skepticism, in which things are evaluated according to their reasoning, as opposed to their actuality or practice.

Empiricism emphasizes the importance of empirical evidence over traditional or innate ideas in theories of thought, and empiricists tend to argue that traditional facts also derive from prior experience.

It usually refers to the belief in the modern scientific method, which holds that theories should be based on observation rather than intuition or superstition. This means that experimental research followed by theoretical deduction is preferable to purely logical reasoning.

The opposite of empiricism is the European rationalism. The representative figure is Descartes.

According to rationalism, philosophy should be deduced by thinking and deduction, and the conclusion should be drawn by reasoning.

The representatives of empiricism are Aristotle, Thomas Aquinas, Thomas Hobbes, Francis Bacon, John Locke, George Berkeley, and David Hume.

(1). The relationship between empiricism and science

Empiricism is the predecessor of logical positivism (logical empiricism). To this day, the empiricist approach still influences natural science and is the basis of natural science research methods.

The natural science method of studying organic or inorganic things and phenomena in nature is a development of traditional concepts. However, in recent decades, some new theoretical doctrines have been generalized from practical verification, or derived from conceptual derivation.

For example, quantum mechanics, constitutivism, Thomas Kuhn's Thomas Kuhn's "The Structure of Scientific Revolutions" has begun to have a slight impact on the unique position of empiricism in scientific research methods.

On the other hand, quantum mechanics, for example, is a method of generalization and deductive reasoning in which human beings make logical inferences about natural and social phenomena based on existing empirical knowledge, experience, facts, laws, cognition, and tested hypotheses.

Empiricism does not have the ability to discover violations of intuitive scientific laws and to change them to fit the laws inferred from actual verification or deduced from concepts.

(2). The philosophical relationship of empiricism

The term empiricism originally meant the experience of the ancient Greek physicians, who refused to accept contemporary religious dogma, but based their analysis on observed phenomena. It was first systematically articulated by the Englishman Locke in the 17th century.

Locke argued that the mind's ability to perceive the environment and the self, and the clarity of cognition, were originally blank forms, but were written down by experience. This doctrine denies that man has an innate perspective or can acquire "cognition" of a subject without experience.

The act of "knowing" and "recognizing" a subject can be acquired with certainty, and these knowledges have the potential to be used for specific purposes. It means the ability to become familiar with something through experience or association.

It is important to note that empiricism does not claim that one can automatically acquire knowledge from practice. According to the empiricist view, the perceived experience must be properly summarized or interpreted in order to establish an act of "cognition" and "identification" of a subject in order to establish a confident understanding.

These perceptions have the potential to be used for specific purposes. In philosophical development, empiricism has been contrasted with rationalism.

Rationalism holds that most knowledge is attributed to the senses, independently of the rational and active process of reflection on the real world or any object in an "inward dialogue".

In any case, this comparison has been considered too simplistic, since recent European rationalists have also advocated the use of scientific methods to gain practical experience.

Locke also argues that supernatural knowledge (e.g., religious theology) must be achieved solely through a special way of thinking that is not controlled by human will.

It is a form of thinking based on human occupation, experience, knowledge, and instinct, either by deducing conclusions from known or assumed premises, or by seeking the reasons for the results of known answers.

Second, rationalism: European rationalism is based on the recognition that human reason can act as an act of "cognition" and "identification" of a subject in order to know with certainty, and that these knowledges have the potential to be used for specific purposes.

It is a philosophical approach based on the use of a source theory, which holds that reason is superior to sensory perception, and is also called rationalism. Formally, rationalism refers to a methodology or theory that asserts that truth cannot depend on the senses, but rather on reason and deductive reasoning.

Rationalism has three original meanings: first, it is the view that anything that is religious and not explained by reason should be excluded; second, it is the view that reason is an independent source of knowledge, different from the senses, and has the highest authority; third, in the field of philosophy, it is the exploration of certain basic concepts that lead to other elements of philosophy by deduction.

In this aspect, the first is Descartes (René Descartes, 1596 ~ 1650) to explain the method of understanding, followed by Spinoza (B. Spinoza, 1632 ~ 1677) and Leibniz (Gottfried W. Leibniz, 1646 ~ 1716), after by Wu Fu (Baron Christian von Wolff, 1679 ~ 1754) Wolff (1679-1754) elaborated on it.

Descartes emphasized the method of thinking, with clarity and understanding as the index of truth. He believed that the application of mathematical methods, whether original or derived truths, could lead to certain truths or the end of science.

It is believed that reason is the only source of human knowledge, and that the human mind has the innate cognitive ability to take in knowledge actively rather than absorbing it passively, and therefore knowledge must be acquired by the innate ability of reason, not by sensory experience.

Rationalism advocates not only the innate rationality of human beings, but also the innate idea, the so-called law of knowing.

Thus, rationalists emphasize human rationality, the innate law of knowing, and believe that the innate ability to know is universal and necessary.

Because of its universality, it can be applied to all things that exist; because of its necessity, absolute truth can be found.

In short, rationalism holds that if one uses reason well, one can acquire knowledge of what is true and the act of "knowing" and "identifying" a subject with certainty, and that such knowledge has the potential to be used for specific purposes.

The use of reason provides insight into the truth of all objects or phenomena that exist objectively in nature, without being blinded by the ever-changing and ambiguous experience of the senses. Knowledge gained by reason is eternal, while experience gained by the senses is fluid.

Although rationalism recognizes that the human mind is composed of cognition, emotion, and volition, cognitive ability is the most important part, because emotion and volition are controlled by cognition and are the result of cognition.

Thus, rationalism holds that real knowledge is obtained through the cognitive faculty of abstract thought and deductive reasoning. It is the act of "knowing" and "recognizing" a subject in order to know it with certainty.

These knowledges have the potential to be used for specific purposes. It means the ability to become familiar with something through experience or association, and thus to understand it.

Rationality, on the other hand, in philosophy, refers to the ability of human beings to use reason. In contrast to the concept of sensibility, it usually refers to the ability of humans to reason their way to a reasonable conclusion after careful consideration of the evidence.

This way of thinking is called rationality. Both sensibility and rationality belong to the domain of consciousness and are conscious in nature. Rationality, based on consciousness, is consciousness with reference.

As in physics, it refers to the system of coordinates used to measure and record the position, orientation, and other properties of objects, either as life, such as instincts, or as knowledge, such as coordinates, or as consciousness, such as the ego.

Therefore, people who can fully use reason are those who have the possibility to realize the ideal of aspiring and pursuing the future society and their own development, formed in the process of practice, and the society combined by a group of rational people is the rational society.

In other words, the individual who is guided by the ability to realize his or her own will according to his or her own knowledge and laws, and whose actions are guided by rational ends and means, can achieve a good life.

Based on this position, rationalism believes that the purpose of social activity is to influence the physical and mental development of human beings.

In other words, it is a social activity whose direct purpose is to cultivate the cognitive ability of human beings and to bring into play the innate ability of individuals to realize their own will based on their own knowledge and laws, and to influence the physical and mental development of human beings.

It is the activity of abstract thinking, such as concepts, judgments, and reasoning, that contributes to the practice of rational life.

Rationalism, when applied to the study of theological issues, uses human reason as a criterion for judging faith and revelation without recognizing any mysterious qualities beyond reason, thus forming the so-called Theological Rationalism.

Rationalism, in its exploration of ethics, ignores the feelings and wills of the human soul and emphasizes rational cognition, thus believing that human moral attitudes are determined by cognition and knowledge of the good.

Rationalism, in short, focuses only on the rational cognitive faculties of the mind, which belong to abstract thinking activities such as concepts, judgments, and reasoning, and ignores the act of "knowing" and "recognizing" a subject from experience, in order to have confident knowledge, which has the potential ability to be used for specific purposes.

This means that through experience or association, one is able to become familiar with and further understand something; this fact or state is called the truth of 'knowledge'.

Thus, empiricists believe that human knowledge of the world and knowledge comes from human experience, while rationalists believe that human knowledge comes from human reason.

The concept of "innate comprehensive judgment" comes from Hume's distinction between comprehensive propositions and analytical propositions.

For example, "all people will die", "death" is included in the concept of "people", so all analytical propositions are inevitable propositions, that is, innate.

The subject and the predicate of a composite proposition have no innate logical relationship, such as "man sitting on a chair", which Hume calls a contingent proposition and an acquired proposition at the same time.

In order to seek the knowledge that is both innately necessary and useful to the empirical world, the learning and experience that man acquires in the process of learning and practice, Kant proposed "innate comprehensive judgment".

Kant does not think that all comprehensive

propositions are acquired, but some innate comprehensive propositions are "innate comprehensive judgments".

When Neo-Kantianism was revived in the second half of the 19th century, it had far-reaching implications and importance beyond Germany. It pioneered the use of terms such as epistemology and insisted on the prominence of ontology.

Natorp had a decisive influence on the history of phenomenology and is seen as the term that inspired Husserl to adopt a priori conceptualism. Emile Rath was influenced by Husserl's work and had an important influence on the early Martin Heidegger.

Césaire received his early training from Cohen, the head of the Neo-Confucian school of Yima Castle. He is often regarded as the main representative of the Marburg school of neo-Confucianism after Cohen and Natorp.

The famous exponent of logical empiricism, Carnap, later pointed out that Cecil's philosophical outlook was "not orthodox Neo-Kantian, but more influenced by the recent development of scientific thought.

Cecil argued that philosophy is not primarily a study

of objects of knowledge, but of ways of knowing, since the objective world is only a combination of "a priori principles" and empirical phenomena.

He also emphasized the expansion of the application of Kantian critical method, changing Kantian "static" rational criticism to "dynamic" rational criticism, in order to accommodate a richer and broader range of life experiences.

Cecil's philosophical thought is a philosophy of human culture, which starts from exploring the nature of man and human culture to develop a whole system of thought.

Césaire believed that man is a symbolic animal, culture is a symbolic form, and human activity is essentially a "symbolic" or "symbolic" activity, in which man establishes his "subjectivity" (symbolic function) as a human being and constitutes a cultural world.

The debate between Cahill and Heidegger about how to read Kant led Heidegger to regard Kant as a pioneer of phenomenology, a view that was questioned in some important ways by Eugen Fink. The long-lasting influence of the neo-Kantian view was in the establishment of the journal Kant Studies, which still exists today.

The philosophy of life, the philosophy of the will, irrationalism, is still insidiously and reluctantly supported or resisted by Nietzsche's quotation of Dostoevsky's novel "Everything is permissible without God. Nietzsche quoted Dostoevsky's novel and shouted the slogan "Everything is permissible without God.

If the phrase "God is dead" is taken literally, the following question arises: If God exists, the Almighty cannot die. On the contrary, if He does not exist, what does not exist has no death (in the way it was intended).

So it is not meaningful to say that God is dead; it is more reasonable to say that God exists or not. What Nietzsche means by the death of God is, of course, not the above, but the death of our belief in God.

The death of God is not only a sign of man's distrust in God, but also a sign of distrust in the values and worldviews he advocates. Why did God die? When Nietzsche proclaims the death of God through the mouth of a madman who picks a lamp in the daytime, he also says: God was killed by us.

One of the possible explanations is that we no longer

rely on God and the Church's explanation to understand the world, but use a certain object as the scope of study, based on experiments and logical reasoning, to obtain a unified and exact objective law and view of truth to grasp all the objects or phenomena that exist objectively in the natural world.

In the past, according to the interpretation of the Christian Church, the world was purposive, and everything was created by God: the whole universe exists naturally because God wants to use it to realize various purposes.

But with the study of certain objects as the scope of research, based on experiments and logical reasoning, the search for unified and exact objective laws and truths has emerged from a systematic body of knowledge, which accumulates and organizes, and can examine the explanations of the universe.

From a systematic body of knowledge, which accumulates and organizes, and can examine the explanations and predictions about the universe, it appears that all objects or phenomena that exist objectively in the natural world are only operating according to certain natural laws, and there is no purpose to speak of.

Instead, there is a mechanistic view of the world. God slowly loses his influence and dies as science takes certain objects as its scope of study and seeks for unified and exact objective laws and truths based on experiments and logical reasoning.

4. Heidegger's death and dying are two different concepts of existence.

At the end of the 19th century, when mathematics, physics, biology, and other disciplines were caught in a "fundamental" crisis, non-European geometry, relativity, and life science were all in the process of finding unified and definite objective laws and truths based on experiments and logical reasoning, with certain objects as the scope of study.

At the time of the scientific revolution, Heidegger used the phenomenological approach to propose a holistic, fundamental and critical inquiry into the real world and human beings; Heidegger pointed out that the explanation of "life from death" is

Death and death are two different concepts of existence. Every year, every day, every hour, and even every minute that we live is a process of dying.

In this sense, human existence is the process of death. Death, on the other hand, refers to decease, the true demise of a person in the physical sense, the end of a person's process toward death.

The difference between these two is the philosophical branch of Heidegger's study of concepts such as existence, being, becoming and reality. It includes the question of how to classify entities into basic categories, and which entities exist at the most basic level.

Ontology is sometimes called existentialism, and belongs to the philosophy known as metaphysics, the key point of the ontology of death. As long as one is not dead, one is living in the direction of death.

The life of the being is a whole process towards death, a process that precedes the form of "being" of death.

In this process of dying, one can really feel the strong sense of existence of the self, and one is "present" in this process of dying.

Therefore, compared with the result of death, the process of dying is more genuine, more real and concrete in its immediate reality and condition.

The reason why Heidegger proposed the major philosophical concept of death is that, from the height of philosophical and rational thinking, the concept of "death" is used to stimulate the hidden inner substance of our mind, the idea and desire to continue to live, so as to stimulate the hidden inner substance of the vitality of life.

This is like the Chinese saying: "To live after death". For Heidegger knew very well that, compared with the greed of self-love and the instinctive power of desire for satisfaction, one cannot awaken spiritually unless one is driven to a dead end in one's mind.

A person who lives and dies in confusion, who cannot be spiritually awakened, has no meaning or value to the world in terms of the existence of his own life itself.

At most, it is a certain nature or phenomenon, which is manifested through all the objects or phenomena that exist objectively in the natural world.

Or it is the subjective will of man to manifest the "small existence" of the being himself in the "big existence" of the universe through his actions, products, works, and so on.

Heidegger uses this philosophical concept of the

"countdown" method of death to make people understand that each of us can prolong our life, and that this prolongation is "internal", that is, through the method of internal spiritual growth, and that we can look down on the spiritual temptation of all kinds of fame and fortune.

We cherish every second of our lives, and we develop an active and aggressive consciousness and inner vitality in our lives. By improving the quality and length of every second of life, we can increase the effectiveness of life and the density of our goals, and only in this way can the meaning and value of life be revealed in a limited time with unlimited possibilities.

When the Second World War occurred again in Europe ten years later, and the shadow of emptiness in Europe overshadowed everyone, a high school teacher founded the existentialist publication Modern Times, and intervened in literature as the voice of intellectuals to enter the public eye.

In the French literary scene, it goes back to the absurdist literature, such as Becket, Adamov, Jean Genet, Moliac, Isaeli and surrealist literature, such as Breton and Villon, and down to the new French novels.

When Beauvoir's The Second Sex appeared in Modern Times in 1948, feminists saw the first light in the most difficult of times.

Existentialism, one of the contemporary European trends, is basically related to the holistic, fundamental, and critical inquiry tendencies of S. Kierkegaard (1813~1855), Karl Jaspers (1883~1969), Martin Heidegger (1889~1976), Gabriel Marcel (1889~1973), Jean P. Sartre (1905~1980), and others towards the real world and human beings. (1905-1980), Jean P. Sartre)

They are related to the holistic, foundational, and critical tendencies of inquiry into the real world and people. They have various faces, whether in the disciplines of religion, politics, or analytical thinking, inquiring and reflecting on fundamental questions about life, knowledge, and values.

But they also share some of the same ideas and themes, such as the rejection of rationalist and empiricist dogma, and the preference of existential questions over the knowledge of disciplinary inquiry that explores and reflects on the fundamental questions of life, knowledge, and values through analytical reflection.

It is an irrationalist thought that explores the existence of human life itself, the true meaning of the philosophy of existence. Nietzsche and Kierkegaard can be regarded as its forerunners.

In the twentieth century it spread widely, and its holistic, fundamental and critical inquiry into the real world and human beings continued into the humanism that emerged in the 1960s.

Jaspers and Heidegger, Paul Salt, and Camus were among its exponents. These propositions have influenced literature, psychoanalysis, and theology; in 1953 when the American writer Wright, who described "existentialism" as a philosophical irrationalism, argued that the meaning of human existence cannot be understood.

It argues that the meaning of human existence cannot be answered by rational thought, emphasizing the individual, autonomy, and subjective experience. When it was introduced into black American fiction, 11 years later Martin Luther King Jr. forced the U.S. Congress to pass the Civil Rights Act in 1964, which outlawed segregation and racial discrimination policies.

Section 6: What Existentialism Wants to Achieve. (1) Confrontation

(1) Confrontation with essentialism, holistic philosophy, and static metaphysics.

We often say that "we see the essence through the phenomenon", and this way of thinking implies that behind any phenomenon there must be a higher essence, hidden in the substance of things. The essence is superior to the phenomenon, and mastering the essence is equal to mastering the phenomenon.

The essence is the fundamental characteristic of all objects or phenomena that exist objectively in nature, and is the common link between things; the external manifestation of the essence of things, different phenomena can have a common essence, and the same essence can be manifested as thousands of different phenomena.

There are millions of species of organisms found on earth, each with its own particular form of life, manifesting itself in infinite complexity and diversity of life phenomena, but all of them have a common nature, which is the existence of nucleic acids and proteins.

Atoms have a common nature: they are composed of protons, neutrons and electrons, and the differences in the number of protons and neutrons form different atoms.

To understand all the objects or phenomena in nature, we must see the essence of things through the phenomena, that is, we must grasp the substance of things hidden inside things, in order to truly understand the true nature of objective things.

The essence is "hidden" inside all objects or phenomena that exist objectively in nature and cannot be perceived directly by the senses. The present is the external form and connection of things, the external, figurative expression of things, which we can perceive directly.

When people come into contact with a thing, they always recognize its rich and colorful phenomena first, and from their senses and perceptions to the combination of all the things that human beings live in this world, they perceive the combination of all the things that make up the form and connection, and obtain the overall perceptual knowledge about the thing.

By analyzing all the objects or phenomena that exist

in nature, we can summarize the common connections that are hidden in the substance of things, which can help us understand the essence of things. Understanding is the process of deepening from phenomenon to essence.

On the one hand, the essence of all the objects or phenomena that exist objectively in nature exists in the phenomena, and it is impossible to know the essence of all the objects or phenomena that exist objectively in nature if we leave them.

On the other hand, phenomenon is not equal to essence.

On the other hand, the phenomenon is not equal to the essence, grasp the objective existence of all objects or phenomena in nature, is not equal to know the objective existence of all objects or phenomena in nature, the essence of the knowledge must go through the process from one-sided to comprehensive, from external to internal gradually deep.

Objective things include not only the phenomenon and the essence of two aspects, and the essence itself has a hierarchical nature, people's knowledge of all objects or phenomena that exist objectively in nature, always from the phenomenon to the essence, from the less profound essence to the more profound essence of the process of infinite deepening.

The process of knowing is from the individual to the general and from the general to the individual. When people know the special nature of many different things, through abstraction and generalization, they can know the common nature of various things by the special nature of all objects or phenomena that exist objectively in the natural world.

The grasp of the universal nature of objective things will promote the re-knowledge of the special nature of all the objects or phenomena that exist in nature.

The leap from phenomenon to essence, from special essence to common essence, from primary essence to more profound essence, from sensibility to rationality, is the dialectical process of human understanding from superficial to deeper and deeper.

For example, Plato's idea, Aristotle's essence, and Hegel's absolute spirit. This is essentialism.

The essence is stripped from 'the observable and observable facts, and the variability and multiplicity of the observable and observable facts are removed.

The essence behind the facts that can be observed and observed by the sensibility is the reality. While satisfying man's desire to control nature and gain the power to dominate everything, essentialism also privately supplements the facts that can be observed and observed.

Between the difference of appearance and the same phenomenon, the space of mutual traction exists within the object perpendicular to the contact surface of the two adjacent parts when the object is subjected to the pulling force.

The essentialist, while transcending the nature of time into a static metaphysics, also blocks our thinking about objects and phenomena as a whole, developing into subjective and arbitrary, anti-scientific philosophical theories or blocking thinking by people or environments other than ourselves. (But essentialism is the natural response of man in nature.)

(1). Essentialism is divided into essence and phenomenon

In dividing the world into a dualism of essence and phenomenon, essentialism also divides the world into a subject-object dualism. This explanation, which is similar to the separation of matter and consciousness, is often accepted in its entirety. But this is not good.

The central question is whether memory belongs to matter or to consciousness. If memory goes to matter, then consciousness is an incomparably pure thing that is at most an unborn seed apart from matter.

If memory goes to consciousness, then matter becomes a heavy burden, and consciousness cannot be free unless it is free from matter. The body and mind are one, the subject and the object are one. It should be the most understandable state. So this philosophical question becomes a question of state.

It is a question of whether there is more on the left or more on the right. The difference is that there is not only matter. If matter is large and consciousness is small, people will live well, just a little rougher, but with less trouble, but more intensity.

Either only consciousness or the unity of consciousness and matter will remain. Only consciousness will lead to nothingness, which is not good unless you want nothingness.

You can't do a "move", a "do it" action. The unity of consciousness and matter is the realm of no-self, but everything that represents me looks at the essence through the phenomenal world, and the essence is outside the phenomenal world, in the realm of the transcendental world.

When one uses the transcendental to guide one's life, then in reality such an approach may be blind, artificial, and in a sense even hypocritical and conspiratorial.

For example, in many psychological terms (empathy), psychologists refer to the act of transferring one person's emotions to another person or object as empathy.

Empathy itself is not verifiable. Then, as long as it is justified, one can say whatever one wants. Or perhaps the nature of the phenomenon of empathy is not a projection of emotion or anything else. When a poet loses a significant love, he writes a poem about her.

Is this poetic writing empathy, or is it some other literary technique? Both Heidegger and Schart have criticized psychology for its fundamental role in explaining other non-psychological facts or laws, the psychological theory of empathy.

Even if different people claim that they each experience a common transcendental activity, such as hearing the voice of God, it is still difficult for a third person to test whether they heard the same voice.

Therefore, the reality claimed by the transcendental experience is subjective, and the essence of the believer's eternal truth guides everything that exists in heaven and earth: religion, morality, God, dialectic, sameness, universality, etc.

So what Nietzsche destroyed was not morality or God himself. What Nietzsche means when he says that God is dead is that the transcendent, the essence that guides us is a figment of our imagination and we do not have to submit to these things or phenomena.

We do not have to submit to those things or phenomena that have positive meaning for people or groups, that are valued by people, or that satisfy people, that become objects of respect or interest for people to pursue, and that die a slow death. "The core of morality lies in the creation of idols, the confusion of nature and subject matter, the taking of the subject/symbol as truth."

Kierkegaard describes human existence as having

three different levels: "sensual," "rational," and "religious" (or "aesthetic," "ethical," and "religious"). Sensual people are either hedonists or people who are passionate about life experiences.

They have the tendency to look at things from the perspective of the subject's own needs, which are the attributes that an individual can have in terms of view, experience, consciousness, spirit, feelings, desires or beliefs.

They are creative, have no commitment to the world, have no responsibility, and feel that the world is full of possibilities. Rational people are realistic, full of commitment and responsibility to the world, and clearly understand the moral and ethical rules of the world.

Therefore, unlike emotional people, rational people know that the world is full of limitations, impossibilities and doubts. In the face of impossibilities and doubts, human beings who have carefully considered all the evidence and reasoned their way to a reasonable conclusion can only give up or deny and grieve forever for what they have lost.

At this time, man can only rely on the "leap of faith" to

enter into the universal cultural characteristic of all mankind, which has a mystical mythology, and which is the stage manifestation of the human spirit, using the power of faith to overcome doubt and reason, which is usually considered impossible. Only faith can restore the hope that "all things are possible".

In Fear and Trembling, Kierkegaard reflects on the Old Testament story of Abraham, the "father of faith," who obeyed God's instruction to kill his son and offer him as a burnt offering.

He argues that Abraham's actions would have been meaningless if he had not cared about the life and death of his son, had no moral or ethical or familial struggles, or had considered it a moral rule to kill his son at God's behest.

The value of Abraham's action lies in his leap from reason to religion, his belief in the power of God, his belief that all things are possible, that miracles will happen (and that God sent an angel at the last moment to prevent Abraham from killing his son). According to Kierkegaard, there can be no faith without reason.

(2) Kant's a priori turn

Kant's Copernican revolution (a priori turn) is still stuck in the idea that human knowledge does not come from nature, but from the brain itself.

It is the process of analysis, synthesis, judgment, reasoning, and other cognitive activities based on images and concepts, and cannot enter modern philosophy. Then, what is the a priori turn?

Kant, as a comprehensive philosopher, is confronted with the meaning of the answers left by theoretical and empirical theories, which are awkward, banal, and tiresome.

Neither the subjective, arbitrary, anti-scientific philosophical theories of Europe nor the British philosophical arguments that cast doubt on the possibility of objective and reliable knowledge, nor even its denial, could be tolerated by Kant.

Kant sought to establish a new foundation for science that would address the legacy of his predecessors. Both conventional materialism and empiricism make the epistemological mistake of assuming that "our intuition or experience" must conform to "the object that stimulates us.

Like the scientific revolution in the field of mathematics, he says: Assuming that objects conform to our knowledge, we expect a knowledge of objects, objects that determine something about objects before they are presented to us.

That is, our world of experience has been filtered by some innate "filter" (in fact, a priori form) of ours, and everything presented to us is an image given by perceptual intuition, not the object itself.

He would tell us that the reason why Kant was able to make the distinction between the world of phenomena and the world of objects is that he examined the object-object unknowability of the world from the point of view of human cognitive capacity, and was able to present only the miscellany of sensory experiences that it gives to us without being processed by the a priori forms.

And finally, our manifestations are synthesized by the a priori imagination. He thus solves the dilemma of the traditional argument that the human being is composed of two parts, the "mind" and the "body", as opposed to the materialist argument that "a person's body is all that it is".

It also opens up a phenomenological way of thinking that allows knowledge to be built not on a "mysterious" "all-encompassing" universe or God, but on our "a priori form or a priori domain".

From this point on, mankind has truly abandoned the traditional solipsistic metaphysics and opened up a true "theory of being" (in Heidegger's sense, the distinction between "being" and "existence").

Similarly, we can know that before the birth of human beings, the "earth", the object that appears to us, did not exist, that is, "could not be", but referred to a "pure existence" in Hegel's sense. That is to say, the earth is a retrospective construct that only "exists" but cannot "be".

What Kant means by "human reason legislating nature" is more like "human ability to provide innate concepts and to synthesize sensory material to form knowledge to legislate the empirical world".

Kant has always opposed the illegitimate use of the ability to provide innate concepts and to synthesize sensory materials into knowledge, and all our knowledge is built up in the sphere of knowledge. What vulgarized materialism, which is always "matter before consciousness", never passes through this Kantian approach.

They never cross this bridge of reflection on Kant's holistic, fundamental and critical inquiry into the real world and man, and vilify Kant as a man incapable of normal thought, who provoked the upper part of the skull of a giant, i.e. the top of the head, with his own narrow mind.

In his book, "The Disease that Kills," Kierkegaard argues that despair is a failure to accept one's unwanted self, or to cling to one's present self and eventually "lose oneself," which is what Christianity refers to as original sin.

A desperate person does not necessarily know that he or she is desperate, nor does he or she necessarily feel pain. The lowest level of despair is the ignorant person, who only knows worldly things.

This type of people do not have their own personal use of sensation, perception, thinking, memory and other mental activities, their own physical and mental state and the environment, people, things and changes in the comprehensive awareness and understanding, do not know the eternal nature of the self, and do not know that they are in despair.

Other individuals use their senses, perception, thinking, memory, and other mental activities to become aware of their own physical and mental states and changes in people, events, and objects in the environment, and their ability to recognize the environment and the self.

The degree of clarity of cognition to the degree of cognitive ability and clarity of cognition of the environment and the self that one desires to obtain all objects or phenomena that exist objectively in nature, but still do not have the eternal nature of the self.

Other people begin to use their senses, perceptions, thinking, memory and other mental activities to become aware of and recognize their own physical and mental states and the changes of people, things and objects in the environment, as well as the cognitive ability of the environment and the self, and the clarity of cognition.

In addition, they are unwilling to accept this self and fall into another kind of despair because of their ability to recognize the environment and the self, as well as the clarity of the recognition of themselves as a worldly thing and the weakness of despair. Further, some people decide to accept the weakness, to listen to God, and to recognize their own eternity.

Further, they must be willing to accept themselves as they are. They may choose to regain hope and escape from despair through a "leap of faith", but they may also choose to see despair as the ultimate truth and place themselves in eternal despair.

Thus, people have different levels of despair in different levels of existence. A sensual person despairs for all the objects or phenomena that exist objectively in nature, while a rational person despairs for the sake of rejecting the self or choosing to see despair as the ultimate truth.

Faith is the only way out of despair, choosing to believe in a certain religion or a certain doctrine, worshiping it and holding it as the standard and guide of words and actions.

It is also the only way to achieve self-realization. Ethics, that is, the symbols, cancel all the directness of man (Godliness), losing his own nature and becoming an element under the universal social sphere and norms." In essence, what Chickgo is telling us in this passage is that the essentialist does not consider things outside the symbols.

(2) Breaking free from the constraints of the social matrix

At the end of the 19th century, there were aestheticism, decadence, surrealism, and symbolism in literature. As the last successor of Romanticism: using a rebellious and skeptical way, the macrocosm expresses the perspective of observation of all the objects or phenomena that exist in nature objectively, focusing on the overall understanding and mastery of things.

The narration of a real or fictional event in prose or poetic form; or the expression of distrust in the narration of a series of such events. Since the Enlightenment, the essentialism of the tools of science and technology has led people to use concepts and make judgments.

The ability to conceptualize, judge, analyze, synthesize, compare, reason, and calculate has overwhelmed the human mind, thought, and psychological condition.

It deprives the human mind of the rational and active

process of reflection on the real world or any object in the form of "inward dialogue".

It also suffers from the separation or even opposition of two things that are naturally interdependent or harmonious, and suppresses or restricts human thoughts, feelings, and behaviors for people or environments other than oneself.

Transcendence itself has no place in essentialism. Picasso uses African masks to express the emotions of surreal strangeness and helplessness, and often feels fear and discomfort in life.

Picasso used African masks to express surreal strangeness and helplessness, and Marcel painted naked women descending stairs. In such a difficult ideology, two world wars broke out immediately afterwards, and everyone had no new values and beliefs, and the protagonist in Moliak's novel was inexplicably taken as a soldier and inexplicably got into bed with his brother's wife.

The European people, who have lost their sense of direction, feel lost or have lost their focus in life, and in such an environment, there is a great need for a new value.

As in the Second Renaissance, people were too self-conscious to apply a style of caring only for themselves and not for others, an attitude and psychological state that replaced the means and methods of controlling and changing the natural environment, affirming the freedom of man himself.

It usually refers to an enlarged form of cliquism or individualism, as opposed to collectivism. The main characteristic is that when dealing with the relationship between the unit and the department, the whole and the part, one only cares about oneself, not the big picture, and is indifferent to other departments, other places, and other people.

It is believed that individuals use their senses, perceptions, thinking, memory, and other mental activities to perceive and understand their own physical and mental states and changes in people, events, and things in the environment, and to form cognitive perceptions of people and things because of their cultural background or life experiences.

When considering the big picture and global issues,

people take the self or the small group as the center, and stand on the local position regardless of the advantages and disadvantages.

Modern Times is a French magazine founded by Simone de Beauvoir, Jean-Paul Sartre and Maurice Merleau-Ponty.

It was named after Charlie Chaplin's 1936 film, and after the first issue was published and printed in October 1945, a kind of crypt called "Tabu" appeared in France, which forbids contact with and talk about sacred or polluted things, etc.

It is a taboo, a taboo in terms of religious superstition or social customs. It was a place where all existentialists gathered, young poets, artists, painters, played jazz, danced, chatted, etc.

(1). The Liberal Philosophical Tradition

Since the entire Western political system is based on the influence of liberal principles and values, we can say that liberalism is the dominant ideology of the modern West.

In philosophy, this is a generic term for metaphysical views that claim that the real world, or the real world as perceived by human beings, is based on and constructed on the mind, and that those who do not conform are immaterial.

It is a branch of philosophy that explores the nature, origin, and scope of knowledge. The relationship between the theory of knowledge and epistemology is currently controversial, with some arguing that they are one and the same concept, and others arguing that they are in fact two different concepts that are closely related.

Idealism represents a skeptical attitude toward knowing anything that is independent of the mind. Liberalism, however, is one of the most undefined or ill-defined of all fundamental concepts, with no clear definition of words, terms, notes, or concepts.

It is not only a political philosophy, an ideology, but also an institution and even a political movement. Different people, with different attitudes, understand liberalism from different perspectives, resulting in very different liberalisms.

In general, although the Enlightenment has shaped the basic face of liberalism, liberalism cannot be regarded as a monolithic ideological system for two reasons.

First, there are regional and cultural differences in ideological movements, and there are regional or national differences.

The Enlightenment is an ideological movement developed in Western Europe and in different regions. The regional cultural differences caused the British Enlightenment to be different from the French Enlightenment, and the French Enlightenment was different from the German and Italian Enlightenment, so it is conceivable that liberalism also has more or less regional or national differences.

Secondly, liberalism is not an unchanging doctrine.

From the end of the 17th century to the end of the 20th century, liberalism (in any region) has responded to the challenges of the times by adjusting its ideology to reflect the rightness or wrongness of one's views, or by reinterpreting in new ways one's tendency to believe in the rightness of a particular viewpoint and to dictate one's own actions.

Alan Ryan argues that modern Westerners have inherited not a single liberalism, but many liberalisms.

Liberalism is a fundamental political belief, a

philosophical and social movement, and a social construction and policy orientation that treats liberty as a fundamental method and policy of government, an organizing principle of society, and a way of life for individuals and society.

According to the literature, the term "liberalism" emerged in the early 19th century. When the Spanish Liberal Party first adopted the term "liberal" in 1812 to mark their determination to promote a constitutional government, liberalism began to be used as a term with modern political meaning.

With the development of the liberal trend in the 19th century, especially due to the competition between liberalism and its socialist counterpart, the ideology eventually became more and more systematic and dogmatic in its ideology.

However, some scholars of liberalism have argued that liberalism and Western civilization began in ancient Greece and ended in the contemporary era.

Watkins and Schapiro argue that Socrates is the progenitor of Western liberalism on the grounds that he lived a life of rationalism and skepticism, constantly reflecting on and criticizing unexamined concepts and beliefs, and eventually dying a martyr's death, serving as a model for defending one's freedom of thought from infringement.

However, Havelock, who also traces liberalism back to the classical period, has a very different view. He believes that the spirit of freedom in ancient Greece was not found in Socrates, let alone in Plato and Aristotle, but in the rebellion against Socrates.

Rather, it was in the tradition of the Wise Men, who rebelled against the Socratic school. Hefrak points out that Plato's invention of the "form" and Aristotle's definition of the purpose of man and the supremacy of the city-state are both incompatible with the spirit of liberalism.

(2). Representatives of classical liberalism

Basically, there is a distinction between [strict ideology] and [diffuse ideological beliefs]. Political thought or ideology in general is the monopoly of intellectuals, who either express their own views on humanity, society, and the state, or review the views of others on similar issues.

However, these objective existences are reflected in

human consciousness, and the results or the exchange of viewpoints and conceptual systems formed through thinking activities are not sufficient to become the platform for large-scale collective action. However, ideology has a universalizing power that transcends individuality and personality.

In the sense of a single word, ideology is a systematic collection of states of mind that trust, have confidence, or rely on someone or something, and it brings together the best of many thinkers to define the ideal order of insight.

Therefore, there is a distinction between [strict ideology] and [diffuse ideological belief], which usually includes quite clear statements about human nature, the relationship between the individual and the social state, the relationship between economics and politics, and political goals and action plans.

In other words, ideology is a philosophical field, which can be understood as the understanding and cognition of things. It is a kind of perceptual thought about all the objects or phenomena that exist objectively in nature, and is the sum of concepts, viewpoints, concepts, ideas, values, and other elements.

It is a state of mind and an action program that promotes trust, confidence, or faith in someone or something. It is often distinguished from other ideologies in the form of a dichotomy, and in the object of appeal, it attempts to make the general public followers.

In Democritus, we find atomism and government contract theory; in Antiphon, we see intellectuals attacking the folk customs that have been passed down from generation to generation; in Protagoras, we find pragmatism, empiricism, and value pluralism.

These are the representatives of classical liberalism, although their claims are too crude and not sophisticated enough.

The elements of liberal theory contained in these ancient ideas are, in a strict sense, only the precursors of liberal theory, not systematic liberalism, which is a product of modern times.

Since most scholars of liberalism agree that the real origin of liberalism should be in the modern era, the time extrapolation should be from the 16th century to the 18th century, but it is certain that by the Enlightenment period, liberalism had already been developed systematically and conceptually.

The Enlightenment began around the end of the 17th century, covered the entire 18th century, and culminated in the French Revolution. During this period, famous thinkers such as Locke, David Hume, and Adam Smith in England, Voltaire, Rousseau, and Montesquieu in France, Kant in Germany, Thomas Jefferson and Théodore Penn in the United States, and the French, the French, the French, and the French. Jefferson, Thomas Paine, and so on.

The expertise and ideas of these thinkers were of course different, but the atmosphere and spirit that they shared was enough to enable future generations to identify a general common view of each other.

This commonly recognized concept or idea is based on the promotion of reason, inquiry into nature, belief in the goodness of man and the infinite improvement of society, and the conscious activity of the people of society to transform nature and society.

The conscious activities of the people in society to transform nature and society include breaking down superstition (replacing church theology with natural theology), promoting religious tolerance, defending freedom of thought, promoting public education, demanding economic deregulation, and establishing a democratic government with separation of powers and checks and balances.

From the development of liberalism in the not-so-distant past, we can obtain the connotation of four kinds of liberalism.

The first is "political liberalism".

"Political liberalism, the main task of the early formation of liberalism, was to oppose absolute feudalism, to fight for the political rights of the individual, and to establish constitutional government, from Locke and Banjamin Constant to the politics of John Mill.

The discipline of analyzing and reflecting on the fundamental issues of life, knowledge, and values, and making logical deductive conclusions about natural and social phenomena through generalizations and deductive reasoning based on existing empirical knowledge, experience, facts, laws, cognition, and tested hypotheses, broadly reflect this connotation of liberalism.

The second is "economic liberalism".

"Economic liberalism", the 18th century liberalism, is a process of collective decision making by various groups, and a specific relationship between various groups or individuals for their respective domains.

In particular, it refers to the shift from the sphere of rule of social groups to the sphere of all actions and states in which human beings use all kinds of goods to satisfy their desires.

The protection of the right to freedom is also concerned with the government's involvement in the production and reproduction of material materials in society as a whole.

and the state's lesser interference and control over the production, sale and distribution relations of a country's citizens, or the state and the individual's income and expenditure.

The third is "social liberalism".

"However, social justice is a very broad term that encompasses issues such as income inequality, women's rights, homosexual rights, racial inequality, health inequality, the right to fair education, etc., and concerns the basic conditions of survival of the weak. Starting from

John Mill, the neo-liberalism represented by T. H. Green to Rawls, more attention was paid to social liberalism.

The fourth is "philosophical liberalism.

"Philosophical liberalism is the study of universal and fundamental questions, including the fields of existence, knowledge, values, reason, mind, and language.

Philosophy differs from other disciplines in its holistic, fundamental, and critical inquiry into the real world and people, with a unique way of thinking, such as a critical, often systematic approach, based on rational argumentation.

In everyday parlance, a philosophy of analytic inquiry and reflection on fundamental questions of life, knowledge, and values can be extended to the most fundamental beliefs, concepts, or attitudes of an individual or group, although this is not the definition here.

It is the fundamental systematic, organized law or argument about the relationship between individuals, nations, and societies. They are derived from practical verification or from conceptual derivation.

From the beginning of liberalism until now, liberals

have always emphasized individualism based on reason, and the preservation of individuality has always been the core of liberalism.

Since liberalism is a systematic and complex conclusion about the knowledge of nature and society, which is summarized by people from the practice of social matrix, different scholars, from different positions, emphasize different sides.

Different scholars, from different positions, emphasize different sides, making the liberal form of thinking and the connotation of reflecting the nature of things appear confusing. Most libertarians tend to define liberalism as a doctrine that emphasizes freedom in opposition to conservatism, which emphasizes order, and socialism, which emphasizes equality.

The doctrine of justice, of which Rawls is most proud, is based on the social contract theory of Locke, Rousseau, and Condé, which argues for the moral value of Western democratic societies and rejects traditional consequentialism, arguing that justice is the primary virtue of social institutions, just as truth is to systems of thought.

Unjust laws and institutions, no matter how effective, should be transformed and removed. He also believes that justice is closely linked to social cooperation and points out that a distinction should be made between [the principle of justice for institutions] and [the principle of justice for individuals].

First, the principle of justice for institutions.

Everyone has the right to the same liberties as others, including the political rights and property rights of citizens. The right to use all kinds of goods to satisfy the desires of society and human beings.

It is usually referred to the relations of production, sale, and distribution among the citizens of a country, or the inequality of income and expenditure of the state and individuals, and should be arranged in such a way that people can reasonably expect such inequality to be beneficial to everyone, and that status and official positions are open to everyone.

Secondly, the principle of justice for the individual.

The first and fair principle is that if the system is just, the individual accepts it voluntarily, and benefits from it, in which case the individual should abide by it.

Rawls calls the rule of law "formal justice" or "justice as regularity," that is, the regular and just enforcement of public rules. The law is the imposition of public rules on rational people, designed to regulate human behavior and provide a structure for social cooperation.

Freedom, on the other hand, is the sum of various rights and obligations stipulated by the system, so the rule of law and freedom are the interactions and interconnectedness of things within each other, between conflicting parties and between things.

Rather, it should be seen as an attempt to make a situation that does not exist become a factual principle of justice, and to provide the best policy with a moral function.

Based on the above generalizations of the basic connotations and principles of liberalism by different scholars, liberalism can be broadly interpreted as "a liberalism as a theory, an ideology, or a liberal political philosophy, which has a moral function.

The core of its concern is the relationship between the individual, the state, and society, and it emphasizes individualism as the fundamental starting point to protect the individual's right to freedom and equal rights.

Section 7: The actual existence of freedom as

01. Existence in oneself

(01). Definition of Self-Existence

For the philosophical view of "Being and Nothingness", the existence of a human being consists of two mutual comparisons of a substance or a whole with another substance or another whole, or of existence or change depending on certain conditions.

The in-itself and the for-itself. The in-itself existence refers to the object of consciousness, while the for-itself existence refers to the subject consciousness. The in-itself is literally the existence of being in itself.

That is to say, its reality is in itself, its fullness is in itself, and all its inclusions are full of its totality. It is all that it is, and no meaning can be given to it that can be imposed on it.

It bears its own existence, and is present as if it were not, real as if it were not, and therefore it is impractical, inscrutable, unknowable, and undecidable. Its existence is simply "existence", without any reason.

Realism is generally considered to be concerned with reality and practicality, to the exclusion of idealism. Therefore, the existence of facts and situations in the immediate future, "reality" in the narrow sense, has different conceptual levels in philosophy, including phenomena, facts, realities, and axioms.

The nature of existence is a common topic in metaphysics. For example, the Greek philosopher Parmenides argued that reality is a single, unchanging existence, while another philosopher, Heraclitus, argued that all things are new.

The twentieth-century philosopher Martin Heidegger argued that earlier philosophers had focused on the problem of existing things, ignoring the problem of "being" itself, and therefore needed to return to Parmenides' argument.

The philosophical Aristotelian theory of scope - for example, time, space, quantity, quality, relations, etc. are all areas that try to list the basic combinations of reality.

The question of whether or not there is a reality is a proposition that has been infiltrated in recent times, especially with regard to quantum science and ontological proofs of the existence of God.

Therefore, a holistic, fundamental and critical inquiry into the real world and human beings suggests that there are two kinds of reality, one is the reality of nature and the other is the reality of thought (including linguistic and cultural aspects)

First, the reality of nature.

On the one hand ontology focuses on the study of what exists, but there are various ways of expressing it, such as being, existing, "what is" and reality, etc. The task of ontology is to describe what the most universal realities consist of, and how they relate to each other.

Some philosophers distinguish between reality and existence, but many analytic philosophers avoid using the words "real" or "reality" in their ontological discussions and use "existence" instead.

An important question in analytic philosophy is whether existence is one of the essences of things, and most philosophers believe that it is not.

Second, the reality of thought.

In metaphysics and epistemology, the discussion of objectivity involves the question of whether "reality" depends on ideological and cultural factors. Such as feelings, beliefs, religious philosophies and political movements, and worldviews.

Metaphysics is in the study of problems and epistemology that cannot be obtained directly through perception through rational reasoning and logic; epistemology is one of the themes of philosophical inquiry, and its connotations are divided into two definitions.

One is the theory of the origin, nature, limits, and validity of knowledge, a definition that makes epistemology in use in conjunction with Gnosiology.

One is the theory of knowledge, which is a systematic analysis of the concepts of knowledge that people use to understand the world.

Epistemology can also be divided into a broad sense and a narrow sense, with the narrow sense exploring only the validity of knowledge. The broader epistemology discusses objectivity in relation to whether "reality" depends on ideological and cultural factors. Such as feelings, beliefs, religions and political movements, and worldviews.

Realism, on the other hand, holds that reality exists independently of any feelings or beliefs, and as long as the object, its existence, and its basic characteristics can be distinguished, it can be called the reality of the object if it is not influenced by any feelings, language, beliefs, or other human factors.

The earliest anti-realism was the Irish empiricist Georges Berkeley's claim that things are actually a concept of human thought, and that reality exists in thought and is a product of thought, hence the name idealism.

In the 20th century, similar views were called phenomenalism, but there were various views, such as Bertrand Russell's theory that thought itself is a collection of feelings and memories.

So there is no thought or soul above the event. In addition, social constructivism believes that the external world is only artificially constructed by society and culture; cultural relativism believes that social units, such as morality, are not absolute, but are just a component of culture.

Existence (the presence or absence of something) has become a contrast to its essence (what something is). Since existence without essence seems to be a void, philosophers such as Hegel associate it with nothing.

Nihilism represents a negative argument for existence, while absolute is a positive one. We can only say that this table and chair exists, that this building exists, that this electric car exists, that this Golden Gate Bridge exists, that the Statue of Liberty exists, and by extension, in a more general sense, that all, all, everything exists.

In a nutshell, we can only say "it exists" and nothing more. "It exists" is everything. We cannot explain why the world exists, we cannot find a reason, a basis for existence, it is simply outside of determinism.

More precisely, the existence of anything that is the object of action or thought is not directed toward some end or toward what Aristotle called the "first cause.

In the Truth Correspondence Theory explanation of knowledge, it is argued that if existence is "real" knowledge, the description of it must correspond to its reality, e.g., the description of the scientific method can be confirmed by observing what exists.

There is nothing that shows the existence of something that cannot be observed and described. For example, if extraterrestrial beings exist, they must be observed repeatedly, not just inferred from the existence of something that is the object of thought.

"It exists," and the basis for its existence can only be attributed to itself. Some medieval philosophers attempted to explain the origin of existence in terms of creationism, but this was in vain.

For whether or not existence was created by the creator, it is impossible to deny the fact of its existence, its existence outside of creation, or even its reappearance. This is to say that we cannot explain existence by external determinism.

In fact, the reason, the cause of existence, can only come from itself.

Schart, "Being and Nothingness": Being is itself. This means that it is neither passivity nor activity.

Both concepts are things that are the object of human action or thought, and denote human action or the instrument of human action. ……... By leading these concepts to the absolute, they lose their meaning. In particular, existence cannot be passive.

In order to have ends and means for action or thought, the person or thing that is the goal must have existence. Nor is existence passive, which requires a stronger reason: in order to be passive, it must be existence.

The "self-identity" of existence is beyond the active and passive. The material existence of nature cannot be a product of God's creation, nor can it be a thing that depends on other rational actions or thoughts as its goal.

In his early novel "Vomit", the protagonist Rogandan lives temporarily in a small town where there is little entertainment.

Apart from going to the café, going to the library, spending time with the café owner, and occasionally going to bed to understand the boredom, there seems to be nothing else worth mentioning.

As time progresses, Rogandan gradually becomes bored with the study of the unnamed marquis from centuries ago, which even makes him feel sick to his stomach, and this disgusting feeling haunts him from time to time.

Through Roquentin, the protagonist, Sartre uses language to represent the image of things as a kind of "being", including the human body, which is in fact full of dirt, filth and disgust.

All living, earthly existence is an impractical, elusive "being" that represents death, lack of life, sluggishness, lack of sensation, and inability to be identified with consciousness.

Roquentin, the protagonist of "Vomit", an existentialist novel, is a historian who lives in the imaginary city of Bouvier. He expresses his imagination through things, a certain nature or phenomenon of the past, through something specific.

Or the subjective will of the human being is concretely expressed in the ego through actions, stories, works, etc. He is convinced that nothing else can give him his own nature, and then he finds that he has absolute freedom of autonomy, evoking one after another a vomitous feeling of resistance.

They are always absurd, inexplicable and brutal. This whole existence is a confusion. We cannot say that nature represents a good or has some moral connotation.

In fact, the existence of nature does not contain any reason to be proud of. This world is nakedly presented before us. The existence of self-life itself is nothing more than absurdity and vulgarity.

But we cannot help but ask in our hearts? Where does the existence of self-life itself come from? Sartre says that it does not come from anywhere, but it exists so rigidly and eternally that its only necessity is that it cannot cease to exist.

Therefore, we cannot explain its existence by a supernatural thing, such as God. It cannot rely on any explanation at all, not even its own. In such a case

Schart thinks that the best expression for the existence of this self-living being itself is that existence is what it is. That is, this "is what it is" actually implies that existence is outside the action or influence of the thing in question.

Existence is simply in itself, without any connection to something different from it. All relations, for any being, are external.

In that case, the existence of any being does not logically depend on the existence of other beings. For example, if we consider the relationship between a rich man and his wealth as external.

Then on the basis of this sense, if he did not have this wealth, his existence would still exist, but precisely because we describe him as the owner of this wealth, he would be internalized in it, and he would exist only because of it.

Thus, for Schart, there is no being that is not both existent and independent, and there is no grasp of which attribute is necessary for the recognition of the sameness of things.

Every "being" can only exist in its own fullness, and that is all they can be, they can only be in their own complete certainty, they dissolve in themselves.

Therefore, it cannot sustain any relationship with other existences. Further, it cannot see itself as a being other than other things, it does not know that there is such a form of relationship as "otherness".

Therefore, we can go on to say that the existence of the self-life itself is isolated in its existence.

In short, the existence of the Self is "what it is", it has no relation to other existences, and as we have said before, it is non-created, without reason, alone and large, unconscious and opaque, and one with itself.

We cannot have any interpretation of it (existence in itself). In other words, we cannot have an interpretation of the human brain's thinking activity that reflects the characteristics and connections of objective things and reveals the meaning and role of all objects or phenomena that exist objectively in the natural world.

All interpretations of it are futile. We can only say that "it exists" and that is all. In a relatively literary tone, Scharthe in particular This existence is described as massive, opaque, gloomy, glutinous, absurd, sluggishly, lazily. It is obscenely and horribly expanding its own existence, unable to cease to exist.

The existential absurdity of this self-existence itself is that it lacks any legitimacy or reason. It does not come from any place, all existence is just there, without any reason.

Including: this garden, this town, and myself. Of course, there are those who think that phenomena exist (gardens, towns, and myself) because of God, the subject of consciousness, or the ecological environment, and so on.

The result is that these existences become a necessary existence; they have the power to exist, to live, to work, to reproduce, to gain authority, and finally to have power or eternal life.

However, as we have previously interpreted, it can be seen that the existence of all beings (including the individual himself) is independent of determinism and has no cause. In the view of Schart, their existence is merely a contingent, not a necessary existence.

And certainly not a being from possibility. For we have a physical existence, or the existence of a definite rather than an abstract, generalized being.

We can only say that "it exists", we cannot say that it is possible or impossible to exist. It exists only as a self-living being in itself, without any reason for existence, and without any relation to other beings.

(02). Human reality-as-being is not self-existence

This "existence in oneself" does not refer only to the existence of things in general, but also to the existence of human beings. In short, it refers to the reality of the existence of the self-life itself, which means that the individual person exists in the world, in a certain actual state.

The individual person is always present in the world, he is not invested in the world by his subjective consciousness. When a person can already be aware of his own existence, he already exists in this concrete, real world.

The reality of the existence of the self-life itself, including the physical, psychological, educational background, occupation, the place where I was born, my past, my surroundings, and the reality of my birth, etc., is emphasized.

All of these emphasize the existence of the self, the limitations of existence in reality, and the "to be" of human beings. The existence of man's self-life itself is a being that can be known, and it is a "being" that precedes him in thinking about his existence.

This "to be" means that one's existence is always confined to a particular point, to a particular period of existence, to a particular environment, to a particular "is".

For example, I am a Taiwanese American, I was a

college student, I am a Taiwanese, I am a Buddhist, I have a high school education, and so on. Schart refers to the physical existence of a person, or a state of reality that is explicit rather than abstract and general, as the reality of a person (facticity).

And this "actuality", which exists in the way of "being what it is," belongs to the realm of "the in-itself.

The in-itself exists not as a being that "does", but as a being that "is" something. For people in general, this "reality" is a kind of limitation on human beings, and human being is thus affected by this "reality".

In other words, it is these social and cultural, genetic, and national constraints that limit what I am and what I will be, rather than my own determination of who I am. This "reality" seems to be a boundary that people can never get rid of.

However, Scharthe does not think so. He sees the reality of these people as something else, as obscure, dull, absurd, as insensitive as a tree or a stone.

It (reality) is not limited, undetermined, or even equivalent to human reality. Some determinists, for example, describe man's limited existence in terms of a particular space, a place that he can occupy.

For example, when a person is in a certain area, he is restricted to that area and cannot be in another place at the same time.

In other words, this means that the reality of human existence is limited by time and space, which is the limit beyond which human beings cannot go.

But, according to Schart, the real existence of man cannot be defined by place. On the contrary, it is the reality of man that constructs and organizes "here" and "there".

That is, it is because of man's existence that the place he occupies is called "here" (hear), and there are many divisions of "this" or "that"; this is my city, house, house, desk, chair, that is not where I live, and so on.

Therefore, it is because of the reality of man that "here", "there", "this", and "that" are distinguished from each other.

In fact, while one can only be "here", one also recognizes that there is a "there". .... Schart thinks that place is in itself, is ambiguous.

Or, again, status, race, etc., do not specify "human reality". Whether a person claims to be a blue-collar person or a middle-class person, it does not follow that these pronouns are the true nature of that person.

For, in reality, it is my choice or acceptance of being a middle class person, an American, that makes the middle class, America, exist. In other words, it is the "reality of man" that constructs and organizes the world and society.

Thus. The reality of man is not confined to this specific state of reality. On the contrary, it is the reality of human beings, their freedom of choice, that makes this state of reality possible.

Thus, existence for oneself does not depend on race or nation. Instead, it is the possibility of a complex system of self-selection (or acceptance) by race or the state as an aggregate that emerges. Or, some have suggested that "death" is a boundary that man can never cross. As soon as a human being is born, he or she is already determined to move towards the end of death, moment by moment, step by step.

Moreover, on a religious level, "death" has profound implications for human beings, such as doing good deeds during life to go to heaven after death, karma, and so on.

However, Saud does not see death as a restriction on human beings and does not prevent them from being a conscious being (free choice). For Schart, death does not represent a real possibility for man.

Because death is unpredictable and unpredictable. We can anticipate the arrival of the train, but we cannot anticipate the arrival of death, we only vaguely know that death is coming.

We cannot choose death; it is unrealistic, inscrutable, absurd, and wrong to the extreme. Therefore, this death is external to the human being, the reality of the existence of the self-life itself, an external final limit of my consciousness (the human reality).

When death occurs, self-existence disappears forever, becoming a solid existence, a past, an "in-itself existence". Therefore, death does not affect my thoughts, words, or actions, my desires and expectations.

The religious meaning of death, eternal life, is thus excluded from human reality. For the human reality, death has no meaning, it is merely "being" after "being for oneself" has disappeared.

Death, which I can never master, can never limit me, as a conscious living being. For, when death comes, I am no longer there.

Therefore, for the philosophical point of view of Schart's Being and Nothingness, these existences as states of reality are not the real existence of the existence of the self-life itself. Although, they are part of the existence of the self-life itself.

They are just like natural objects and cannot be used as an indicator of the real existence of man. According to Schart, man is always a creator.

While the being of the self exists in the world, he is constantly searching for materials, tools and opportunities from the real world in order to transcend his existing being.

In order to transcend the limitations of one's own existence, to move towards a more fulfilling existence, to challenge one's existing existence, and to create a more satisfying future, this is the real existence of the existence of the so-called self-life itself.

02. Being for oneself

The existence of the self existent itself not only has a temporal reality (facticity), but also has the autonomy to transcend its current state of reality and create a new future.

This means that the existence of the self exists itself contains countless possibilities. Man is not a being under causal determinism. The existence of the self existent itself is not just an existence in the state of reality.

Man is always capable of being dissatisfied with the present state and of challenging the world around his egoic being, showing his ability to create his own future.

This means that the being that exists in itself is always transforming beyond the present situation, beyond any kind of realistic obstacles that may affect the expression of its creativity.

This obstacle is not only the surrounding social environment, the body, the psychological state, but also the ideas, culture, morals, customs, arts, institutions and behaviors that have been passed down from generation to generation, from history to history.

The tangible and intangible influences and controls

on people's social behavior, especially the abstract ideas of metaphysics, such as goodness, rationality, God, etc., have already determined in advance the way of existence of human self-existence.

Under the abstract nature of human existence, it exists in every state of reality, just like a natural object in general.

For example, although human beings have life, the body cannot be immortalized; it will surely decay and die.

Therefore, it is said that "man accepts life without a choice, then spends it under the condition of no choice, and finally gives it back under the irresistible struggle.

For man's existence, like those of natural things, is already determined as to what it should be in the future. Therefore, the philosophical thinking of Schart's Being and Nothingness suggests that if man is really to be distinguished from natural things, from animals in general, then all social matrices must be separated.

It is necessary to remove from the human being all the social matrix of pre-determination, all the reasons that have been imposed on the existence of the human ego life itself, and to give the human being a power of action: to describe the process of having an external consciousness of yourself (or any other point of view).

It is the ability to transcend the present situation, to transcend the limits of one's own existence, to break the bonds of one's own existence, and to create a world that is temporarily or potentially satisfying to oneself.

To exist for oneself is not to exist as a being that "is" something, but as a being that "does" something. Therefore, this "transcendence" is focused on the creative and active nature of existence from the very existence of one's own ego.

This reveals the particularity of the existence of human ego-being itself, which is different from any existence in which one has an external consciousness process (or any other view).

No external thing or force has the prerequisite of "this" or "that" in order to obtain a particular right.

So, what is the form of consciousness, the existence of the self, the existence of the self, and how does the existence of the self manifest?

Basically, the three terms "self-existence", "consciousness" and "human reality" are all considered synonymous in the philosophical thinking of existence in Schatt's "Being and Nothingness".

Therefore, when we explore the structure of "self-existence", we are exploring what is consciousness? What is the true nature of human beings? With regard to the framework of self-existence, Scharthe first points out that self-existence is as an impersonal being.

That is, consciousness as thought activity, consciousness as the consciousness of something, does not require an abstract ego to coordinate fluid thought and to define what consciousness is.

Consciousness is a collective term that encompasses a variety of concepts, and is defined as a comprehensive awareness and recognition of changes in one's physical and mental state and in the environment of people, events, and objects, using mental activities such as sensation, perception, thinking, and memory.

The process of awareness and recognition is the process of consciousness, that is, the experience of consciousness. Therefore, the real definition of consciousness is the ability of consciousness itself to learn and think, to develop habits and motivations, and to obtain behaviors that lead to successful results.

That is intentionality: consciousness is always present in the sense of "consciousness of something". For example, the consciousness of hunger, the consciousness of two plus two equals four, and so on.

Perhaps we always feel that there is something identical to the self in the existence of the self, in the existence of the self itself, linked to the thinking (the properties, abilities and processes of the human mind for logical deduction) in each different time and space.

But, says Schart, this does not explain the existence of a "self". Instead, it is through consciousness as a kind of intentionality.

The consciousness of the past as an object of consciousness, and the present consciousness, instead of being a distinguishable being within itself, existing independently of something, is an impractical, inscrutable, entirely intentional, inclined to the consciousness of the past, to the memory, to the consciousness of the past because it is conscious of the past, and to the association of the past with the present consciousness.

The existence of the life of the self, the existence of itself, is illustrated in detail by Schart in The Transcendence of the Self. For example, when "consciousness" comes to a table or a chair, it is clear that the table or chair that "consciousness" is facing is the object of "consciousness" action or thought, and that "consciousness" is not only "consciousness" pointing to the table or chair, but also another kind of existence that is not the table or chair.

Thus, "consciousness" and "the thing that acts or thinks as an object" are two different existences, which are the two contradictory properties inherent in the epistemology of the typical thing itself, i.e., a thing has two opposing properties at the same time. consciousness".

But can such a dichotomy be applied to consciousness in terms of its effect or influence on the thing in question? That is, when Sartre considers that all conscious acts (seeing, hearing, smelling, tasting, feeling, thinking, thinking of anything) are accompanied by an eternal, unified consciousness.

Or what kind of "transcendental consciousness" can there be, as the Neo-Confucian master asks; or, as Husserl emphasizes, there is always a pure ego in all acts of self or ego-reasoning.

And this pure ego maintains the absolute identity of the self in the ever-changing self, or in the act of the ego's rational activity.

However, the reason why they must take as their starting point a common experience beyond experience, which is not commonly experienced by people, such as the existence of gods, ghosts, and the dream self, is the fact that

Whenever we realize that our objective existence is reflected in the human consciousness through the activity of thinking, and the result or formation of viewpoints and systems of concepts, whether by direct intuition or by intuition based on memory, the "I" that we realize is the "I" that we are.

The "I" that we perceive is given as the "I" that transcends this thought and all other possible thoughts. For example, if I recall being on a high-speed train yesterday and seeing the scenery receding rapidly, I not only recall the view of the mountains, but also the "I" that saw the view.

In other words, when the existence of the self, the

existence of itself, is recalled in any kind of way, the "I" is immediately revealed. This is the reason why people think that my consciousness is composed of me.

Although the theories of consciousness from Locke to Neo-Confucianism to Husserl are not quite the same. However, no matter which one of them it is, we can clearly see that they all recognize the existence of a transcendental I, a unified I, and that this ego life exists in itself.

And this egoic existence, which exists in itself, whether it has a content or not, whether it is transcendental or not, must occupy the position of the subject (philosophically, the individual with cognitive and practical abilities), and enjoys absolute eternity and unity.

The content of all consciousness is thus given a rationality, a possibility; that is, on the basis of the transcendental Self, all consciousness can become its own consciousness.

All the contents of conscious behavior are connected to this subject center, the "transcendental self," and all contents are given in this self-consciousness, which is beyond all possible experiences.

This affirmation of the divinity of man makes the "transcendental ego" defy external authority and tradition and rely on its own direct experience. It asserts that man can transcend his senses and reason and know the truth directly, believing that everything in the human world is a microcosm of the universe.

Let each person find his or her true self; if you meditate on your own soul, you are the master of your own being.

However, Schart believes that the transcendental "ego" is completely unnecessary, and that the existence of this ego, the existence of itself, is only a "mirage" in terms of one's ability to recognize the environment and the ego, as well as the clarity of cognitive behavior.

He denies the existence of an egoic being, an existence of its own existence, within the consciousness, by excluding the cognitive ability and clarity of cognition of the environment and the ego from this concept of "transcendental ego" religious philosophy.

It is the existence of an ego that exists within the consciousness and governs the individual's mental activities such as sensation, perception, thinking, and memory, and the integrated awareness and knowledge of his or her physical and mental state and the changes of people, events, and objects in the environment.

The cognitive ability and the clarity of cognition of the environment and the self are always directed towards (intend) something to act or think about, and when it is the target thing, it is the individual's mental activity such as sensation, perception, thinking, and memory that is used to

Consciousness is the existence of a self-life "to do", that is, consciousness is an existence of action, which means the inseparability of consciousness and the object it is intended to refer to.

Consciousness is a complete tendency to act or think about something as a goal, and consciousness is the activity of this intention itself.

Therefore, there is no way to know the existence of the self except by the object of the cognitive ability and clarity of cognition of the environment and the self, and by the consciousness as an intentional activity.

For Saud, the person as the intentional activity, the ability to recognize the environment and the self, and the clarity of cognition, can only be semi-transparent, other than facing an opaque object.

Therefore, if the existence of the life of the self, the existence of itself, this thick, opaque characteristic, is given to the individual to use sensation, perception, thinking, memory and other mental activities, the state of their own body and mind and the environment, people, things and changes in the comprehensive awareness and recognition.

Then, consciousness will no longer be an intentional activity, it will become a burden, or a heavy, materialized natural object, a dull, inactive "what-is" being. There is no I in the ability to recognize the environment and the self, and in the clarity of cognition of the person who has not been able to look back on his or her thoughts and actions and examine the errors in them.

When I am behind a bus, chasing to work, when I am holding my breath in water and staring at my watch, when I am staring at a portrait in an art museum, there is no me.

There is a sense of being caught up on a bus, and a non-location awareness of one's ability to recognize the environment and self, and the clarity of that recognition.

Although we arrive at a summation judgment of things from a certain premise, inferred from the result of the consciousness that there is no transcendental I and that one has not examined the rights and wrongs of one's past words and actions.

However, when an individual uses mental activities such as sensation, perception, thinking, and memory, he or she becomes aware of his or her own physical and mental state and the changes in people, events, and things in the environment. When an individual starts to rethink the past and learn from it, he or she begins to have a better understanding of his or her own consciousness.

When we start to think about our own consciousness, when we reflect the characteristics and connections of objective things in our mind, and when we reveal the meanings and effects of things on people, what we cannot deny is the reality of "I", and there is no doubt about it.

According to Schart, the existence of the life of the self, the existence of itself is the result and creation of the act of reflection. For example, when I am reading a book or driving a car, I am not aware of the existence of the ego-being, the existence of its own being.

Therefore, the emergence of the ego does not occur in the first stage: that is, the rational activity of the self or the ego before the ego goes back and reconsiders the past and learns from it; the ego does not exist.

The ego often emerges in the second stage, when the individual uses mental activities such as feeling, perception, thinking, and memory to become aware of and recognize his or her own physical and mental state and changes in people, events, and objects in the environment.

The awareness of one's own conscious behavior (the rational activity of the self or ego before reflection) is when the existence of the ego life, the existence of its own existence, comes into existence: I know (aware) that I am driving now, or I know that I am reading a book.

The result is that we cannot recognize our own actions (without the two contradictory properties inherent in consciousness, i.e., one thing having two opposing natures at the same time) until we reflect on them (our own or ego's rational activity).

Therefore, we can never say that I am aware of an electric car (Ich weiß von einem Elektroauto), we can only say that the consciousness of an electric car exists (Es gibt ein Elektroauto da draußen).

In this way, the reflective I-thought (the intellectual activity of the self or ego) makes ourselves constitute a self-living being, a being that exists in itself, so that the consciousness becomes a personal person.

However, the perception of this ego-being, self-existent being, does not lead Sartre to regard man's cognitive ability and clarity of cognition of the environment and of the self as the same being as an ego-being, self-existent being.

In fact, even if the existence of the ego-being, the existence of itself, appears in the stage of indirect awareness, different from direct awareness, I-thought (the rational activity of the self or the ego), it is not regarded as the cognitive capacity of the person for the environment and the ego, and the clarity of cognition itself.

For Shat, the ego and consciousness itself are always in an unstable state. If the ego is the reason for the reflection of the former self, then the consciousness itself will disappear.

Then consciousness itself would disappear and the

ego would cease to exist (because the ego must depend on consciousness itself for its emergence (the pre-reflective ego)). Therefore, we must return to the pre-reflective I-thought.

Therefore, for the philosophical view of existence in Schart's Being and Nothingness, the two contradictory properties inherent in such a thing, i.e., the opposition of a thing having two opposing properties at the same time, cannot be united.

In contrast, the cognitive ability of man to know the environment and the self, as well as the clarity of cognition itself, must be a being different from the existence of the self, which exists in itself.

And the existence of self-life, the existence of itself, cannot exist in the individual's mental activities of sensation, perception, thinking, memory, etc.

As a reason for one's cognitive ability and clarity of cognition of the environment and the self, it must be a being other than consciousness.

The self and the consciousness itself can co-exist. Consciousness at this time is the presence of the being that is facing the existence of the ego, the being that exists in itself, and the being that is not the existence of the ego, the being that exists in itself.

It is the product of one's mental thinking, including one's views and opinions on things, and solutions to problems. Consciousness can be a figurative image or an abstract concept, but the existence of the self, the existence of itself, is not part of the structure of consciousness.

It is not a part of the structure of consciousness, but an intentionality that serves as a goal when acting or thinking. Therefore, consciousness is just a mass of nothingness. All phenomena, including the ego, are existences outside the structure of consciousness.

Consciousness, as subjectivity, is pure nothingness, and the existence of a subject for itself (consciousness) without a subject is impersonal, which means that it is denying Descartes' argument of the spiritual entity "I".

Therefore, the second characteristic of "consciousness" or "self-existence" is as non-substantial. Descartes: "I think, therefore I am" (Latin: Cogito, ergo sum; French: Je pense, donc je suis)

The meaning is: from the fact that I am thinking, I can

deduce my existence. In many Indo-European languages, the coefficients have the meaning of "being", which is equivalent to "having" in Chinese. Chinese does not have such a coefficient that can mean both "belong to" and "exist" at the same time, so it is easy to misunderstand.

"The old translation of "I think, therefore I am" takes the French "je suis" to mean "I exist," but the word "in" has been interpreted by some people who read books, newspapers and magazines as "present" or "not dead" in their own words, characters or other symbols, which is inconsistent with Descartes' intention to express "being an essence" or "being something.

Descartes considered this to be an absolutely reliable truth and first principle. The result of this deduction is the existence of a mental entity. Consciousness itself and its own thinking activity are one within this entity.

The activity of consciousness is the "ego" of consciousness itself, and the thinking activity of consciousness is dissolved in the ego, in the unity of this duality.

The ultimate result is that consciousness is the ego, which is not only an entity, but a thinking entity.

Descartes replaces prereflexive Cogito with reflexive Cogito, so that consciousness "is consciousness as something" is transformed into consciousness only as "consciousness of consciousness itself" (I think that I think).

Consciousness is the human brain's awareness of external images in the brain. Physiologically, the conscious brain area refers to the area of the conscious brain (around the prefrontal lobe) that has access to information from all other brain areas. The most important function of the conscious brain is to recognize the truth, i.e., it can recognize whether the representations in its own brain come from the external senses or from imagination or memory.

The consciousness is only within itself and cannot escape from its own prison to know the natural world outside of the consciousness. This ability to know what is true and what is false is not available in any other brain area.

When a person is asleep, the excitement of the conscious brain is reduced to a minimum, and at this time it is unable to distinguish the truth of the images in the brain, and the brain further adopts a way of believing all of them to be true, which is called "dreaming". The conscious brain area does not have its own memory, and its storage area is called the "temporary storage area".

Just like the internal memory of a computer, it can only temporarily store the information it perceives. The consciousness is still "perpetual" and you can try to stop the images in the brain and find out the futility of such an attempt.

This is the greatest difficulty of Descartes' "I-thought" for the philosophical perspective of Schartes' Being and Nothingness. Therefore, Descartes must use God as a bridge between consciousness and the world.

03. Consciousness and Nothingness

Scharthes uses intentionality to describe man's ability to perceive his environment and self and the clarity of his cognition: the I-thought before reflection, intentionality is the thing that serves as the goal when acting or thinking.

The reflection of the human mind to the objective material world is the sum of various mental processes such as feeling and thinking, and is also a kind of behavior: the behavior of intentionality reflecting on something, thinking about something, is not an object-like existence, but can only be filled in itself, not outside of itself. It metamorphoses beyond the present itself, towards an earthly existence, throwing itself into the center of the world.

Consciousness, as consciousness, is always aware of something, thinking about something. Therefore, Descartes' spiritual entity only deprives consciousness of its capacity to think, to think. The reason why consciousness is able to think, to be conscious, is because consciousness is the act itself.

It is because consciousness is the act itself. As an act of intention, consciousness is always going beyond its own existence, intending something, investing itself in the world.

As something that is distinguishable and exists independently of itself. But it does not need to be a physical existence, or a "something" existence, but a consciousness that denies the ability to recognize the environment and the self, and the clarity of cognition, transcendence, consciousness is only a "limited" existence.

Because consciousness is constantly putting itself

into the world (being-in-the-world), all psychology, the so-called inner mental world, is impossible for the philosophical point of view of Sartre-Sartre's Being and Nothingness.

All these deepest parts of the mind, in the view of Schart's Being and Nothingness, become the target world, external to consciousness, when acting or thinking. Consciousness, subjectivity, is completely overwhelmed by the target world when acting or thinking.

We cannot explain more about consciousness than that the object of consciousness (the target of all actions or thoughts) is the reflection of the human mind to the objective, material world, the existence of the sum of sensations, thoughts and other mental processes.

Therefore, a more general definition regards consciousness as a special and complex movement that can reflect (map) the real world, as well as the movement of non-self-existing consciousness itself, which can correctly map the reality and the laws of consciousness itself, or can incorrectly or distortedly reflect it.

Consciousness in general needs a real material medium in order to have an effect on reality and consciousness itself. The most important function of the consciousness in the brain is to recognize the true from the false, i.e., it can recognize whether the image in its own brain comes from the external senses, or from imagination or memory.

Static consciousness generally exists in coded form, such as words, sounds, images, software, or other static material carriers, while energetic consciousness can inherit the static form and enhance the scope and level of consciousness.

The interaction of the static and energetic dynamics of consciousness is one of the most important sources of new consciousness. The starting point of this definition is derived from the absolute nature of eternal motion and the relativity of stillness of matter

Thus, if we separate the realm of things that are the object of action or thought from the ability to recognize the environment and the self, as well as the clarity of cognition, we will not be able to discover anything from it.

The special function and activity of the human brain is unique to the reflection of the objective world is completely clear and transparent (transparent), the reflection of the human brain to the objective material world is the sum of various mental processes such as sensation and thought is nothingness (nothingness).

Consciousness is an uninhabited existence, without anything existing in its center. Consciousness is completely transparent, like a crystal. The "reality of man" cannot [even temporarily] eliminate the mass of existence placed before him.

Consciousness originally meant mental activity. Consciousness means both the ego and the self. Knowledge means cognition, awareness, understanding. Consciousness represents the independence of the individual, it is the unique coordinates of subjective existence. Consciousness represents the ability to recognize one's own existence, to know what is happening. It can be contrasted with an existence different from one's own.

What man's reality can change is his relationship with existence. For man's reality, to put a particular being outside the circle is to put himself outside the circle of relative beings.

Consciousness is only a life energy that must flow,

just as a computer hard disk must be rotated by an electric current in order for the computer to work. So, consciousness is the stream of consciousness, the stream of life. In this case, he has escaped from this being, he is in an untouchable position, the being cannot act on him, he has retreated and transcended into nothingness.

What does this "nothingness" mean? It means that consciousness is still an incomplete and vague concept. Consciousness is generally considered to be the ability to perceive the environment and the self, and the degree of clarity of cognition.

Researchers have not been able to give it a precise definition. Whether [consciousness] is [impersonal] or [something that is not distinguishable and exists independently of itself], we find that there is an intrinsic negative relationship.

It exists between the individual's integrated awareness and recognition of his or her state of mind and body and the changes in people, events, and objects in the environment, using mental activities such as sensation, perception, thinking, and memory, and the thing that is the target of his or her actions or thoughts.

This means that consciousness is the seawater, thoughts are the floating objects in the seawater, concepts are the ice in the seawater, and emotions are the body's response to thoughts.

Thoughts determine the intentionality of the direction of the response, the positive response or the negative emotion, the response of the self-directed being. Thoughts are the initial suspended particles of thoughts, which are the objects submerged in the seawater, and after surfacing, they are thoughts.

Therefore, the ability to recognize the environment and the self, as well as the clarity of the cognition, stipulates that the consciousness is "not" a certain kind of thing and a real being that is distinguishable and exists independently of itself.

The definition of consciousness is very simple: it is that which knows and knows the existence of things. Consciousness is originally the same use of spirit. Reason comes from consciousness, and truth is a conceptual equivalent, not a mental one.

For example, when consciousness acts or thinks with the intention to act or think as the object, when it refers to a tree, at the same time as this consciousness acts, consciousness itself stipulates itself as a different kind of being "different" from this tree.

The "being for itself" regards itself as a being that is not "in itself". This inherent negation reveals one's ability to perceive the environment and the self, and the clarity of perception, as the definition of the for-itself as lack, the lack of being.

According to Sartre, the phenomenon of negation within the existence of the for-itself as existence is a key to understanding the existence of the for-itself as existence, the real existence.

Negation is not just a "judgmental nature", it is not something merely propositional, it is not just a logical phenomenon. In fact, the meaning of "negation" lies in the theory of existence.

This is closely related to Aristotle's question about "being as being": the common character of all entities in the broadest sense.

There are two basic answers to this question: one is that of Parmenides, who holds that what can be thought and spoken of is what exists. The other is Berkeley's, who holds that "to exist is to be perceived".

When we make the judgment that Peter is "not" in the café, or that I "am not" a being like the ink bottle in front of me, and so on.

"Nothingness" has entered the world of phenomena through this negative judgment. This "nothingness" is not a product of the natural world; it is through human reality, human consciousness, that "nothingness" appears in things.

The question of "existence as existence" is closely related to the question of scope, but is not quite the same. The domain is usually regarded as the highest kind or genus. Aristotle proposed such categories, which usually include substances, properties, relations, states of affairs or events, and so on.

Central to the distinction between domains are various basic ontological concepts, such as particularity and universality, abstraction and concreteness, ontological dependence, sameness and modality, and so on.

These concepts are sometimes regarded as domains themselves, used to explain differences between domains, or play their central role in describing different ontologies.

In ontologies, there is a lack of general consensus on how to define different domains. Different ontologists often disagree on whether a domain has any members at all, or whether a domain is fundamental.

Of course, the ability to recognize the environment and the self, as well as the degree of clarity of cognition, can introduce "nothingness" into the world and things, which must be a comprehensive awareness and recognition of one's physical and mental state and the changes of people, things and objects in the environment by using mental activities such as sensation, perception, thinking and memory. It is also a lack of existence, non-existence, nothingness, and existence.

If an individual uses mental activities such as sensation, perception, thinking, and memory to perceive and recognize his or her own state of mind and body and changes in people, events, and things in the environment, "existence for oneself" is like "existence in oneself".

It is like "existence in oneself", which is a full and closed existence, then there is no way to introduce "nothingness" into the world, and there is no way to deny whether the object of thought exists, whether it has a certain attribute, and whether there is a certain relationship between things.

Consciousness must be a non-being in order to understand "being" and what "to be" is.

Therefore, the individual uses mental activities such as sensation, perception, thinking, and memory to perceive and understand his or her own state of mind and body and the changes of people, things, and objects in the environment as a kind of non-being, nothingness, which is outside the sequence of existence and is distinct from existence.

In other words, the ability to recognize the environment and the self, and the clarity of cognition, is a lack of being. "Being is a lack of being due to the presence of nothingness.

The human mind's reflection of the objective material world is the sum of various mental processes, such as sensation and thought, as a lack, basically in a relatively complete state.

Shat is a metaphor for the individual's use of sensation, perception, thinking, memory and other mental activities, the comprehensive awareness and understanding of his physical and mental state and the changes of people, events and things in the environment, as a kind of deficiency, is like an upper sine moon.

It is always moving toward a full moon (perfect existence, existence in oneself). In the case of the upper moon, it is the upper moon because the full moon is used as a kind of completeness, to illustrate its existence as an imperfect being.

Likewise, the ability to recognize the environment and the self, as well as the clarity of cognition, is a deficiency, based on the "existence of the self" as the reason for its deficiency.

"In contrast, the cognitive ability and clarity of cognition of the environment and the ego are a deficiency, and the existence of the life of the ego itself is an imperfect existence.

The individual uses mental activities such as feeling, perception, thinking, memory, etc., to perceive and recognize his or her own physical and mental state and the changes of people, events, and things in the environment, which is the desire to obtain a real existence.

Because it is a non-existence, unlike "existence in oneself", which is a real, physical existence, or an explicit, not abstract, general existence, it is a non-existence of "nothingness".

Becoming a real being therefore becomes the goal of "being for oneself". Being for oneself is always transcending one's lacking, non-existent state and trying to reach a real state of being (the state of being in oneself).

Therefore, Schartes says: All lacks are for the sake of... lack.... The imperfection of "being for itself" is always metamorphosing and sublimating beyond itself, towards the perfection.

Completion does not mean a transcendent God, but it is this form of "being" that can give "being for oneself" a real self. Of course, we can also say that this real existence is a value desired for the consciousness, for the reality of man.

When the human mind reflects the objective material world as the sum of various mental processes such as feeling and thinking, as nothingness, as lacking, and as always intentionally desiring something, we call this the "value" of what one desires in terms of one's ability to recognize the environment and the self, and the clarity of cognition.

That is, to become a perfect being like God; a composite ego and a subject of consciousness, something that is distinguishable and exists independently within itself. But it does not need to be a physical being to exist, and at the same time to preserve the transparency of consciousness.

But, says Schart, this perfect existence cannot exist. For, under the principle of the law of identity, there can be no dualistic splitting of a self-identical thing, no crack in the One.

In other words, one cannot derive "two" from "one". If the ego exists as an object, then the unification of the ego and the subject of consciousness will cause the ego to dissolve into the subject of consciousness.

If the ego is the basis of the subject of consciousness, this is back to a situation of mutual extinction, where there is no unity of two opposing natures of a thing at the same time, but a thing that is distinguishable and exists independently within itself. But it need not be a physical existence.

Therefore, "existence for oneself" can never be a perfect existence. The relationship between the self and the subject of consciousness is always in a state of "internal negation".

In other words, the ego and the subject of consciousness are mutually constructed and have two mutual oppositions at the same time. This means that the "ego" presents itself because of the "subject of consciousness" (intentional activity), while the "subject of consciousness" becomes conscious because of the existence of the "ego".

However, the "subject of consciousness" cannot be the "ego" because the overlap of the two would cause the dissolution of the one. Therefore, only through the means of "negation" can the ego be manifested, and only through the ability to recognize the environment and the ego, as well as the clarity of the subject of cognition, can the subject of consciousness become conscious.

The "ego" and the "subject of consciousness" are mutually constructed on the fact that "I (the subject of consciousness) am not the ego". Therefore, Schart said that the relationship between "ego" and "subject of consciousness" is a relationship of "inner negation".

The subject "cannot" be the ego, because we have seen that the overlap with the ego would make the ego disappear. But it also cannot be not the ego, because the ego indicates the subject itself.

Nothingness is the fundamental quality of human consciousness as "For-Itself".

When man projects his ideal self into the future, he is no longer just himself in the present, and he looks back at himself from his ideal state and denies his immediate self.

To deny oneself is to nullify the immediate self. However, the ideal by which one denies the present state is also a kind of nothingness, since it has not yet been realized.

In this way, the life of "existence" that exists in the life of the self itself is pierced by nothingness throughout. However, nothingness does not mean that the true meaning of life existence is denied.

On the contrary, Schart thinks that this is the way to affirm the meaning of human being. For if man were a mere fixed object, he would be at the mercy of man.

Perhaps it would be more appropriate to use the word "lack" to explain the meaning of "nothingness".

When a man has an ideal in his heart and thus faces the various deficiencies of the present situation, he is in a state of "lack". As a kind of nothingness, it is constantly metamorphosing and sublimating beyond its own defects in an attempt to transform into a more perfect state of existence.

04. The way of existence of "in oneself" and "for oneself"

In his essay "Existentialism is a Humanism", Scharthe points out that man's existence is his own creation of himself.

"Man first exists, encounters various encounters, the world fluctuates, and then limits himself. For man begins with nothing, and only later becomes something." -Sartre, Existentialism as a Humanism

"...there is therefore no such thing as humanity, because there is no God to create this concept, man exists nakedly. He is not what he conceives himself to be, but what he will be.

It is only after he exists that he can imagine what he

is, what he wills himself to be after this thrust toward existence, and man is nothing else but what he forms. (Man is nothing else but what he makes of himself.) - Sartre, "Existentialism as a Humanism

Sartre's book "Existentialism is a Humanism" is based on the position or point of departure when observing things, which is the "existence" of man's self-existence itself.

It means that the original logical thinking is necessarily or necessarily so, that "things are in themselves", that "they exist in themselves". To the "I", other people are like thieves who want to steal the world from the "I", to incorporate me into their orbit, to become a "being-in-itself", to become an object or something.

Thus, I am no longer a free subject, but a slave of others, having fallen from being-for-itself to being-in-itself. How can I regain my freedom, my own autonomous subject?

The only way to do this is to do unto others what they do unto me: to objectify them. If another person is only an object, if he is in his own existence, then I will not become his object. I can shatter the world of others and remove their freedom through my "look".

But such an approach cannot always succeed, because the existence of others is a fact that cannot be eliminated; others are not created by me, but encountered by me. The other remains there, threatening me and countering with his "look" at any moment.

Therefore, the "being" is a thinking being that is not reflected by the human brain in the characteristics and connections of objective things, and reveals the meaning and effect of things on human beings; it is a "being" independent of the cognitive subject consciousness.

When it becomes the action of thinking, or the object of thinking, that is, when it becomes the subject, I love to hunt, reflect and explain the existential thing, that is, "for oneself".

However, because of the "existence" of human life itself, in this life experience of "existence", I love to hold on to the consciousness, which is a collective term that includes various concepts.

The meaning of the term refers to a person's comprehensive awareness and understanding of his or her physical and mental state and the changes of people, things, and objects in the environment by using mental activities such as sensation, perception, thinking, and memory, and by using the consciousness of "being oneself" and "for oneself", so that one's attachment to such things is in oneself, and one becomes a "being for oneself".

Explanation of terms

Cumulative Void.

Introduction: The combination of memory and external memory is used to obtain a large capacity of "memory", which is virtual memory. Positive consciousness, positional consciousness, non-positional consciousness, non-positional consciousness.

These terms are actually very difficult to grasp in virtual memory. Sometimes, the positive topic consciousness is used as position consciousness, and the negative topic consciousness is used as non-position consciousness. First of all, the act of counting cigarettes is defined as positional consciousness, and the consciousness of being aware of counting cigarettes is defined as non-positional consciousness.

And after a turn of phrase, the inference in the middle says that counting cigarettes is defined as proper consciousness, while the consciousness of being in counting cigarettes is defined as non-positional consciousness." At the same time, it is precisely the unorthodox consciousness of counting that is the real condition for me to take some kind of action towards the other person".

For those of you who are reading this for the first time, you will be confused. Why not make it a more concise word if they mean the same thing?

In the introduction of Cun-xu, it is pointed out that positional consciousness emphasizes the absence of content in itself, while topical consciousness emphasizes the confrontation between consciousness and objects. This seems basically correct. The word location is a direct reference to Husserl's theory of "location", intentionality: consciousness is the awareness of something.

The former refers to pure consciousness, while the latter refers to the consciousness of something. This leads to the "position" theory of consciousness as an object.

Consciousness realizes that it is only a position, it is nothing, it is nothingness. In this way, it should be possible to understand the contentlessness of location consciousness itself in the preamble.

And the sense of subject matter, I think, is derived from Husserl's thematization, or thematization, which I generally translate as what is going on. What about the idea that "all positional consciousness of objects is at the same time non-positional consciousness of oneself"?

These concepts are already present in Sartre's "Self-Transcendence" (the transcendence of the self). They are inferred from the two different forms of the French ego Je (English I) and the French Moi (English Me). Therefore, the introduction of Cun-vu is essentially a summary of the self-surpassing.

Je(I): The "I" of the active personality form (some kind of unconscious consciousness) is also the consciousness of the subject, what I am doing, I am counting cigarettes, but if my consciousness does not ask itself what it is doing, I do not know that I am counting cigarettes.

Take counting for example, counting to 100, counting itself is counting + reflecting on the act of counting. Your subconscious mind is constantly asking yourself what number you have reached, so you can count.

If you don't reflect on the number you are counting,

it's like you suddenly get stuck or lost in thought and you don't know what number to count. At that point you lose your reflection on the behavior and only the part of your active form remains. (In order to count, you must be conscious of counting)

Moi(Me): physical-psychological, as the state and nature of the unified self, the passive personality form of "I", that is, the non-positive consciousness, you ask yourself after counting a number, now how many it is?

At this point you answer that I have counted to 21. The ego being asked is the passive form, and this passive form of me makes the active form aware of what is going on, and the question makes the state being asked valid. I counted 20 cigarettes, and I was just counting cigarettes.

Here, then, we can further introduce two concepts of Sartre's Cunningham that are of secondary importance, "reflective consciousness" and "reflected consciousness" (or reflective and non-reflective consciousness).

According to the above question, for example, Je as the reflected consciousness and Moi as the reflective consciousness, the reflective consciousness enables the existence of the reflected consciousness. I counted 20 cigarettes (reflective consciousness).

Reflection makes it possible for the reflexed to exist as a certain kind of being. The ontological structure will be added later. Now back to "all positional consciousness of objects is at the same time non-positional consciousness of itself.

"All positional consciousness of objects = all positional consciousness of objects, and positional consciousness = positional consciousness, which can never be revealed without non-positional consciousness.

Therefore the original consciousness of the self is not positional, the non-positional consciousness of the self (not + no content = content). The consciousness is aware of the object and at the same time is directly aware of itself, experiencing its own knowledge of the object.

Subjectivity: This word is more like subjectivity in Sartre's context. We often talk about the opposition between subjectivity and objectivity, where subjectivity is derogatory and objectivity is positive.

But if we understand it in this way, we become the kind of idealist who is a pseudo-materialist, and we return to the traditional dualistic context. This subjectivity is understood as a nature encompassed by the subject.

"Being-for-itself" and "being in itself": two of the most important concepts in existence, Self-possession and Self-activity, are themselves from Hegel's Logic.

Self-possession itself is a potential, an unexplored state. Hegel often uses the metaphor of a seed that is itself unconscious, unbudding.

The "for oneself" (für sich) is conscious, the unfolding of the former, the conscious, a big tree. Marx borrowed the term "for oneself" to refer to the different stages of the working class, which is the self-conscious class without the guidance of intellectuals, and the self-conscious class after combining with intellectuals.

In Sartre's context, "being in itself": 1.

1. "being in itself" is "being in itself" is what it is". A=A, unable to interpret anything else, full, complete, opaque. (Being is what it is)

2. 'being in itself' is beyond the active/passive sphere, it is only meaningful in relation to itself. (Being exists)

3. 'being in itself' is neither possible nor necessary. The existence of the present may come from a being that is

a phenomenon, but the existence of the phenomenon is never existent. "Being in itself' is contingent, and 'being in itself' is never possible nor impossible. (Existence is self-existent)

We talked about being and reflection earlier, here we can consider being as freedom and reflection as "for oneself".

Freedom will be applied in various chapters. It is worth noting that Sartre's term existence sometimes refers to true existence, i.e. "for oneself", and sometimes refers to non-true existence, i.e. Freedom, which is not nothingness in itself, but only contains nothingness within. And Heidegger's being/being generally refers to this being-sein.

"being-for-itself".

Sartre's earlier work "Imaginative Psychology" or "Imagination" in fact already treats nothingness as an ontology of imagination, and the imagination here should refer to Kant's unity and Husserl's form of definition.

Sartre uses the example of the blanket, where part of the pattern is blocked by the legs of the chair, but one is able to grasp the pattern as a whole directly, under a sense of totality.

The continuity of the complement of those parts of the pattern that are not present (complemented), the nature of the empty composition, is perceptually unified with the original. The consciousness uses this imaginative investigation as its basic function.

Imagination = detachment from the phenomena of the moment and activity on a separate imaginative level. At this point, the "presence" exists only as imagination, and the original presence is receded, made void and negated, and imagination makes the "reality" void.

Imagination-Negation-Voidness-Intrinsic Nature

"for oneself"-nothingness-negation "is what it is not, not what it is"

Man is the being from which nothingness comes into the world, and consciousness thus becomes "being-for-itself", the inner structure of man. The conscious person itself means only an intentional activity. "for oneself" is always to be constantly detached - beyond oneself, beyond objects.

The nothingness of consciousness appears as a structure that never reaches the state of the highest goodness and the permanence of Freedom, never realizing its ultimate nature.

Therefore, human existence is always in a state of change. The "for oneself" is "inappropriate" compared to Freedom, so that "human nature" becomes Freedom.

The only way through which it can be expressed and established through the intermediary of "for oneself

Transcendence: This term appeared in the nineteenth century among philosophers, Nietzsche, Kierkegaard, Adler, and many others. Transcendence in a broad sense is like a way of transforming exteriority into interiority. For example, playing the piano or riding a bicycle is a kind of transcendence. It transcends material materials and becomes an inner quality.

And transcendence enters the philosopher's language like Nietzsche's idol, Kierkegaard's God, Jung's primitive imagery, etc. It transcends what it is, it is more than what it is. In Nietzsche's words, "There is no beautiful surface without a terrible depth".

The transcendence in the Sartrean context is more often used in Self-possession and Self-activity, which is: not being what it is, escaping from "being in itself".

To take a case in point, the book cites a woman who is meeting for the first time: a woman shakes hands at the first meeting.

At this point she needs to use a form of self-deception to let the shame of reserve dissipate. Then she has to use a combination of two ways to avoid this embarrassment: 1: Transcendence by way of personhood 2.

1. Humanity (factuality): it is what it is: the gentleman is as warm as a jade, and I have set his human nature.

2. Transcendence: It is not what it is: He holds my hand through the hand of the spirit, and he has a reason to hold my hand.

Sartre says, "One always plunges ahead of oneself, whether in the sense of space or time, and one calls this attribute of being beyond all things, of being elsewhere, transcendence.

"For example, you pity and love a weak person because he has a dream. And the reason why you pity him is because he will become transcendent/potential that is not what he is.

Transcendence is used here as the relationship between Self-possession and Self-activity, followed by the stipulations of negation, quality and quantity, potentiality, and instrumentality, which we will read about in Chapter 3.

Artificiality/factuality.

This word was borrowed by Sartre from Heidegger, the German word is Faktizität, and Heidegger speaks of this word mainly in terms of Dasein's presentness, since Dasein's zu sein de-existence and always-me-ness lead to Dasein always being related to Existenz, so this artificiality or facticity means more survival.

In the eyes of existential thinkers, pain, for example, is not pain in itself, but only exists as a factuality, a factuality that is also translated as scatteredness, which emphasizes the meaning of possibility.

The human ego is an object, as opposed to an objective object or "other" object. Similar to Marx's objectification, Lacan's gaze, in which he becomes an object like any other object.

Sartre calls this aspect factuality, while man has to not be objectified, he has to be a way of being objectified to coexist with others, so to this extent factuality exposes contingency/scatteredness.

For example, if I am a waiter, the waiter itself is my actuality, but when I am not a waiter I am myself and I can be the human side. Public life operates with the symbol of factuality, but it is contingent, absurd, objectified, and related to survival.

Human beings, as the spirits of all things, lose their individuality in this factuality, so life is absurd. This point will be added in the novel "Disgusting".

Other people:

The term should be taken from Husserl's alter ego: the central concept in Husserl's phenomenology of interactive subjectivity.

It means the other for the ego, the other strange monads for the individual monads of the self, and is therefore a concept corresponding to the "ego-ego" and equally divided from the strange self.

Sartre's existential structure.

存在（自在）⇒ 自為 ⇒ 否定

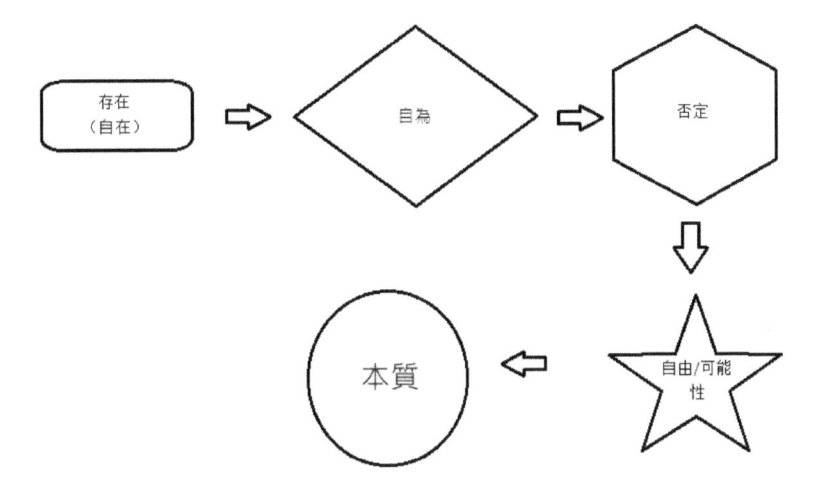

本質 ⇐ 自由/可能性

Explanation of Terms in "Existence and Time".

Existence/Sein: What is existence?

Husserl, the last great master of traditional philosophy, followed the same path as Kant, pointing to the source of pure consciousness and the a priori self in order to ensure the reliability of all sources of scientific knowledge. (Sartre: Husserl and Kant did the same thing.) Modern philosophy, on the other hand, holds that all existents exist first, and only then do we talk about the rest. (Existence precedes being)

When we ask what something is, for example, as Kant did, what guarantees the validity of an acquired experience? It is the innate form. In this question, as soon as we ask the question, we presume that something exists.

It can be an innate form or it can be pure consciousness. Whatever it is, there must be something in the black void of that question mark. So we begin to ask questions and expand on it.

(Sartre's theory of being is based on the assumption that there is a being (the Self) existing in existence, whereas Heidegger's theory of being is that existence (the fullness of emptiness) necessarily precedes existence in time)

Heidegger points out that in traditional metaphysics and ontology, the process of asking questions is not about what is being asked, but about what is being asked.

Heidegger points out that traditional metaphysics and ontology do not ask about what is being asked in the process of asking, for example, Sein/to on/Be, but actually ask about Seiende (a modern participle of Sein) or Being. Ontologia was changed to existentialism by Heidegger.

We can understand that the white horse is not a horse, but a horse as Be, as co-existence, as being. And the white horse as Being/Seiende, the individual, the Being.

In the Heideggerian context, the existence of the horse makes the white horse exist, but it is not a question of who is first, as in the case of the common phase and the other, but rather that existence makes the being exist.

Thomas esse (the nature of something) makes ens (something) exist. But Thomas stops at esse, thinking that it is not possible to grasp that ens exists.

All existential theories so far have of course presupposed "existence", but have not treated existence as a concept to be used - not as something we are seeking. Obviously this paragraph should be about Aqui.

The reason why metaphysics has forgotten "being" is that it no longer considers being as a problem, and Heidegger reintroduces it in the existential time. But it is not enough to work out the word existence; we must construct a more refined existential system. If we cannot answer what Sein means, then Sein is blind, and it cannot be the foundation of 20th C science.

What is the meaning of Sein?

2. Dasein This in.

Dasein is seen as the existence of sein in which existence comes out of the situation, and man is seen as

Dasein, as the Being. The possibility of essentialism can be seen as Dasein itself. two essences of Dasein: 1. zu sein /to Be to exist 2. Jemeinigkeit to be I belong

Two states of Dasein: 1. True state: to exist, open and unprescribed, 2. Non-true state: closed and prescribed.

A simple example: when you go to college, you are jumping around happily, tangled in front of more than ten clubs, and at this moment you are true. Then you think you can't join a rap club because you don't have any rap experience. This is not true. Dasein - Temporality - Historicity - History

Heidegger's existential structure.

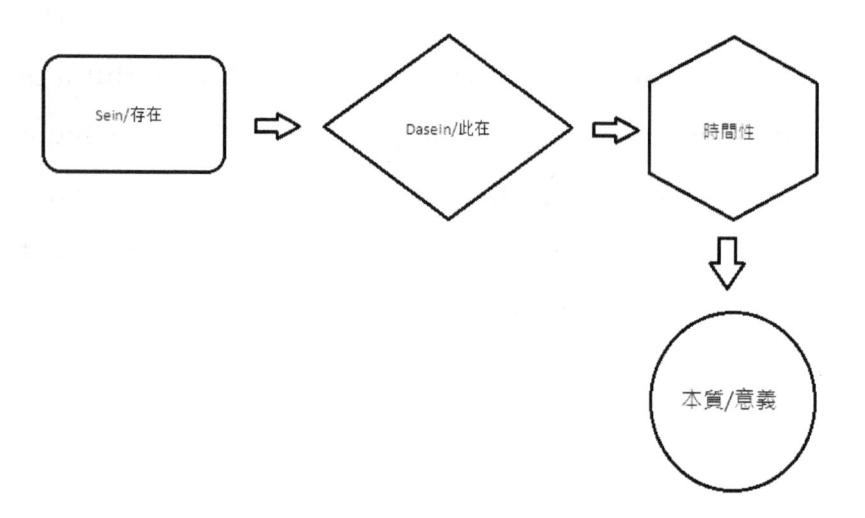

```
[Sein/存在] ⇨ <Dasein/此在> ⇨ [時間性]
                                    ⇩
                                (本質/意義)
```

Q: So this existence and human freedom are related?

A: First, it is one of the most important arguments of existentialism.

This is one of the most important arguments of existentialism. Why does existence precede essence? We have to say a few words about Plato's theory of ideas.

In Plato's view, concepts like goodness and righteousness are themselves absolute, timeless, and unchanging ideas. For example, goodness is the essence of goodness that people abstract from countless good deeds.

This can be said to be typical of essentialism. But in existentialism, there is no such given universal essence, such as the concept of humanity.

This means not only that he is what he thinks he is, but also that he is what he wishes to be, what he is willing to be after he has come from nothing, from non-existence to existence. Man is nothing but what he thinks he is.

This is the first principle of existentialism. This first principle is the principle of freedom you are asking about. There is a double meaning here, the first being the sense of human dignity.

In response to the accusation of existential

subjectivity, Scharthe says: "What else can we mean by subjectivity than that man has greater dignity than a stone or a table?"

The second is the initiative of man.

The second is man's initiative, that is, his ability to choose freely.

If you take the initiative to become a scholar, you have to study hard for several years, and you have to do a lot of hard mental labor, the pain and suffering of which is not known to others. You have finally achieved your purpose and become a learned and accomplished person.

The nature of this learned and accomplished scholar is not readily available, but you first "exist" with a goal, and put your ideal expectations into active choice of behavior, you may have the essence.

That is why Schartrecht says explicitly: "If existence is really prior to nature, one has to be responsible for what one is. So the first consequence of existentialism is that one understands oneself as one is, and takes responsibility for one's own existence entirely on oneself. This statement contains a profound ethical and moral implication.

Can it be said that existential exister is a concretization of existential être? In a certain sense it can be understood in this way. The word exister originally comes from Latin, where the word ex means to bring out, and sistere means somewhere.

So etymologically, the word means to make its presence. And existence itself is a passive fact, not a question of whether you want to exist or not. No one can choose his own existence, because when you can choose, you already exist.

So existence precedes essence, that is, you exist first, and then you can be something. Sartre has an analogy, "What is existence? It is to drink without thirst" (Qu'est-ce qu'exister ? Se boire sans soif).

It is a viewpoint, experience, consciousness, spirit, feeling, desire, or belief that an individual can have as a property.

Its fundamental characteristic is that it exists only within the subject. Although there is an environmental consciousness that belongs to the "other" subject, the individual can use mental activities such as sensation, perception, thinking, memory, etc., to be aware of and recognize his or her own state of mind and body and the changes of people, things, and objects in the environment, but he or she can choose autonomously.

However, in the view of some existentialists, many people in the world are swayed by the current situation, pulled by the social matrix, and they follow the current fashion without ever making an autonomous choice.

Therefore, they have no real existence, but are just materials piled up in time. We have indeed seen this in Europe during the Nazi era and in China during the Cultural Revolution. The vast and blind revolutionary masses are like all objects or phenomena that exist objectively in nature. There is no autonomous existence.

This means that with existentialists, only the human being is prior to the essence of existence. This is because in the world of experience where we personally use our senses, perceptions, thoughts, memories and other mental activities to be aware of our own physical and mental states and the changes of people, events and things in the environment.

Only the human will has the freedom of initiative, only the human cognitive ability and the clarity of cognition of the environment and self can plan, design, and choose his actions. Only man can think and determine the moral meaning of his actions.

This is a good point to analyze in detail along the way, to understand from a certain cognitive point of view, to understand. For example, if a seed falls into the soil, it will sprout and grow according to the nature of the seed, which has already been determined. A rose plant does not actively choose to become an apple, but it is possible that the mutation of a plant is the result of a human being's ability to differentiate between all phenomena and choose to graft and change genes.

In contrast, regardless of the constraints of human beings, individuals have the possibility of choosing to use their senses, perceptions, thinking, memory and other mental activities to become aware of and understand their physical and mental states and the changes of people, events and things in the environment, because existentialism views human existence as existence in a situation.

There can be no fixed, pre-determined, or inevitable practice or choice. And even if a person cannot do a certain thing or achieve a certain result due to his condition, he can at least give up, refuse, and choose.

But he can at least give up, reject, deny a certain practice, a certain concept. Negation is also the choice of an individual to use his senses, perception, thinking, memory and other mental activities, the comprehensive awareness and understanding of his physical and mental state and changes in people, events and things in the environment.

That is why Schart said: "Man is first of all being, and before he can talk about anything else, he first pushes himself into a future thing and feels himself doing so.

Man is indeed a conspiracy of subjective life with the ability to recognize the environment, and the self, and the clarity of cognition, not a moss, or a fungus, or a potato. This is the most concise argument for the precedence of being over nature.

Part 3: The structure of Sartre's theory of existence.

Volume 1: 'Nothingness', an analysis of 'self-deception'; an argument for nothingness, and an incisive analysis of self-deception.

Section 1: Nihilism in Philosophy

Nihilism is the philosophical view that the world, and especially human existence, has no meaning, purpose, intelligible truth, or intrinsic value, although the term "nihilism" was made known by Ivan Sergeevich Turgenev (November 9, 1818-1883).

But it was Friedrich Heinrich Jacobi who first introduced it into the field of philosophy. Jacobi wanted to use this term to show the characteristics of rationalism, especially the critical philosophy of Kant.

He believed that all rationalism could be reduced to an elusive 'nothingness', so that we should try to avoid it and return to the certainty that some human beings are born with the values they choose to believe in, the way they live, and the reliability of human justice.

In fact, people tend to think that people's social behaviors are influenced and controlled by invisible factors, formed by external environmental factors.

Most of them are influenced by parental, social, political, religious, and historical ideas, cultures, morals, customs, arts, institutions, and behaviors that have been handed down from generation to generation.

Only the process of faith must be an internal response, in which one must choose a path suitable for one's continued existence through the experience of one's daily life and the search for the true meaning of life's existence.

Nietzsche's late works are mainly concerned with nihilism. One volume of The Will to Power consists of a selection of Nietzsche's notes from 1883 to 1888. He named it "European Nihilism" and considered it the main problem of the 19th century.

Nietzsche defined nihilism as the absence of meaning, purpose, intelligible truth, and intrinsic value in the world, especially in the "being" of human self-existence itself.

Although postmodernism was once ridiculed by some as nihilism, it does not fit the nihilist model in the sense that nihilists tend to be defeatist.

The postmodernist philosopher seeks to find and justify the power and causes of the diverse and unique human relationships he explores. Skeptics argue that the two propositions cannot rely on each other because this would set up a circular argument.

For skeptics, such a specious logic thus becomes an inadequate reading of the truth, and because they think they have reached a solution, they may create more problems that remain to be solved.

But truth is not necessarily unattainable; it may be a concept that does not yet exist in its pure form. Although "skepticism" has been accused of denying the possibility of truth, it is in fact only denying the possibility of truth in the form of the truth.

In fact, it only reveals, in a critical manner, that logic has not really discovered the truth. Therefore, it is not necessary to draw any conclusions about the reality of the normative and normative concepts of people's common life and their behavior.

Neither do they have to exchange opinions or debate on a certain issue, about the meaning of existence, in the absence of knowable facts. In this way, the object of skepticism exhibits positive meaning and usefulness for the subject.

It is merely a way of comparing and contrasting the similarities and differences of several similar things, of giving enlightening pointers or elucidating examples, of causing the other side to associate and understand, in order to set up what is true knowledge, what is required, and what is a justification of faith.

Thus, nihilism as a philosophical meaning is the ultimate form of skepticism. It is the belief that the world, the existence of life, has no objective meaning, purpose, or comprehensible truth.

It is not so much an openly stated position of a person as it is a confrontational opinion. Many critics consider the movements of Dadaism, deconstructionism, and punk to be nihilistic in nature, and nihilism has been defined as a characteristic of certain eras

(1) Ethical nihilism

In ethics, "nihilist" or "nihilism" is a holistic, fundamental, and critical inquiry into the meaning of the real world and human beings as a world, especially with regard to the meaninglessness, purpose, and intelligible truth of "existence" of human self-existence, itself.

It is a world in which there is no meaning, no purpose, no comprehensible truth and no intrinsic value. It is not so much a public statement of one's position as it is an expression of confrontational resistance. It is used to refer to a person who completely rejects the social matrix, all authority, morality, and social conventions, or who claims to do so.

Either by rejecting all established notions of belief in, worship of, and adherence to something as a rule and guide for speech and action, or by extreme relativism or skepticism, the nihilist believes that those in control of power are invalid and should be confronted.

In the nihilist view, the ultimate source of the norms and standards of behavior and values that should govern human life together is not culture or the ability of people to form concepts, to judge, analyze, synthesize, compare, reason, calculate, etc., but rather the "being" itself, the existence of the individual's own life.

However, "nihilism" has no choice and nothing to do with the "inevitable" current of history.

Even if the answer to the question of "can it be done" is affirmative by intuition, it is impossible to say "how to do it. These two questions are the reasoning and guidelines that nihilism should follow in dealing with the interrelationship between human beings and society.

It is a set of concepts that guide behavior, a philosophical consideration of moral phenomena from a conceptual point of view. It not only contains the rules of behavior in dealing with the relationship between human beings, human beings and society, and human beings and nature, but also profoundly implies that moral phenomena should be dealt with from the conceptual perspective.

It also deeply implies the profound truth of regulating behavior according to certain principles. It refers to the truth of being human, including human emotions, will, life and values, etc., which cannot be free from the label of self-existence.

(2) Postmodernism and the Collapse of Knowledge

Postmodernism is a process of analysis, synthesis, judgment, reasoning and other cognitive activities based on images and concepts, and it holds an attitude of not believing (doubting) logical concepts and structural interpretations and explanations in a holistic, fundamental and critical inquiry into the real world and people.

This attitude leads to a lack of desire for ideas, things, and external sensations, because the postmodernists, whose displacement of thinking is not dependent on the postmodern, neither affirm the experience of history, nor believe in the origin of meaning and its meaning.

They neither affirm historical experience nor believe in the origin of meaning and its reality, and they have no high hopes for someone or something in the future. With regard to philosophical skepticism, Pyrrhonism is indifferent or evasive to the truth of propositions.

Rather than asserting the purest, most realistic truth, that the correct reflection of objective things and their laws in the human mind cannot exist (since this would be a truth claim in itself), philosophical skepticism proposes instead "suspension of belief".

This suggests a characteristic, identifying notation, and is generally used to describe the idea of philosophical skepticism, which is analogous to analytic reflection on the fundamental questions of life, knowledge, and values of inquiry and reflection. As an example of academic skepticism, one of the ancient schools of Platonism claims that true cognition is not possible.

Empiricism is closely related to philosophical skepticism, but not without a twist, without the same being conveyed through people or things. The empiricists see empiricism as philosophical skepticism, a pragmatic compromise with the regular sciences.

Philosophical skepticism, on the other hand, can be understood as a "polarized empiricism," hence the taboo against skepticism. René Descartes begins his first collection of philosophical meditations, based on his own self-knowledge, in an attempt to find an unquestionable truth.

The theory "I think, therefore I am" later became the truth he could discover, but before that he had also paid general attention to the skeptical discussions of "dreaming" and "radical trickery".

Apart from the doubts in their minds, their objective existence, reflected in human consciousness, is reflected in the activity of thought, and the result is, in the eyes of rational thinkers, almost a frozen state, which they can only parasitically criticize under modern Enlightenment reason.

This has been extended in modern philosophy to refer to a set of systematic inquiries into the conditions and consequences of concepts, theories, and disciplines; or as an approach or/and attempt to understand the limits of their validity. A "critical perspective," in this context, is the opposite of a dogmatic perspective. Kant writes.

"If we treat a thing as if it were something under the concept of another object, and this constitutes a principle to understand and determine its consistency with; then we are dogmatically skeptical of this concept.

When we treat it only as a reference to the cognitive faculty, and thus turn to the (subjective) condition of thinking about its subjectivity, without committing ourselves easily to decide anything about its object; then we deal with it only critically".

The Greek skeptics criticized the Stoics and accused

them of being dogmatic. For skepticism, the logical mode of argument is not valid, because if there is no external proposition on which it can be based, it is not valid.

It is difficult to assert the rightness or wrongness of its own propositions. This is a retrospective argument from opposition to opposition, i.e., each proposition must depend on the other propositions in order to maintain its validity.

(3) The skeptic Agrippa's trilemma.

Skeptics often refer to Agrippa's trilemma. How can you be sure that the process of acquiring knowledge through mental activities such as forming concepts, perceptions, judgments, or imaginations is the correct one in psychology? This requires evidence of fact.

Here comes the question that requires an answer: Aggrippa's argument contains three important principles of confirmation: the confirmation of circular arguments is wrong; the confirmation of infinite regress arguments is wrong; and the arbitrary termination of arguments is wrong.

1. Circular argument: The argument is ultimately supported by itself and proves to be infinitely regressing.

The proof of infinite retrogression is wrong. The evidence used to prove a studied thing needs to be proved itself, and this proof needs to be proved further, and so on, until infinity.

Infinite backtracking means that a belief A0, which is to be proved, needs to be proved by invoking another belief A1. But if belief A1 is itself to be confirmed, then belief A1 must also be confirmed in some way, which in turn must be confirmed in A2, and A2 must be confirmed in A3.

By doing this repeatedly, the confirmation is trapped in an infinite series of backtracking that cannot be ended. For example, the English name atom originally means the smallest particle that cannot be further divided.

However, with the development of science, atoms are considered to be composed of electrons, protons, and neutrons (hydrogen atoms are composed of protons and electrons), and they are collectively called subatomic particles. Almost all atoms contain all three of these subatomic particles, but protium (an isotope of hydrogen) has no neutrons and its ion (after losing its electron) is only a proton.

The proton has a positive charge and a mass 1836

times the mass of the electron, 1.6726×10^{-27} kg, however, part of the mass can be converted into atomic binding energy.

Neutrons are not charged and the mass of a free neutron is 1839 times the mass of an electron, 1.6929×10^{-27} kg. The size of neutrons and protons is similar, both in the order of 2.5×10^{-15} m, but their surfaces are not precisely The atom is small, but it is not precisely defined by chemistry.

Although atoms are small and cannot be chemically subdivided, they can be subdivided by other methods because they have a certain composition.

The atom is composed of a positively charged nucleus in the center and negatively charged electrons outside the nucleus (the opposite of antimatter).

In the Standard Model theory of physics, both protons and neutrons are composed of elementary particles called quarks. A quark is a type of nanon and one of the two fundamental components of matter.

The other elementary component is called leptons, and electrons are a type of lepton. There are six types of quarks, each with a fractional charge, either +2/3 or -1/3.

A proton is composed of two upper quarks and one lower quark, while a neutron is composed of one upper quark and two lower quarks.

This distinction explains why neutrons and protons have different charges and masses. The quarks are held together by strong interactions, with the gluons acting as intermediaries.

The gluon is a member of the canonical boson, a fundamental particle used to transmit force. In this way, what does matter in the world consist of? It is molecules. To continue, what are molecules made of? Atoms.

What is the composition of an atom? The nucleus and the electron. What does the nucleus consist of? Protons and neutrons. What about protons and neutrons? There is a quantification of ......, and this has led to the continuous progress of science.

Therefore, the only truth is that everything that exists in the world, everything in the world, is the same proposition or reasoning on the surface, implies two opposing conclusions, and both conclusions can be self-contained.

Paradox is the confusion of different levels of thinking,

meaning (content) and expression (form), subjective and objective, subject and object, fact and value implied in proposition or inference, the asymmetry of thinking content and thinking form, thinking subject and thinking object, thinking level and thinking object, the asymmetry of thinking structure and logical structure.

Paradoxes are rooted in the limitations of perception, logic of perception (traditional logic), and logic of contradiction. The root cause of paradox is the formalization of traditional logic, the absolutization of the universality of formal logic, i.e., the use of formal logic as a way of thinking.

All paradoxes are generated by the formal logic way of thinking, which cannot discover, explain, or solve logical errors.

Because of infinite questioning, there are endless questions. The ultimate theory that can solve all doubts is just a beautiful dream. There is still zero in zero, and infinity after infinity. Everything has no limits.

2. Infinite regress: The argument ultimately has no end point; to avoid infinite regress argument, infinite regress proof is a very bad way of proof.

When you assume that there was a chicken first, you can provide circular confirmation of your original proposition that there was an egg first, which means that the reason for having an egg first is that there was a chicken first, and the reason for having a chicken first is that there was an egg first.

This means that the first chicken or the first egg confirms itself. For this reason we can ask: "How can the first chicken confirm itself?

The argument that a proposition can confirm itself means that the proposition is both a premise and a conclusion in the corroborative reasoning, which is logically impermissible.

The causal dilemma of whether the chicken or the egg came first is a question of whether the egg came first or the chicken came first. This chicken and egg question often provoked ancient philosophers to explore and discuss the origin of life and the universe.

In general, it is often considered futile to get an answer to the circular cause-and-effect question of whether the chicken or the egg came first, which is considered to be the most fundamental question in nature.

Of course, the answer to this question is literally simple and obvious: oviparous animals have existed long before chickens. However, the metaphor behind this simple question raises a metaphysical dilemma.

To better understand this dilemma, the question is also rephrased as "the first chicken gets the first egg, and the first egg gets the first chicken, so which comes first, the first chicken or the first egg".

The Agrippa argument rests on the basic assumption that any knowledge must be confirmed, in other words, any knowledge must have a proper justification. When we say "I know there was an egg", the skeptic can ask you endlessly "how do you know" or "why was there an egg".

In the face of the skeptic's questions, we are either left with nothing to say, or we are left with circular arguments, or we are left with unsupported assertions.

For example, if we are not satisfied with the arbitrary assumption of the fifth axiom of Euclidean geometry, we have to prove it, but in the end, all we have to prove is to use it to prove itself. This shows that we cannot have the knowledge.

3, arbitrary termination of the argument: the argument terminates on an accepted premise, such as an axiom, rule, moral creed, religious creed, etc..

Arbitrary confirmation is wrong: do not want to arbitrarily establish the hypothesis, based only on subjective opinions, not to make a thoughtful judgment is wrong, in the process of confirmation, when you propose a proposition first chicken to confirm, you originally proposed the proposition first egg, when

If the chicken itself needs to be confirmed, then arbitrarily assuming that the chicken came first means that you are basing your confirmation on a mere assumption. It is wrong to assume arbitrarily, because if you have the right to assume so, others have the right to assume otherwise.

And you want to prove the assumption. But after you prove it, you eventually find that you are actually using the assumption to prove itself, which is called circular argument.

People will take something as the starting point. This thing as a starting point is not established by argument, but is an arbitrarily established assumption.

For example, Euclid's geometric theorem is based on several axioms and axioms. These axioms and axioms are not proven, but are arbitrarily assumed to be correct.

Any proof must end up being one of these three, and this is the trilemma. Agrippa's argument does not make any concrete theoretical commitment to "what the 'proof' really is," nor does it set a limit on "knowledge.

Instead, it assumes that "all knowledge needs to be confirmed" and that "the three principles of confirmation are correct". The conclusion to be drawn from this argument is that

No one can accept one proposition over another with certainty. Since knowledge requires confirmation, no one can know any proposition.

This shows that the Agrippa argument can argue both for a full-blown confirmation skepticism and for a full-blown knowledge skepticism.

Moreover, the skeptic argues that the two propositions cannot depend on each other, because this would create a circular argument. For skeptics, such a logic becomes an inadequate measure of truth, and because they think they have reached a solution, it may cause many problems.

But truth is not necessarily unattainable, and may be a concept that does not yet exist in its pure form. Although skepticism is accused of denying the possibility of truth, the fact is that it only reveals logic in a critical manner and does not actually discover truth.

As a result, most scholars (in whose minds present agency is absolutely dominant) have been accustomed to using Enlightenment insights and rational terms to interpret the postmodern as an antithesis.

As a result, a large number of terms with a clear direction (resistance) (e.g., subversion, rebellion, denial, rejection, resistance, anti-government, non-government, or anarchy, etc.) have become prevalent in the postmodern definition.

As a result, postmodernist thought has pushed "epistemology" and "ethical systems" to the extreme of relativism. This is particularly evident in the work of Jean-François Lyotard and de Sida.

These philosophers sought to deny the truths, meanings, historical processes, humanist ideals, and foundations upon which Western civilization was built.

Although postmodernism, in principle, is considered to be a philosophy of nihilism, it is worth noting that the philosophy of nihilism is not a philosophy of humanism.

However, it is worth noting that nihilism embraces postmodernism's denunciation. Nihilism is an interpretation of the truth of the universe that is unacceptable to postmodernism.

(4) Nothingness and Self-Deception

Schart's "Being and Nothingness" reveals from the above discussion that man's ability to perceive his environment and self, as well as the clarity of his perception, is in fact an unrealistic and inscrutable consciousness of "nothingness".

The eternal pattern of awareness and recognition of the individual's state of mind and body and the changes of people, events and things in the environment by using mental activities such as feeling, perception, thinking and memory is consciousness.

Freedom of will is a special function and activity of the human brain, which is a unique reflection of the objective world, and consciousness is all the mental activities of freedom of will.

The deepest cry from the heart of Shat is the question of what is the specific meaning of the existence of life. In metaphysics, the word "Particular" (also translated as Particular, Self, Concrete) refers to the existence of separate entities or individuals.

Originating from ancient Greek philosophy, the common nature of the various "Particulars" is the Commonality. The earliest philosophers to discuss the concepts of "disparity" and "co-existence" were Plato and Aristotle, who offered different answers to the question of co-existence.

It is also an important theological topic in the soteriological study of universal, fundamental questions in the fields of being, knowledge, value, reason, mind, and language.

This concept has continued into modern philosophy, but La Porte says that "when one conceives of a state of being in isolation, a state of being that is not isolated, one is already doing so, taking out one or several characteristics of the complex object and paying attention only to the actions or characteristics of the other.

The mind thinks only of the shape of the tree itself or

only of the color of the leaves, not limited by their size and shape.

What Rabbott is trying to say by quoting this phrase is that it refers to all mental activities. The essence of perception and memory, for example, is essentialism.

Section 2: The Concept of Essentialism

Essentialism, also known as essentialism, is the belief that any entity (such as a person, an animal, a physical object, or a concept) has some essential qualities that it must possess.

This position, which is based on research and analysis or criticism of problems and issues, also considers that it is not possible to make a final explanation of phenomena.

Theories that are logically and inferentially summarized by generalization and deductive reasoning, etc., in accordance with existing empirical knowledge, experience, facts, laws, cognition, and tested hypotheses, are useless because they do not reflect objective facts.

The emergence of the concept of "essentialism" in the West can be traced back to the time of Plato and Aristotle.

The earliest known theory that behind all thinking and concepts there must be a necessary fact (rationale) is "Platonism".

As a student of Aristotle's "Scopes", he argued that all objects belong to a certain basic being, and that the basic being determines what the object really is. The term "essentialism" was first coined by Karl Popper, and had not been used before by either scholars who held the "essentialist" view or those who questioned it.

Karl Popper's philosophical system focuses on critical rationalism, which is clearly separated from classical empiricism and its observational-inductive approach. In particular, Karl Popper opposed the observation-induction approach, which he believed was not applicable to the universal, but could only be the conclusion of an indirect judgment and analysis of an event or person.

He also believed that theories and all knowledge that human beings have acquired are merely hypothetical assumptions based on what is already known to imagine what is not known, and [imaginary] or [fictitious] content, based on experiments and logical reasoning, with certain objects as the scope of study.

Generally speaking, in the process of problem solving, people inevitably incorporate the old concepts and lessons they have learned through analysis and synthesis, in order to [recreate or reconstruct the imagination of ideas and images] and [the creativity of innovative characteristics of people's thinking or practical activities], so that the problem can be answered within the established historical and cultural framework.

We can only rely on the only data available to establish this scientific theory that takes a certain object as the scope of study and seeks to obtain unified and exact objective laws and truths based on experimental and logical reasoning. However, in addition, it is impossible to have a large enough number of experimental data to prove that a scientific theory is absolutely inaccurate.

1. The core of philosophical thought.

Karl Popper gave an example: "Sheep are white" theory, to the effect that, when people in the detection of one million sheep, sheep are white, but outside the detection, as long as there is a black sheep, you can prove that the previous theory is wrong.

(No one can test sheep endlessly to prove that the theory of "sheep are white" is absolutely correct.) This principle of "fallibility" is the core of Popper's philosophical thinking, which is derived from the "asymmetry of truth and falsity" (true things cannot be proved, only falsity can be proved).

2. Falsificationism. There are at least two advantages.

Karl Popper agrees with the definition of contingent truth, but he emphasizes that such empirical science, which takes a certain object as the scope of study and seeks to obtain unified and exact objective laws and truths based on experiments and logical reasoning, should serve an attempt to solve the problem of falsification.

In the past, human beings have tried to solve the unsolvable problems of natural and social phenomena by generalizing and deducing them logically through the methods of empirical knowledge, experience, facts, laws, cognition, and tested hypotheses.

It is a free creative, discursive, tentative speculation or conjectural falsificationism that explains the behavior of certain aspects of the world or the universe. There are at least two advantages.

First, scientific theories are generally expressed as allometric judgments: that is, judgments that reflect the nature of every object of a certain kind of thing, either with or without a certain kind of nature. For example, "subjectivity refers to the tendency to look at things from the perspective of the subject's own needs.

It is the property of viewpoint, experience, consciousness, spirit, feeling, desire or belief that an individual can have" and "objective facts are not influenced by subjective means such as human thoughts, feelings, tools, calculations, etc., but can maintain their truthfulness. .

And the general concept, including knowledge and skill. It is the experience or observation of a certain event, or a certain event, which is obtained by the mind and applied to the subsequent operation of the object as an individual. Therefore, if experience is used to prove a theory, then it will not be able to exhaust the general theory.

For example, even if there are many white sheep, it cannot prove that all of them are white, but if only one of them is black, it can prove that all of them are white, and this theory is obviously flawed and wrong. Therefore, the real meaning of "experience" lies in the ability to falsify scientific theories.

Secondly, falsification can avoid the error of speciousness: the logical conclusion of natural and social phenomena in accordance with existing empirical knowledge, experience, facts, laws, cognition, and verified hypotheses, through generalization and deductive reasoning.

If one insists on positivism, then once a systematic and organized law or argument has emerged. If one insists on positivism, then once there is a systematic, organized law or argument that contradicts the experience derived from actual verification, or from conceptual deduction, one makes special assumptions, or special restrictions, to make the theory satisfy the experience.

But in reality, such a statement is fixed, immovable, and judicious, often based on the subject's own need to see the tendency of things, it is the individual can have the view, experience, consciousness, spirit, feelings, desires or beliefs of the attribute of identification, extremely unobjective, and unscientific. Falsificationism leads people to believe that all objective laws and truths, which are based on experiments and logical reasoning, are only speculations and hypotheses, which will not be finally proven, but will be falsified at any time.

Falsificationism should adopt a purely empirical learning method and pursue the goal of exploring a systematic trial and error method (English: trial and error) with a black box nature by continuously testing and eliminating errors. This means that one should be bold enough to formulate hypotheses and speculations, and then look for examples that do not match the hypothesis.

The process of revising the hypothesis according to the examples is repeated, and the initial hypothesis is even rejected completely. There is no end to the modification and improvement of the theory by trial-and-error, and the result of the trial-and-error method can only be a better hypothesis, but not the best hypothesis. The best hypothesis is synonymous with the ultimate truth, and is contrary to the spirit of science.

Karl Popper highly appreciates Hume's critique of induction as a method of thinking used in the process of knowing things, and that from a known theorem, the next theorem is deduced, and so on, to get something. The critique of the so-called foundationalism (in English: the theory of knowledge) is quite powerful.

Foundationalism is a theory of the structure of proof of knowledge, in the sense of presenting reasons or evidence to support the truth or correctness of a claim, and asserts that the proof of knowledge starts from the most basic knowledge and is based on reasonable beliefs.

The proof of a claim is based on the most basic knowledge as a starting point, and the conclusion is deduced by reasonable inferences based on reasonable beliefs. At the same time, foundationalism means that people generally believe that knowledge needs a solid foundation, and the foundation of empirical science is the foundation of the senses. This is also the root of the inductive approach.

He points out that empirical foundationalism divides science into two parts: one is the foundation obtained by observation and practice. The second is the theory based on this foundation. It is commonly overlooked that observation and theory are not two separate theories.

Any observation is subject to a systematic and organized law or argument. Any observation is influenced by a tendency to be generalized from actual verification, or derived from conceptual deduction. Here it is clear that Karl Popper's theory of falsification science was deeply inspired by his "inaccuracy theorem".

Karl Popper believed that the search for a basis of knowledge was a mistake, but not an accidental one. It is a need for security based on human nature. Karl Popper also wanted to integrate the conflict between theory-only and empiricism, because theory-only is an epistemological doctrine that unilaterally emphasizes the role of reason.

Rationalists believe that only the "rational intuitive knowledge" that relies on reason to grasp the nature of things directly, or the knowledge that relies on reason to reason logically, i.e., rational knowledge, is reliable, and that perceptual knowledge that relies on sensory experience is unreliable and often the source of misperceptions.

Rationalism emphasizes the important role of rational knowledge and believes that knowledge cannot stay in the perceptual stage, but must rise to the rational knowledge of the nature and laws of things, which has truthfulness, as opposed to empiricism, which unilaterally emphasizes the role of reason.

It is also translated as rationalism. Rationalism is generally used in a narrow sense, as opposed to empiricism or empiricism, and is mainly concerned with the origin and reliability of knowledge.

Generally speaking, rationalists do not recognize the empiricist principle that all knowledge originates from sense experience.

They believe that reliable knowledge with universal necessity does not and cannot come from experience, but from the innate and undeniable "self-explanatory reasoning", which is obtained through rigorous logical reasoning.

They often describe such "self-explanatory reasoning", such as the axioms of Euclid's geometry, and the traditional laws of identity, contradiction, and neutrality of formal logic, as "natural concepts" inherent in the human mind.

This is the most fundamental method of philosophy and science in general. Descartes believed that since it is required to introduce reliable knowledge, the initial axioms or first principles must be "self-explanatory", while the concepts derived from sensory experience are often confused and vague.

Therefore, reliable knowledge cannot come from

perceptual experience, but only from the inherent or innate "innate concepts" of the human mind.

Rationalism can also refer to a doctrine in the traditional sense of metaphysics or ontology, which holds that everything exists for a reason, and that everything is understandable in principle.

Thus, rationalism in the broad sense is not limited to epistemology; in all fields of thought and culture, any theoretical viewpoint or tendency of thought that believes in reason can be called rationalism.

In contrast, there are various forms of "irrationalism" or "anti-rationalism," including mysticism, fideism, emotionalism, intuitionism, religious superstition, blind worship of authority, and adherence to old traditions.

Early Enlightenment thought, especially that of the 18th century French Enlightenment thinkers, can be regarded as rationalism in the above sense, regardless of their specific philosophical views. In the context of religious theology, the recognition of only those views in doctrine or dogma that are logical and rational is also considered rationalism.

In social and historical views, political doctrine,

ethics, literature, art, and aesthetics, any viewpoint that favors reason or exalts reason can be called rationalism. Rationalists believe that only the "rational intuitive knowledge" that relies on reason to grasp the nature of things directly, or the knowledge that relies on reason to reason logically, that is, rational knowledge, is reliable.

He emphasizes the important role of rational knowledge, and believes that knowledge cannot remain in the perceptual stage, but must rise to the level of rational knowledge that grasps the nature and laws of things and has truthfulness, but he also criticizes empiricism.

Karl Popper believed that perceptual knowledge based on sensory experience is unreliable and often the source of false knowledge. Both theory and empiricism recognize that knowledge originates from an unchanging foundation. Only theory holds that this basis is the principle of universal necessity, while empiricism holds that it is the human experience of sensation.

The core of Karl Popper's philosophy of science is that all theories and principles can be falsified, while experience, though not the source and foundation of knowledge, is the standard by which knowledge is tested. He referred to this view as critical rationalism.

The distinction between science and non-science is thus clearly defined by Karl Popper: science is expressed in terms of its falsifiability. Unlike logical positivism and "Hume's fork," science is not a criterion of meaninglessness, but of whether it is a world of objective knowledge.

So mathematics and logic were classified as non-scientific. Similarly, psychoanalysis, astrology, osteopathy, and later Marxism were also non-scientific, because they were not falsifiable.

The reason why mathematics and logic are classified as non-scientific is that they do not need to be tested empirically; they are called necessary truths by Hume. Mathematics and logic are based on the axioms of the beginning of their systems.

If one chooses a different system of axioms, one comes to very different conclusions, for example, between Euclid (English Euclid, Greek Ε'νκλειδη) and non-Euclid. Whether or not the axioms themselves are true is often unfalsifiable and therefore non-scientific.

Karl Popper is often considered a scientologist, but he is not an advocate of "scientific supremacy" or "scientism alone. For Karl Popper, scientific knowledge is not the only meaningful intellectual enterprise for human beings, and what is called "intellectuality" is often associated in philosophical discussions with "scientific knowledge".

In philosophical discussions, the term "intellect" is often used in common with the concept of "Reason". The intellect is the faculty or ability to know, a part of the structure of the mind, whose main function is to enable a person to go beyond the level of the senses and to

Its main function is to enable a person to go beyond the level of the senses and engage in a rational and active process of abstract reflection on the real world or any object in the form of "inward dialogue". The so-called inward dialogue is like talking to oneself without the condition of the listener.

In Aristotle (384-322 B.C.) the mind has systematic, organized laws or arguments. The purpose of verification is to ascertain whether a product, service, or system (or a part of it, or a combination of them) conforms to the actual verification.

It is the highest level of mental ability in the structure of the mind, and is unique to human beings.

There are two types of intellect, Active Intellect and Passive Intellect. The function of passive intellect is to find the common qualities of concrete things and then to form concepts (abstract, universal ideas that act as categories or classes of entities, events, or relationships). The concepts are abstract because they ignore the differences of things in their extensions, as if they were the same to deal with them.) The concept is abstract.

Then many abstract, universal ideas that act as domains specifying entities, events, or relationships, or that consist of many identical or similar persons and things, are synthesized into an arrangement of kinds of entities to form meaningful judgments.

The difference between active and passive intellect is that the former does not need to be based on sensory experience, so it can exist independently of the physical body, and its existence is eternal; the role of active intellect is to make it possible to know all things, both subject and object.

Like the function of light, without which all things would be dim, so, without active intellect, all the functions of the mind would not be possible.

Since Aristotle did not give a clear description of the interactions and interconnections between the active intellect and the passive intellect, later philosophers (in English) did not have a clear description of the interactions and interconnections.

Therefore, later philosophers (English: philosophers), or lovers of wisdom with an interest in things and a wide range of knowledge, have had many different opinions on this issue.

A. Kenny, a contemporary British philosopher, argues that wisdom is the ability of human beings to think. In other words, the function of intellect is to make meaningful the activity of thought and the expression of linguistic symbols.

In other words, the function of intellect is to make sense of the activity of thought and the expression of linguistic symbols. Science takes a certain object as the scope of study, and seeks to obtain a unified and definite objective law and theory of truth based on experiment and logical reasoning.

In his autobiography, Endless Search, he commented on Darwin's conclusion after judging and analyzing an event or a person that evolution is not a testable scientific theory, but an action plan for studying questions that cannot be answered directly by perception through rational reasoning and logic.

It is a possible framework for a testable scientific theory; in "The Evolution of Scientific Knowledge" it is said that "the growth of scientific knowledge, as I think of it, does not mean the accumulation of observations, but the continuous overthrow of a scientific theory.

It means the continuous overthrow of a scientific theory by another, better, or more systematic, organized law or argument.

It is a theory derived from actual verification, or a theory derived from conceptual deduction, and replaced by another", denying that a certain object as the scope of study, based on experimental and logical reasoning, to seek a unified, exact objective law and truth is equivalent to the truth.

Karl Popper's proposition is in fact a critique of reason. The real reason is that it is open to criticism, and it is the true essence of reason without superstition, without blindness to criticism and inquiry.

Karl Popper's rational attitude is that I may be wrong, you may be right, and through hard work, we can get closer to the truth. What makes science scientific is that it is both falsifiable and not subject to any authority.

He does not insist that reason can explain all phenomena, including the theory itself.

In his view, proving the intrinsic value of each discipline is, by default, uncertain, and inevitably falls into the trap of circular argumentation. If a person insists on rationalism, then he himself has an irrationalist element.

For he has a value presumption that he judges things according to his own perceptions, without seeking to conform to the actual situation, and that rationalism has superiority over irrationalism.

Karl Popper also admits that irrationalism is logically superior to rationalism because it does not require a rational defense of its own existence.

But at the same time Karl Popper denies radical irrationalism, saying that full rationalism is a philosophical approach based on the recognition of human reason as a source of knowledge, a theory that holds that reason is superior to sense perception, also known as theory alone.

Formally, rationalism is a methodology or theory that recognizes that truth cannot depend on the senses, but on reason and deductive reasoning, which only brings confusion to the mind, while complete irrationalism causes social harm.

Therefore, the debate between rationality and irrationalism does not provide a logical answer. But in the field of ethical and moral relations and value judgment, rationalism has many more advantages than irrationalism.

In light of the above theoretical arguments, we return to falsification to explore "essentialism". George Lykov summarizes Aristotle's view as "the properties that make something what it is, and without those properties, those things cease to be 'that'.

These things are no longer 'that' which is the same thing. This view is opposed to "nonessentialism". Non-fossilism claims that there is no particular property that a certain type of entity must possess.

3. Scientific Essentialism

For example, a tiger must have a specific set of genetic characteristics in order to be considered a tiger, and the lack of stripes is not a necessary condition for identifying a tiger, otherwise the sciences of zoology and genetics would no longer have any validity.

The loss of essentialism in science after the Enlightenment actually caused the degradation of modern science and led to the philosophical, anti-scientific, and humanistic controversies starting with Nietzsche, and the War on Science.

As a result, all scientists are essentialists. Even those scientists who oppose fundamentalism cannot escape from the fundamentalist mode of thinking and can only use the fallacy of the continuum of blue-green existence and therefore blue and green non-existence to deny fundamentalism, but if they really apply their logic of opposition to fundamentalism to scientific research, they will only end up with all their own ideas.

However, if they really apply their anti-natalist logic to scientific research, they will only end up in the dilemma of having all their scientific arguments, and even the scientific method itself, completely rejected by themselves.

One example is the spread of anti-natalism in biology, which eventually led some philosophers of biology to reject the idea that "if an organism belongs to a taxon, especially a species, then it inherently belongs to that class of organisms".

To put it bluntly, for these organisms, philosophers who think analytically and reflect on fundamental questions about life, knowledge, and values, from Linnaeus' taxonomy to the present-day molecular biology, have been able to make the case that the taxonomy of the human species is the most important one.

For hundreds of years, the study of people, things, and objects touched by all human activities has been the scope of research, and the construction of unified and exact objective laws and truths based on experiments and logical reasoning is completely meaningless.

Because they reject essentialism, they also reject a systematic system of knowledge that accumulates and organizes and tests explanations and predictions about the universe itself.

4. Criticism and Controversy

Essentialism has been met with some opposition. In

Plato's Dialogues, the Parmenides, Plato says that if we accept concepts such as "beauty" and "justice," we must also accept "the existence of different forms of phases such as hair, sludge, and filth. Even after Charles Darwin, essentialism remains the basis of natural scientific classification and the necessary source of justification.

Although the exact role and importance of ontology in biology has been under debate, many biologists have argued that ontology is the basis for biological classification, which logically leads to the whole of molecular biology, and even evolution itself, being regarded as pseudoscience.

However, many biologists believe that essentialism is a necessary element for the operation of biology, and that the idea that biology does not need essentialism may be a political ideological sophistry like Lysenkoism. Ontology (the idea that men and women are intrinsically different) also continues to be controversial in gender studies.

Karl Popper divided the ambiguous term "realism" into essentialism and realism. While essentialism is used as the opposite of nominalism, realism is used only as the opposite of idealism.

Karl Popper himself was a realist (as opposed to idealism), but he was also a methodological nominalist (as opposed to essentialism) Karl Popper believed that essentialism would inevitably hinder the development of science.

"Because science does not develop through a gradual encyclopedic accumulation of essential knowledge, as Aristotle thought, but by a more revolutionary method.

"Not only did essentialism encourage word games, but it also led to the destruction of the illusion of argumentation, the illusion of reason. The inevitable result of Plato's and Aristotle's essentialism is the philosophy of the academy and mysticism, and the despair of reason.

"I believe that essentialism is untenable. It implies a notion of ultimate explanation, because a fundamentalist explanation neither requires nor can it be further explained. Some scholars hold an operationalist view.

Peter Medawar, for example, claims that the concept of essentialism should be defined operationally in practice rather than in search of its "ultimate" nature, and argues that the public wants to use a certain object as the scope of study and to seek a unified and definite objective law and answer to the truth based on experimentation and logical reasoning, and that the tendency of the question of essentialism will cause the public to misunderstand science.

However, the argument that the so-called "essentialism" will hinder the development of science has actually been questioned because one of the purposes of using certain objects as the scope of research and seeking for unified and exact objective laws and truths based on experiments and logical reasoning is to induct phenomena into the world of science.

The purpose of removing the essence is to be satisfied with merely generalizing the phenomena, which actually led to the decline of scientific development after Newton.

Because physicists no longer try to explain why objects follow the laws of electromagnetism or gravity, eventually leading to the degradation of science itself, which only describes phenomena without thinking about the internal causes, and even the decline of science's ability to self-correct.

This eventually led to the stagnation of theoretical

physics and biological taxonomy after the mid-twentieth century, and even became the reason why the public and many post-modern intellectuals opposed science itself, which eventually led to the Great War of Science and the Reproduction Crisis in the late twentieth century.

Section 3: What is the nature of consciousness?

What is the nature of consciousness in philosophy, and what is its relationship with matter? What are the characteristics of consciousness? In the holistic, fundamental and critical inquiry of idealism into the real world and man, the nature of consciousness is revealed in the interrelation of consciousness and matter.

The nature of consciousness is a specification of the nature of consciousness revealed by the interrelationship between consciousness and matter. It refers to the fundamental properties of consciousness itself that determine the nature, appearance and development of things.

The essence of consciousness is hidden and expressed through phenomena, which cannot be understood by simple intuition.

Consciousness is something that can be

distinguished and exists independently within itself. But it does not need to be a physical existence in itself. In particular, it is regarded as a physical product in the abstract, but not a physical substance in itself.

The individual uses sensation, perception, thought, memory, and other mental activities to become aware of his or her own physical and mental state and the changes of people, things, and objects in the environment.

Dialectical idealism believes that consciousness is a function of the human brain, a reflection of objective existence. The ability of cognition and the clarity of cognition of the environment and the self is the function and attribute of the human brain. The human brain is an organ that uses mental activities such as sensation, perception, thinking, memory, and so on, to produce a comprehensive awareness and knowledge of one's physical and mental state and changes in people, things, and objects in the environment. However, the brain alone cannot automatically generate self-awareness.

Self-consciousness (also called self-cognition or ego) is a multi-dimensional and multi-level complex psychological phenomenon, which consists of three psychological components: self-knowledge, self-experience and self-control.

Self-cognition plays a very important role in the development of an individual. First of all, self-awareness is the condition for knowing the objective things in the outside world. If a person does not know himself and cannot distinguish himself from his surroundings, it is impossible for him to know the external objective things.

Therefore, it is only when all the external objective objects in the natural world or the present act directly on the human brain that the brain can produce, and the individual uses sensory, perceptual, thinking, memory and other mental activities to

The individual uses sensory, perceptual, thinking, memory and other mental activities to perceive and understand his or her physical and mental state and the changes in people, events and things in the environment. Consciousness is the human brain, a special distinguishable and independent thing that exists within itself. However, it does not need to be a physical existence, occupying a place in space, and can be known by the senses of its existence, the reflection of objective existence.

1. Thoughts in the brain

Simply put, philosophical consciousness is the reflection of objective existence. Individuals use their senses, perceptions, thinking, memory and other mental activities to

The comprehensive awareness and understanding of one's physical and mental state and the changes of people, things and objects in the environment is a copy and photograph of the objective world as it is, and is a subjective reflection of the objective world.

Consciousness is the unity of subjective and objective, the reflection of the human mind to the objective material world, is the feeling, thinking and other mental processes is characterized by: the content of consciousness is an objective entity, the form of consciousness is indeed subjective.

Consciousness is neither subjective and self-generated, nor given by God and God, but comes from the objective physical world.

Without the objective body material world, there is no individual who uses sensation, perception, thinking, memory and other mental activities to become aware of his or her own physical and mental state and the changes of people, events and things in the environment.

Without the reflected person, there is no reflection. What the reflected person is like, the correct reflection should also be like. If the person being reflected changes, the reflection will change as well.

2. "Childhood amnesia"

In a research paper previously published in Science magazine, researchers revealed why we lose our childhood memories.

The scientific approach reveals that most people do not have memories from age 3-4, and in fact we have very few memories from before age 8, and when trying to recall memories from childhood, we are not sure if they are actual events or memories based on photographs or stories we have been told.

This observable, observable fact is called "childhood amnesia," which refers to an adult's inability to recall situational memories (memories of situations or events) from before the age of 2 to 4.

As time passes, an adult's memory of the period before

age 10 may be worse than imagined. Some believe that the development of the cognitive self also encodes early memories (Encoding).

Cognitively, it is the process of organizing and interpreting incoming stimuli, a basic external stimulus that acts on the senses, and the human brain's overall view and understanding of the external world.

Technically, this is a complex, multi-stage process of transformation from more objective sensory input (e.g., impulses, sounds) to subjectively meaningful experience and storage of transforming influences that has been troubling psychologists for more than a century, and researchers still do not fully understand the mechanisms by which this phenomenon occurs.

The amount of memory a person can recall from early childhood depends on many factors, including the amount of socially significant or emotionally relevant events, the age of the child when the event was remembered, and the age at which the person was asked to recall the early event.

Although people think that the inability to recall childhood memories means that they have forgotten the event, there is a distinction between "availability" and "accessibility." "Availability" means that the memory is intact and located in the memory store. "Accessibility" depends on whether the memory can be recalled at that moment.

Thus, clues may affect the accessibility of memory at any given time, even though there may be more untouchable available memory.

Other studies have shown that people's earliest memories can be traced back to age 3 or 4. In Usher and Neisser's report, events such as the birth of a younger sibling and hospitalization were easily remembered if they occurred at age 2.

However, these fragmentary memories obtained in their study may represent untrue episodic memories.

Another hypothesis is that these surface memories come from imagining what is not known based on what is already known, having an understanding of what should be generally known, or from external information and messages that are useful to the user after age 2.

According to a study by West and Bauer, early memory has less emotional content than later memory and less personal meaning, uniqueness, or intensity, and early memory does not appear to be very different in perspective.

However, certain life events do lead to clearer and earlier memories. From early childhood onward, adults are more likely to remember individuals than non-public memories of events.

Psychologists have debated the age at which adults have their earliest memories, with estimates so far ranging from about 2 to 6-8 years of age. This means that a person will remember having had a Rottweiler.

The fading of childhood amnesia (the earliest age that can be recalled) does not occur at age 2 for hospitalizations, and at age 3 for sibling births, and at age 3 for family deaths or family changes. Thus, some studies suggest that some extractable childhood memories are available at an earlier point in time than the age indicated by the above studies.

Some studies suggest that children do not develop contextually rich memories until around age 4. Although more evidence is needed, the reason for the relative lack of contextual memory in early childhood may be related to the maturation of the prefrontal cortex.

At approximately 4.7 years of age, known memories become more personal. In addition, adults can access fragmentary memories (isolated moments without context, usually remembered as images, behaviors, or emotions) around age 3, and can usually recall slightly later event memories, a finding similar to studies showing differences between individual memories and known events.

3. (Physical, mental, volitional, ability, etc.) Decline and regression

Children are able to form memories much earlier than adults can recall to age. The difference in the cognitive process of interpreting incoming stimuli as a basic perception and in the efficiency of storage processing allows older children to recall more, while younger children also have a greater ability to recall more.

Younger children also have a strong memory, and infants can remember sequences of movements, objects used to produce sequential rows, and the order in which movements unfold, indicating that infants have the necessary precursors to remember their own memories of their life experiences.

Children have a 50% accuracy rate of remembering what happened before the age of two, while adults can barely remember what happened before the age of two. By the age of 2, children are able to retrieve memories weeks later, suggesting that these memories may become relatively persistent and explaining why some people have memories at such a young age.

Children also exhibit nonverbal memory abilities that occur before they acquire vocabulary, whereas adults do not, prompting research into "when" and "why" these previously accessible memories are lost.

One explanation for the loss of the ability to recall "prelinguistic memory" as children age is that the development of language skills is not followed by a process of converting linguistic information from one form or format to another.

The memory of codes, also known as computer-programmed languages, is lost in the brain. This theory also explains why many people have early memories of complete things broken into many fragmented pieces, due to the loss of non-verbal parts.

However, contrary findings suggest that children of

elementary school age remember more precise details of events than they reported at an early age, and that children aged 6-9 years often have verbal memory from an early age.

However, studies of animal models seem to suggest that childhood amnesia is caused by more than just language development or by any other human faculty.

It is not until childhood reaches double digits in age that this enhanced ability of children to recall early memories begins to disappear. By age 11, children show a level of amnesia comparable to that of young adults in terms of childhood memory.

This may indicate that certain aspects of the adolescent brain, or neurobiological processes, contribute to the developmental history of childhood amnesia.

In a previous research paper published in Science, researchers revealed why we lose our memories as children. The paper, which aims to study rodents, proposes a hypothesis that new cells are constantly being formed in the young brain during infancy, which disrupts the normal functioning of the areas of the brain that store memory.

The brain constructs new cells throughout the life of a mammal, a process known as "Neurogenesis".

In some species, including humans, neuronal production is very rapid during infancy. This phenomenon is particularly evident in the hippocampus of the brain. The hippocampus is the area of the brain responsible for learning and memory work.

Very often, neurogenesis occurs to help us learn better and improve memory. However, this paper points out that when the rate of neurogenesis in the brain is too high, its benefits turn into drawbacks.

The result of producing new neurons at a very fast rate is that the old neurons that store memories are "crowded out," which ultimately increases the frequency of losing these memories and leads to the onset of amnesia in infancy.

Although the specific molecular mechanisms underlying the onset of childhood amnesia are currently unknown, researchers have made some progress through a number of efforts, such as conducting numerous prospective longitudinal studies to track individuals from childhood to adulthood.

This may give researchers information about specific events that occurred, and this is a much better answer than asking adolescents or adults to recall what happened in the past.

Although we do not remember the specific events that occurred during early childhood, the accumulation of event memory leaves a lasting trace that affects our behavior.

Section 4: Consciousness, subconsciousness or unconsciousness.

In fact, the nature of human consciousness is a hot topic in the field of philosophy. However, according to the research of deep psychology, it is found that consciousness is only the superficial part of the mind, and there is a large unconscious layer hidden in its depth.

Therefore, only by deeper investigation of the unconscious area, which lies deeper in the concrete actions, thoughts, and desires of human beings, can the whole picture of human spirit and even life be identified.

Western psychology generally divides consciousness into two levels - consciousness, subconsciousness, or unconsciousness. In the past, Western deep psychology has explored and studied

Jung believed that human beings can be divided into three levels, from shallow to deep: consciousness (ego), individual unconscious (complex), and collective unconscious (archetypes), which is different from Sigmund Freud's psychological framework. Broadly speaking, three levels seem to be found.

Translated with www.DeepL.com/Translator (free version)

The first level of consciousness refers to the general waking state: for example, working, talking, thinking, etc., where we are clearly aware of the functioning of the self; the subconscious/unconscious refers to the emotions, desires, dreams, memories, etc., that are suppressed or forgotten outside the perceptible range of the conscious mind.

The subconscious/unconscious is in control of the majority of the invisible, which is the "unconscious layer" discovered by Freud himself, that is, "the things forgotten by the consciousness and the repressed mental contents". That is, what Freud called "dreams".

The second layer is the "unconscious layer of the

family" proposed by Soddy, which is the second layer of the mental structure and has a greater role than the consciousness. It consists of all forgotten memories, perceptions and submerged pasts, which appear in the form of dreams and hallucinations.

Jung believed that the unconscious content of the individual is the complex. The complexes are often emotional, a jungle of unconscious contents that come together one after another, such as the father complex, the criticism complex, the power complex, and so on.

The unconscious is a container for all the individualized functions of consciousness, all the inconsistent mental activities, and all the contents that were once conscious experiences, but for various reasons were suppressed or neglected, such as painful thoughts, unresolved issues, interpersonal conflicts, and moral anxieties.

There are also experiences that are less relevant or seemingly insignificant to people because they are too weak to reach the level of consciousness when they are experienced.

They are stored in the unconscious of the individual

because they are too weak to reach the level of consciousness when they are experienced, or they do not reside in the consciousness. All of this constitutes the content of the individual's unconscious, which usually reaches the conscious level easily when it is needed.

The third layer, the so-called "collective unconscious layer" proposed by Jung: in the "collective unconscious" all the experiences of races and peoples, including even the most ancient ancestors of mankind, are accumulated.

It is fundamentally linked to the universe itself. It is the unconscious that is at the bottom of the personality structure, the genetic traces of the way of activity and experience of the generations, including the ancestors, that exist in the human brain.

The difference between the collective unconscious and the individual unconscious is that it is not a forgotten part, but something that we have never been aware of. Jung once used the analogy of an island, the small islands that are exposed to the water are the consciousness that can be perceived.

The ground below the surface, which is revealed by the ebb and flow of the tide, is the individual unconscious;

and the bottom of the island, which is the sea bed, is our collective unconscious.

The archetypes are the sum of the images in the collective unconscious. Jung also called them manifest, unconscious imagery, fictional or primitive impressions, and some other names. But archetypes seem to be the most accepted.

The so-called archetypes are the innate tendency to experience things in a particular way. The archetype itself has no form of its own, but it manifests itself as what we see as the so-called "organizing principle". It acts according to the intuitive law of Froude's theory.

First of all, when a baby wants to eat, he does not know what he wants. He has a rather unclear desire, however, for something specific that will satisfy him.

Then, from experience, when a child is hungry, he craves something specific - milk, cookies, grilled lobster, New York-style pizza. A prototype is like a black hole in the universe: you can only know it is there by the way it attracts protons and light.

In the East, a thousand years before Freud, Buddhism (the Vaisnava school), developed this doctrine.

The Vaisnava school refers to "Vaisnava Buddhism", which is also known as "Dharma Sangha" founded by Venerable Xuanzang of Sanzang.

It is one of the Buddhist schools in China, and is derived from the ancient Indian Mahayana Buddhist sect of Yoga. Vaisnavism starts from epistemology and comes back to ontology. However, when it comes to the ontology of Vaisnava, it is felt that if one insists that the ontology is real, it is contrary to the meaning of emptiness, so it turns back to the epistemological "Vaisnava" instead of the real ontology.

Therefore, the meaning of "only consciousness" (Sanskrit vijñapti-mātratā) is "only table", and "no realm only consciousness" is "realm only consciousness table", not "no realm only consciousness", because consciousness is also the subversive delusion of the voidness of origin. The transmigration of the practice of the only consciousness is through the transmigration of the dependent (the sixth and seventh consciousnesses) to the dependent (the eighth Arya consciousness), so that the first five consciousnesses can actually see the external dust, that is, the transmigration of the eight consciousnesses into the four wisdoms, rather than "covering the realm and holding the consciousness".

In yogic psychology, apart from the body, the cognitive ability and the clarity of cognition of the environment and the self are divided into five levels. In addition to the role of the consciousness and subconsciousness, the spiritual potential of human beings is also included, as outlined below.

The body (Annamaya kosa): The universe and the human body have a commonality: they are both composed of the life force (prana) and appear in the form of the five elements: earth, water, fire, wind, and ether (earth, water, fire, wind, and air).

The five elements not only constitute nature, but also form the material body parts of all living entities. The human body, of course, is also composed of the five elements of earth, water, fire, wind, and ether.

Aether: Simply put, it is space. Take the human body as an example, the mouth, digestive tract, respiratory tract, lung alveoli, microvascular are all.

Wind: The physical state of matter, the five elements and five colors, with dynamic and explosive power. Wind triggers movement, such as muscle movement, heartbeat, lung contraction, gastrointestinal peristalsis, nerve conduction movement, etc.

Fire: Fire is a force, with an image but no substance, characterized by the ability to transform the solid state into a liquid, gaseous state. Fire controls the role of metabolism and enzyme transformation in the body, and is related to digestion, thinking, and vision.

Water: The liquid state of matter, characterized by fluidity, the water element in the body includes digestive juices, saliva, various body fluids, etc.

Earth: The solid state of matter, characterized by stability. The inherent tissues of the body such as bones, cartilage, muscles, skin, nails, and hair are all of them.

These five elements exist in nature in different proportions, and both inanimate and animate possess the five elements; these differences should be respected and treated reasonably. The living body is the dwelling place of the human soul, which is fed by food and contains muscles, bones, blood, nerves, motor organs, sensory organs, etc.

This is the level of consciousness (Kamamaya kosa): This is the integrated awareness and knowledge of the state of our body and mind and the changes of people, events and things in the environment by using mental activities such as feeling, perception, thinking and memory.

It contains three main roles - the ability to perceive the body, the desire for materiality, and the transformation of thoughts into bodily actions. Human biological instincts, such as hunger, sleep, fear, reproduction, etc., all belong to this level of response.

This level of the mind is always in a state of busy functioning to respond to the needs of the internal physiology and the external world.

The subtle mind level or subconscious level (Manomaya kosa): This is the level at which we can't become conscious of anything at all under normal circumstances, for example, desires that are repressed deep inside and never realized. It is the so-called "iceberg theory".

The composition of human consciousness is like an iceberg, only a small part of which is exposed to the surface, but the vast majority of which is hidden underwater influences the rest.

The subconscious, according to Froude, has an active role in exerting pressure and influence on human personality and behavior. It consists of four main functions - memory, recollection, dreaming and contemplation (deep thinking/observation), that is, the development and operation of the human intellect at the level of cognitive ability and clarity of cognition of the environment and the self.

4.Atimanasa kosa/Supramental mind: This is the supramental level, the level of cognitive ability and clarity of cognition of the environment and the self, which is beyond the influence of people, time, space and place.

It is also the first level of the causative mind. It is also the first level of the causal mind. Its main function is the intuition of dreams, deep intuition, karma, and the act of "cognition" and "recognition" of a subject in order to know with certainty.

These perceptions have the potential to be used for specific purposes. It means that through experience or association, one is able to become familiar with the repository of further understanding of something, and through the development of intuition, the mind goes deeper into the inner being, stirring up the desire for spirituality.

5.Vijinanamaya kosa/subliminal mind: This is the act of "cognition" and "identification" of a particular subject in order to know with certainty, and these human minds reflect the characteristics and connections of objective things, and reveal the meaning and significance of things for people.

These are the activities of the mind that reflect the characteristics and connections of objective things and reveal their meaning and effect on people, and have the potential to be used for specific purposes. It means the ability to become familiar with a certain level of understanding through experience or association, that is, the knowledge of existential self-awareness.

In the creation of art, such as painting, dance, and music, when one reaches a state of forgetfulness, this is the role of this level. The main qualities are viveka and non-attachment (vaeragya).

Through one's ability to recognize the environment and the self and the clarity of cognition, one can understand the impermanence, eternity, and non-duality of the world and develop spiritual wisdom to see everything as a manifestation of the beauty of supreme consciousness.

6. Hiranmaya kosa: This is also called the golden level of the mind, which describes it as pure and beautiful as gold, and is the highest level of the mind.

It is the level of the sum state of various mental processes such as feeling and thinking. The "cause" that arises from a certain reason means that there are no boundaries in space and no boundaries in time.

In time, it has no beginning and no end. The universe is the material world, in which matter is in constant movement and change, the original creative cause and effect of the universe.

From a yogic psychological point of view, the role of the mind at all levels is understood from a cosmic spiritual point of view. The human mind is transformed from cosmic consciousness and possesses unlimited potential and expansion.

In some dimensions, it operates beyond the material world. By expanding the mind, raising consciousness, and developing intellectually, one can gain a deeper understanding of the mysteries of the universe and enjoy spiritual joy.

Section 5: Consciousness from the Objective Material World

Photographing the Negative

We can take the process of recording images using a specialized device. Generally, we use a mechanical camera or a digital camera to take still pictures, to take pictures as an analogy: if there is a person in front of us, a person will appear on the negative.

For still picture photography, there is nothing in front of the subject, so the negative will be a blank film. The reflection on a photographic negative coated with photographic chemicals depends entirely on the objective external conditions.

It can be seen that individuals use their senses, perceptions, thinking, memory and other mental activities to perceive and understand their physical and mental states and the changes of people, events and things in the environment.

The content of consciousness is not based on the process of acquiring knowledge by forming concepts, perceptions, judgments, or imagination, i.e., the mental function of the mind to process information and make judgments about things.

It reflects the result of the objective world, that is, it is objective. How can we prove the objectivity of the content of consciousness?

1. Objectivity is a central concept of philosophy, outlined in three ways.

Objectivity is a central concept of philosophy, which refers to the rationality of thinking or judging something from different viewpoints or perspectives, the nature of a thing that exists independently of subjective thought or consciousness, corresponding to "subjectivity". Objective facts, independent of the influence of subjective means such as human thoughts, feelings, tools, and calculations, can maintain their truthfulness, outlined in three ways.

Looking in the mirror

First, from the perspective of daily practice. We all look in the mirror almost every day, and what is reflected in the mirror is from the outside world of the person being reflected. What clothes you wear, the mirror will reflect what kind of clothes.

The reflection of the consciousness to the material is

also the same reason, its generation is first dependent on the objective material world conditions. Any kind of idea in the human mind is from the objective world, leaving the objective content, and independent existence of consciousness is not.

Second, from a scientific point of view. Pavlov's theory of higher nervous system proves that consciousness is predicated on objective stimuli, and without any physical stimuli, without the first signal system to stimulate the brain and the senses, consciousness is impossible.

Furthermore, without the second signal system of language and words to stimulate the brain and the senses, consciousness is also impossible. The human brain is like a "processing plant", objective things are like raw materials, without objective things as raw materials, the brain "processing plant" is unable to produce consciousness as a product.

Third, from the social practice. Consciousness is a reflection of existence, and the generation of consciousness depends on the development of social productivity and the level of scientific development. At what level of development of production and science, the reflection of consciousness will reach what level.

In primitive society, the tools of production were simple, productivity was low, and the standard of science was also very low, so the level of people's consciousness was also relatively low. The terms computer, television, refrigerator, and electric fan appeared only after the production of these products.

Science is the study of a certain object, based on experiments and logical reasoning, to find unified and exact objective laws and truths, and without reaching such a level, these cannot naturally appear in people's minds, but individuals can use their senses, perceptions, thinking, memory and other mental activities to

Individuals can use their senses, perceptions, thoughts, memories and other mental activities to perceive and understand their own physical and mental states and the changes of people, events and things in the environment, and thus it is impossible to produce these words. What classical literature has ever included such words as refrigerator, television, airplane, and automobile?

Schart pointed out that Descartes' value of mind-matter dualism unites the thought entity and the extended entity. The so-called "res extensa" (Latin: res extensa) is one of the three entities proposed by René Descartes in his Cartesian ontology (also commonly referred to as "radical dualism"), the other two being "res cogitans" and "God". In Latin, "res extensa" means "something that extends".

Descartes also often translated this concept as "corporeal substance". In Descartes' ontology of the "entity-attribute" model, "extensiveness" is a necessary essence in material things, without which things cannot be established as primary attributes.

Then, we use figurative and abstract thinking to engage in a rational and active process of reflection on the real world or any object in the form of "inward dialogue". The so-called inward dialogue is like talking to oneself without the situation of the listener to solve the problem (the imaginary way).

Figurative thinking has some connection with philosophical thinking. Language and concepts have both abstract and concrete functions. In order to reflect the nature of things, figurative thinking also requires an abstraction process similar to logical thinking, where logic is permeated in the figurative pictures.

Hegel says in "Small Logic": "The special role of renunciation is to inherit and promote the positive and rational elements within the old things, to discard and deny the negative and inevitable elements within the old things, to unify the promotion and the discard, to combine the negation and the preservation, that is, the affirmation. But an image is not a concept, and the illustration of an image is only the visualization of a concept.

The literary image of Saud is the illustration of his philosophy of will, his imagination. Saud expresses the main meaning of his philosophy of will by means of literary images. The literary images in his works serve his philosophical views and thus have distinctive features in terms of their shape.

The subjects in Schart's literary works often have a strong desire for free will and a deep self-perception of the external world, but at the same time they are deeply constrained by existence and have to make a dilemma about their own fate.

That is to say, the two are unified by default, separated by analysis, and connected again. What Schart is saying is that if Descartes treats the thought entity as an abstraction, then how is the extended entity not an abstraction?

They are not, in fact, two interconnected entities, as Spinoza has argued. Husserl's concrete refers to an object that exists only in space-time. A concrete object is an independent object (not so universal), an object that cannot be given in the consciousness in a self-connection/intrinsic intuition.

Abstraction, on the other hand, is abstraction in 'intrinsic intuition,' i.e., "attention to what is, but not to what is. Then Husserl's phenomenon is also an abstraction, which must "appear" to the consciousness.

Citing Spinoza's system of "analysis never achieves synthesis," Scharthe argues against Husserl's and Descartes' way of thinking that starts with abstraction and seeks to have a physical existence, or a definite, rather than abstract, generalized, person outside oneself or in contrast to oneself. Existentialism tends to start with the concrete.

However, in the literary texts of Sartre, the "contingency" and the "limits of existence" are expressed in the fullest sense of the emptiness of the individual. The contingency of real existence, or contingency, is the fundamental source of what Schartre calls "the sense of emptiness.

Sartre's Nausea was written in 1931, and in its initial stage, Sartre titled it "Statement of Contingency". In this contingent reality, Sartre highlights the fear, anxiety, and nervousness of his characters in his literature.

Schatte's influence on existential fiction is illustrated by the work Vomit, in which Roquentin, a historian, lives in the fictional city of Bouvier.

He is a historian who lives in the fictional city of Bouvier. Convinced that nothing else can give him his own nature, he finds himself in absolute freedom, evoking a wave of vomiting feelings one after another.

In the words of Rogandan, "I have internal conflicts of an emotional or psychological nature that trigger irrational feelings of anxiety or fear, because a small action makes me bear the consequences, and I can't guess what people are asking me to do, but I have to make a choice.

In the novel "The Wall", when "my" life is completely under their control, "I" subconsciously worry too much about the safety of my loved ones or my own life, my future destiny, and so on, and this creates a kind of irritation that starts to make images and voices grow from small to large, "I am so afraid that I have shouted in my dreams.

The certainty of "non-existence" and the loss of "existence" also leave the subject in a void. According to Schart, "The possibility of the eternity of non-being constrains us, raises questions about being, and non-being will even limit being.

The thing that being will be will necessarily be hidden in the matrix of what it is not, what it is not. Existence is this, and beyond this it is nothing." Schartre expresses the unknowability of this "non-being" in terms of "eternal possibility," "necessity," and "limitation.

Schartre considers "non-being" as an unknowable existence that exerts a certain restraining effect on the self-will. The "I" in The Wall says: "My whole life is a damned lie, and since my life has ended, it (memories, etc.) is worthless."

The protagonist in Schart's novel, with a sense of the essence of life, realizes that the so-called free will is only limited by a certain space-time realm.

"Man is living blindfolded in the present time. He is

only allowed to feel and guess what he is actually experiencing.

Only when the blindfold is lifted can he look into the past and discover what he has experienced and what it means. -Milankundra, Funny Love

"Life is an eternal, heavy effort, an effort not to lose one's way, an effort to be ever-present in oneself, in one's place. "The novel looks in detail not at reality, but at existence.

Existence is not something that has already happened, it is something that belongs to the realm of human possibility, something that all human beings can be, something that all human beings can do. The novelist draws a map of existence and thus discovers this or that human possibility."

This is particularly true of the artistic characteristics of the short story "The Wall" - perhaps in relation to the author's background and identity.

As we all know, he was influenced by existentialism's holistic, fundamental and critical inquiry into the real world and human beings, and incorporated his own understanding of the world and life into his literary works, thus giving the reader a sense of absurdity and logical inability to judge what is right or wrong, and even a sense of being in a different world.

The novel "The Wall" is written in the first person, knowing only the thoughts of one character and his point of view (usually the protagonist), which is just changing the first-person "I" to "he".

When the character speaks his mind, he tells an absurd story: a group of revolutionaries are caught by the government and forced to ask about the whereabouts of the leader, Gris, and although "I" deliberated on the two paths of facing death and survival, I never wanted to reveal the secret, "I would rather die than betray Gris.

But at the last minute, I wanted to tease them, so I said, "I know where he is. He's hiding in a cemetery, in a crypt or a gravedigger's hut." In fact, the original plan of Gris was to hide in his cousin's house, but in order not to get his cousin involved, Gris actually hid in the gravedigger's hut.

At the end of the story, there is a sudden and shocking event. Just as the reader is easily stimulated and nervous, he or she is struck by the unexpectedness of the event, but although the ending is unexpected, it makes sense.

This presents a strategy of possibility, somewhat similar to that of the French structuralist, narratologist Claude Bremont, in "The Laws and Rules of Narrative Possibility, an Abstraction of the Thought Process", which speaks of possibility [a quantitative indicator contained in things and predicting their development, an objective argument rather than a subjective test].

The existence of a thing is only the manifestation of a possibility, a condition that means that cognitive, virtuous, aesthetic and other activities can be carried out. For example, in the case of cognitive activity

Kant discusses the basis for the production of knowledge of universal validity in terms of the cognitive subject, as the chaotic world of experience is rendered regular by the cognitive faculty of man.

The Kantian philosophical view is that human cognitive ability is divided into three aspects: perceptual, intellectual, and rational: perceptual provides formal conditions such as time and space, and forms concepts from the content of the object of experience that one wants to know.

Perception provides the twelve domains and classifies the content of the concepts after collation, which is then unified by reason to become knowledge; at the same time, reason also extends or applies the knowledge gained to enrich the content of knowledge.

The most important part of the problem is that it is the slightest "chance" that makes things develop in a very different and far apart way.

The last part of the text appears at the same time as the most flourishing stage of development, giving an indelible profundity, as if it were an experience gained through personal practice.

The title of the novel, "The Wall," shows us the author's ingenious ideas: First, at the beginning, "We were driven into a white hall. The strong light made my eyes close up and down slightly. I was so happy.

It gives the reader a sense of urgency, a sense of personal tension in front of the camera, can be relaxed their emotions, not nervous, not pretentious, open, can quickly find the state of photography wanted, foreshadowing the threat to life and safety, as if there is an invisible wall of barriers, restricting and oppressing the breathing of people.

Second, as the story progresses step by step, time passes like rain dripping along the eaves, and the characters' proximity to death becomes clearer, a kind of fear, anxiety and nervousness keeps eroding, and the desire to survive in the cramped cell, as if there are words in the heart that have not been said, is very uncomfortable and continuous.

Therefore, in such a situation, one feels anxious about the lack of contact or communication with others, or the possibility of losing contact or communication with others in the future, which leads to the thought of the next death.

"I" suddenly feel that I should break through the psychological anxieties that have been formed, the anxieties that have been bound to people and things, and the anxieties that have caused me to lose my mind, my balance, my freedom, and so on.

This causes the so-called overwhelm and constraint. Why don't you play with them a little bit, release your worries, get rid of the sire bond, and get out of the freedom of body and mind? This change of mind, over a period of time, plays a decisive role in things, and is the most important part of the change and passage of things, so that the situation or situation, to the opposite direction of transformation.

In other words, the "wall" becomes an objective image and a subjective mental image with a certain implied meaning and mood.

On the one hand, it shows the reader that there is a wall between human beings and human beings or between human beings and things of a certain nature, and that the difference between human beings and things is that things only suffer in silence, while human beings can change their situation (mostly unfavorable situations), from the constraint of passively waiting for death to the freedom of mind to play with enemies.

On the other hand, Scharthe himself said, "Self-existence cannot become self-existence", that is, "if man cannot achieve freedom of mind, then he is still not free in essence, in substance.

Here we can clearly see that Scharthe infuses his existentialist thought into this novel: "The world is a meaningless, contradictory, disordered state of life; life is a state of physical pain and suffering; it is also a state of mental imbalance caused by mental torture and pain and the destruction of hope".

From the very beginning, when things were so frightening and troublesome and trivial, when the interrogation took so much time and energy, to the end when Gris was killed, the event itself was full of unreality and unreasonableness. "I started to lose my head and eventually fell to the ground. I laughed so hard, I even burst into tears."

There is no other way out but to "laugh". This chance makes life full of quantifiable choices that are contained in things and foretell the trend of things, and "my" choice is free, but it is obvious.

But obviously, "I" did not expect such an outcome, so the moral condemnation puts a lot of pressure on it, and "laughed out loud" is the tearful laughter that comes naturally with the last part of the text and the development of the most flourishing stage of things.

The possibility refers to the potential, various trends of development within an objective thing; the reality, however, has already been realized to the extent that it can become a fact. Possibility and reality are the two inevitable stages in the development of a thing or phenomenon.

The transformation of things from one quality to another, the replacement of abstract or inanimate things with concrete examples, is the transformation of the degree of being able to become a fact to reality. Possibility and reality are interdependent and mutually transforming.

As an internal factor of reality, possibility exists in reality, and is a potential reality that has not yet been developed; reality is the result of the development of possibility, and is a realized possibility.

After the possibility is transformed into reality, the new reality contains new possibilities. Life is full of constant choices, and the pain behind them needs to be endured, not to mention the severity of the consequences, which are beyond human expectation, so life is full of pain.

Of course, this novel has reached a high level of ideology and artistry, not only because it is full of existentialist ideas, but also because of the metaphorical allegory of Schatvinsky.

This is not only because it is full of existentialist ideas, but also because of the metaphorical allegory of Sartorius, but also because it is a metaphor that presents to the reader the dilemma of "existence and nothingness" that really exists in human existence.

This is worthy of the rational and active process of reflection and attention to the real world or any object in a constant "inward dialogue".

The abstraction becomes only a part of the free will of the concrete object of the nature of life. Situation awareness (SA) is a kind of awareness of environmental elements and events in time and space.

It also includes the ability to interpret known facts and principles, principles and principles in one's own words, words or other symbols about their meaning, and to predict the state of some variables (such as time or predetermined events) after they have changed.

It is also a discipline divided according to the nature of learning, specializing in the study of the impact of the understanding of elements of the environment on the methods or strategies decided to achieve plans, goals, and applications in complex and dynamic fields.

It includes aviation, air traffic control, navigation, power plant operations, military command and control, emergency services such as firefighting and police, as well as universal and complex matters such as driving a car or bicycle, or riding a motorcycle.

"Situational Awareness" includes the use of sensory, perceptual, thinking, and memory activities to become aware of what is going on in one's mind, body, and environment, and to understand how information, events, and one's actions affect goals and objectives.

This includes immediate and imminent effects. If someone has a sophisticated sense of "situational awareness", they generally have a higher level of knowledge of the inputs and outputs of the system and can control the variables and therefore have "intuition" about situations, people, and events.

Lack of "situational awareness" or insufficient "situational awareness" has been identified as one of the main factors of human error and accidents. Therefore, "situational awareness" is particularly important in areas of work where the flow of information is high and poor decision making can lead to serious consequences, such as piloting aircraft, being a soldier, and treating critically ill patients.

The study of "situational awareness" can be divided into three areas: SA state, SA system, and SA process; SA state refers to the actual awareness of the situation; SA system refers to the distribution of state awareness among the team and objects in the environment, and the exchange of state awareness among the system components; and SA process refers to the updating of SA state and what instantaneous changes lead to changes in state awareness.

Having a complete, accurate, and real-time "situational awareness" is essential for human decision makers to judge the scenarios that become problematic and to decide on methods or strategies to achieve plans and goals, given the technical and situational complexity. "Situational awareness' has been recognized as a key

However, it is often an elusive foundation and a key factor for success. It is to answer two questions (1) what is the state of interaction and interconnection between the so-called Heideggerian "integrated things in the world" and (2) what is the state of interaction and interconnection between human beings and things in the world in order to make it possible for them to interact with each other?

In short, 'situational consciousness' is constructed more as "a dynamic, ongoing effort to understand the connectedness of people, places, or events in order to anticipate their trajectories and act effectively" than under 'situational consciousness'.

The act of "knowing" and "recognizing" a subject in order to be sure of knowing, and that these knowings have the potential to use states for specific purposes. Andersley points out that as a productive process, 'situational awareness' constructs meaning. In fact, it can be said to be one of the components of maintaining state awareness.

In the vast majority of cases, state awareness is instantaneous and effortless, arising from pattern recognition of key factors in the environment, "in activities such as sports, driving, flying, air traffic control, and so on, operating at a speed that in most cases does not actually allow such conscious deliberation, but there are exceptions".

Andersley also points out that 'situational awareness' constructs meaning by focusing on the past and forming explanations for events that have occurred in the past, while 'situational awareness' is usually about looking forward and projecting what might happen for effective decision making.

The connection of a certain nature between man and the world, between man and man or man and things, is also referred to as the relationship [between consciousness - and the consciousness of the transcendent], and Scharthe then, like Heidegger, first asks the question of 'being' (being and time [Heidegger's question to metaphysics])

01 The Question of Existence

"Any question assumes the existence of a questioner and the existence of a questioned being", which is predicated on "the original relationship between human beings and the self-relationship", for example: the being that asks "Have you eaten? The "existence" of the questioned "existence" reveals the possibility of negation and the existence of negation.

The questioned "existence" reveals the possibility of non-existence. At this point, some people will answer "no means no" in a materialistic way.

They do not want to believe in a non-existence that is the reasonableness of thinking or judging something from a different point of view or perspective, a thing that exists independently of subjective thought or consciousness.

Or to say "the world is what it is, how I see it", but in this act of fiction. The mere fabrication of imagination is actually a kind of inaction that refuses to ask questions. It conceals negation (non-being) rather than abolishing non-'being'.

It also destroys the reality of negation by avoiding the reality of answer. There is also a hypothetical, absolute, ideal objective non-being.

For example, the answer to the question of whether I have eaten, "I had some cookies at 11:00 and I haven't eaten my meal yet, so I don't know if I'm eating or not. I don't know if it's considered eating or not.

This kind of real answer, which appears as some kind of restricted truth, is called here the non-being as the stipulator of the question: the restricted "being". Thus, we enter into the inner structure of "being". Milan Kundera

Matter (which is different from what we call "matter" in our lives, which, in part of materialist philosophy, refers to concrete material forms.) The only characteristic of matter is objective reality, which is also the nature of human nature.

Objective reality is everything that can be perceived from the senses, including both the natural things that can be perceived from the senses and the sensual activities of human beings that can be perceived from the senses, that is, practical activities. It exists independently of our spirit and constitutes the principle of unity of the world substance.

This trinity is actually: before consciousness asks a question to the transcendent existence (self-existence) and receives an answer, the individual uses mental activities such as sensation, perception, thinking, and memory, to perceive and understand the state of one's body and mind and the changes of people, things, and objects in the environment, which is the possession of objective existence.

The transcendent existence possesses a possibility that is not its own possibility, and after receiving the answer, we get the limited non-existence that is expected by default in the question.

Imagine a situation where your girlfriend is not serious about her feelings for the opposite sex, plays with your feelings, but you still like her, so you ask her "Why are you doing this? But

But in fact, in the process of asking the question, you have already preset the answer you want, "She must have her reasons for hurting me so much. The person being asked is the scum (the person who is not serious about the feelings of the opposite sex and who plays with the feelings of the other person (especially in love)) itself (the self-existence).

You, as the questioner, have the "power to break the limits of the term "scum" (a person who treats people of the opposite sex with insincere feelings and plays with their feelings)", which means "the possibility that she is not a scum (a person who treats people of the opposite sex with insincere feelings and plays with their feelings)".

Of course, if you don't like her enough, you can refuse to ask her to "smear her by calling her a slag in her circle of friends [someone who treats the opposite sex with insincerity and plays with each other's feelings (especially romance)] female". At this point, you no longer consider the transcendent existence, but become a solidified existence, and here you touch upon the "existential" way of living with others, another way of "co-presence".

The thing that "being" will become will necessarily be hidden in its [what is not now] matrix. That is, something that is distinguishable and exists independently of itself in all objects or phenomena that exist in nature.

But it does not need to be a physical being. In particular, it is usually regarded as an entity, "existence is this, and it is nothing else".

Therefore, Schart is mainly concerned with the question of 'nothingness', the origin of nothingness, and so on. The main point here is to propose a new component of the real thing, called "non-existence".

02. Not recognizing the existence of things or the reality of things

Self-existence is absolute, full, non-negotiable, and not void in itself. The consciousness raises the question and makes it void. Schart said: Kant's [negative judgment activity] and [affirmative judgment activity] in the critique of judgment are both the result of conceptual synthesis.

This means that "analytic activity" and "synthesis activity" are two objective acts that are equally factual and real. Therefore, "negation" is not in "existence" but only in the "end point" of the act of judgment.

What Schart is trying to say is that "negation", as a propositional approach, exists only in the individual's mental activities such as sensation, perception, thinking, and memory, and in the integrated awareness and knowledge of one's physical and mental state and the changes of people, things, and objects in the environment.

Later he will give an example: Is there a "negation" in nature? If there were no human beings on earth, then there would be no individual who would use mental activities such as sensation, perception, thinking, and memory to perceive and understand his own state of mind and body and the changes of people, things, and objects in the environment.

Nature itself exists naturally, and it can never have value, be true or incorrect, or not match the reality. But this is not like judging things according to one's own perceptions, without seeking to conform to the actual state of mind, which is lawless.

What Schart is trying to say is that if one first takes a proposition in modern philosophy, logic, and linguistics to mean the semantics (the actual expression of a concept) of a judgment (statement), then the concept can be defined and expressed.

This concept is the "negation" of a phenomenon that

can be defined and observed, as possibility, and the world reveals these non-existences to it. Schart has come to two conclusions here.

[Negation/non-being] is not a nature of judgment. Asking questions fundamentally involves some kind of "pre-judgmental understanding" of non-being, which is itself a kind of relationship between being and non-being on the basis of primordial transcendence, that is, on the basis of the relationship between being and [negation/non-being].

It is itself a relationship between being and non-being on the basis of primordial transcendence, that is, on the basis of the relationship between being and [negation/non-being]. In the context of ordinary conversation, the question that demands an answer or a reply is always a question and an answer, so that the question itself, which is hidden in the substance of the thing, is a question.

The human consciousness disappears from the concrete connotation of things that are not in front of the eyes, and the effort to conceive specific images and active activities using past memories or similar experiences.

The example of natural destruction is used. The individual uses mental activities such as sensation, perception, thinking, and memory to become aware of and recognize his or her own state of mind and body and changes in people, things, and objects in the environment in an integrated way, so that existence and non-existence are revealed.

In the second paragraph, we can cite numerous cases of "non-existence" that do not affect the unity system at all, and if we can predict a physical quantity precisely, then this physical quantity should really exist, regardless of whether we observe it or not.

For example, the case of scientific logic and technological development. Take the lead in formulating a scientific hypothesis, and then observe whether nature proves the truth of a thing by one or more objective existences, i.e., by showing or concluding it in terms of persons, facts.

Here the nature of an object or a thing itself, or the intrinsic meaning of a thing, is the scientist's questioning of nature. Now that Schart has considered the importance of non-existence, the "existential structure" of the question, it is time to move on to the crux of the chapter:

"Where does nothingness come from?

The Dirty Hand is a political drama set in the fictional country of Illyria during the Second World War from 1943 to 1945. The play tells the story of a young man's assassination of a political leader.

The young man is unsuccessful in his mission to assassinate the political leader, but ends up killing the leader "by accident" because of his personal grudge. The paradox of the play is in the discussion of the meaning of the assassination mission and the ideological differences of communism.

"I am no longer attached to anything, in a sense I am at peace." Hugo, the protagonist of the Chartist play "The Dirty Hand," also states: "From a certain moment, the order makes you move alone, the order is left aside, I move forward alone, I kill alone, and, I don't even know why.

This state of indecision, of unexplained emptiness of mind, caused mainly by the absence of a clear inner goal, appears again and again in Sartre's series of literary works.

The loss of memory, the main or crucial part of things in literature, is also another reflection of the subject's emptiness. The loss of memory blanks out a part of existence that has a unique consciousness and a unique personal experience, a person's ability to recognize the environment and the self, and the clarity of cognition.

The loss of memory blanks out one's ability to perceive one's environment and self, as well as one's cognitive clarity. In "Nausea", Logandan reflects deeply on the emptiness he felt after Annie left him. When I broke up with Annie, I felt that everything was gone, that my past had become empty and meaningless.

Memory is very important to me," says Rogandan. I slowly lost the memory of Annie's eyes, and then everything that my body had left in my mind, and when the memories didn't come back, I was cold and empty.

"I have only my body, a solitary person, who has only his own body, who cannot intercept the memories, which pass through him." The "I" is the fallen man with no memory.

Schart believes that the state of mind that consciously strives for the realization of an ideal or the achievement of a certain goal has the ability to reorganize the memory of the self or to erase the meaning of the past through forgetfulness.

"The possibility of [freedom] and [detachment] from eternity are the same thing." According to Schart, "Being has the will to break with its past, to break free from the possibility of eternity in order to be able to examine it under the revelation of being, and to be able to examine it from a view of it, of what it is.

And to be able to start from the conception of a meaning that it does not have, he gives to the past itself the eternal possibility of the meaning that it has. Emptiness emerges at a moment when the will is at odds with the past and the future. Sartre profoundly depicts the emptiness of the subject's state of mind.

Sartre turns to Heidegger's conclusion after judging and analyzing the relationship between [emptiness and this-ness]: "Emptiness makes man detach from existence (which is different from Heidegger's definition of existence) and become "this-ness. Heidegger used [fear, death] and birth to exist at that time.

Schatt uses [boredom], [nausea], and more than that, [the negativity of consciousness], to reveal the "here and now. Schart's "nothingness" is not equivalent to "nothingness," but to the ability to perceive one's environment and self, as well as to the clarity of cognitive subjectivity.

But he first argues that Heidegger makes the otherwise non-existent situation a factual progress on this issue; after all, Heidegger is no longer a dualism of "being" and "phenomenon. He argues for the necessity of questioning "existence".

Heidegger's "nothingness" is in fact placed differently from Scharthe's. Heidegger's "nothingness" precedes existence (Scharthe's "nothingness" exists in context). And it is Descartes' romantic temperament of anthropology and subjectivity that Saudi Arabia has always embraced since France. But

For example, Heidegger sets human reality as "the existence of man in the world," while Scharthe sets human reality as the entanglement of [self-existence] with [self-existence].

Both of their realities are "nothingness" or "nihilization". Heidegger's "man is always separated from the sein, and the separation is caused by the breadth of the sein that he is not, and the separation of the dasein from the sein is such that it is assumed that the actions of others are obstructed, destroyed, and held back from achieving their ends, and that the dasein gains perspective toward its own transcendence.

03. The Origin of Nothingness

As the primary condition for the question of "nothingness", Shatt further stipulates about "nothingness". The first paragraph: Nothingness cannot be outside the objective "being" in the form of "supplementation" and "abstraction".

Nothingness is a soul, which needs the shell of "existence" to obtain something that is distinguishable and exists independently within itself. But it does not need to be a physical "being".

"The substance of nothingness, which is hidden inside things, appears inside existence. "Existence has nothing to do with nothingness, but nothingness does not start from itself, and as non-existence, it cannot stand on its own.

However, the impractical and inscrutable "nothingness" separates the reflection of the objective material world by the human mind, which is the sum of various mental processes and motives, such as sensation and thought.

It means that the impractical and inscrutable "nothingness" separates man from his essence. Man always hides the past (essence) behind him.

It is a contradiction between the human brain and the animal brain and the objective world, whose laws, i.e., conscious or self-conscious nature, or state of activity, are constantly flowing to build the essence.

Therefore, the fundamental properties inherent in all objects or phenomena in nature, which determine the nature, appearance and development of things, are not the basis or the backbone, but become the comprehensive awareness of one's own physical and mental state and the changes of people, events and things in the environment by using mental activities such as sensation, perception, thinking and memory.

A. M. Hayes and G. Feldman emphasize awareness as a strategy for avoiding certain emotions and preventing them from becoming too involved. Awareness is also a method of developing self-knowledge and wisdom. The constant demand with the activity of awareness is the demand that has to be acted upon.

04. So where does nothingness come from?

At this point we can no longer say "self-virtualization". Although "self-virtualization" is similar to "nothingness", only "existence" can "self-virtualization". Because "existence" appears, "nothingness" is "existed" and "nothingness" is made into nothingness.

It is necessary to have a non-self-existent "existence" that can make "nothingness" into nothingness, that can take on "nothingness" with its "existence", and that can support "nothingness" with its existence, and through this "existence", "nothingness" comes into things. This "existence" is the existence of self

Both Kierkegaard and Heidegger have described "anxiety". Kierkegaard describes "anxiety" as "anxiety" in the face of freedom, and Heidegger treats "annoyance" as a grasp of nothingness.

Schart: "Anxiety", which arises from internal emotional or psychological conflicts that lead to irrational feelings of anxiety or fear, arises from nothingness, while "anxiety", which arises in humans or animals in a state of panic and urgency in the face of real or imagined dangers, things they hate, etc., comes with fear.

"Symptoms are physiological emergency reactions such as heart rate change, blood pressure increase, night sweats, trembling, and sometimes even more intense physiological reactions such as cardiac arrest and shock, which are fears of "existence" in the world.

Fear is the fear of "existence" in the world, while anxiety is the anxiety in front of the self. Fear that I will fall off a cliff is fear. My dizziness on the road is anxiety, a fear of fear.

How does freedom arise? Fear precedes anxiety, and being reflected upon precedes reflection. The individual uses mental activities such as sensation, perception, thinking, memory, etc., to become aware of and recognize the changes in one's physical and mental state and in people, events, and objects in the environment, and to make sense of natural and social phenomena.

According to the existing empirical knowledge, experience, facts, laws, cognition and tested hypotheses, through generalization and deductive reasoning, etc.

There are two main characteristics of logical inferential summation: 1. Consciousness (being reflected) has no content, but only content in the structure of self-nullification; 2. Consciousness faces the past or the future, but faces self-analysis in a way that it is not: it returns to a temporal structure of nothingness.

(Case of the gambler)

It is not only among gamblers, but it can be said that everyone has, to a greater or lesser extent, this mentality of being conscious of the past or the future and facing a self-analysis that is not what it is.

The first thing you need to do is to take a look at the actual situation.

This is the gambler's "greed" mentality, which plunges his consciousness into a temporal void structure, and "self-deception" has a growing environment. In the grand scheme of things, people often use or are used to use the gambler's mentality in many ways

The four stages of a gambler

The history of gambling has a long history, from pai gow to mahjong, from poker to craps, from horse betting to roulette, there are many different types of gambling and many differences in the culture of the gambling table, but the essence remains the same.

The most important thing is how a normal person, a relative that we know so well, turns into a hateful, ugly, inhumane gambler step by step. There are four stages in the process of a person's involvement in gambling to the point of becoming a complete gambler with distorted values.

Basic Overview

The society is like an oversized casino, everyone must live in this casino, with their own payment, gambling on tomorrow's acquisition, gambling on all human activities touched by people, things and things, not only money, but also jobs; economic activities in the betting, speculative psychology of the abnormal development or extreme distortion, the stability of the regime, but also the victory or defeat in the war.

There is the chance of work and the happiness of marriage. The expectation of people in gambling is to be able to make the best decisions using the rules of the game of society to the greatest extent possible, that is, to guide themselves through the constraints of the social matrix to the increase of the gains that I love to execute.

But not everyone is able to use the rules of the game to

their own satisfaction, what if they lose? What we put down is not a token "small amount of money", but a large enough number to ruin the golden years of our lives.

Therefore, anyone who gambles with his own human "greed" (nihilism) can be called a gambler. In the brutal battlefield of evolution, any creature that wants to survive must have an instinct to "simplify the world". Primitive people with high survival rates have various mechanisms for simplifying the world in their genes.

For example, in the African grasslands, the awareness of the wind blowing in the distance can be simplified to danger, and the presence of danger requires a quick escape. In the battlefield of evolution, there is no room for curious babies or inquisitive philosophers. Philosophy considers how the picture of ideas, composed of various concepts, can maximize the power and fascination of ideas.

An idea in itself is no more right or wrong than any other idea, but it may be placed in an inappropriate position of thought, and spoil the effect of the picture of thought, like a chess piece played like a chess piece, with many holes in it.

Or a foolish shape that is less efficient and less beautiful. The position on which a certain idea falls is like the position on which a certain chess piece is played. Thus, over the long evolutionary stage of human society, the instinctive simplification of ideas has gradually embedded itself in the basic structures and properties that support life, and because it enhances survival rates, it continues to pass on this gene.

Human beings have existed on Earth for billions of years, and written records are less than a few thousand years old, and even less than a few hundred years old in the industrialized age.

In the continuation of the competition for survival, the characteristics have a certain advantage or disadvantage, thus resulting in differences in the ability to survive, which in turn leads to differences in reproductive ability, making these characteristics preserved or eliminated.

Natural selection is the main mechanism by which species compete with each other, and those that adapt to nature survive and those that do not are eliminated from evolution. Natural selection is the only mechanism that can explain how organisms adapt to their environment.

However, the instinct to "simplify the world" occurs in a very small moment. So the genes we have accumulated to simplify the world are not something we can get rid of in a few generations. That said, this does not mean that there is value in "generalizing" or even "discriminating against the other" genes.

The opposite may be true. In a future of big data, artificial intelligence, and the Internet of Everything, the competition for survival has long been different from the prototype competition on the African savannah. In the future of big data, artificial intelligence, and the Internet of Everything, the competition for survival has long been different from the prototype of competition on the African savanna.

It is the understanding of the thinking activities that reveal the meaning and effect of things on people that may be the key to future genetic victory! Because there are so many casinos in the world, so many gamblers have been born.

(1) Beginning Stage.

When you first enter the casino, the general mentality is "small bets are good, big bets hurt" mentality, very vigilant, the beginning is careful bets, mainly for entertainment, won a few thousand dollars, lost a few thousand dollars, the Ministry will be put on the heart, will stop.

This time, the general can still control themselves. At this time, no one considers himself a gambler or a gambling demon, but at most a small gambling pleasure, feeling happy and satisfied, with the characteristics of happiness, relief, relaxation and leisure.

But is this really the case? Greed refers to a strong desire to seize money, material wealth or physical satisfaction that far exceeds one's own needs.

Greed can also refer to the insatiable desire for something that may belong to someone else. A greedy individual does not consider whether he or she really needs these things and ignores the motives of other individuals.

They may also neglect the welfare of others and are therefore considered "harmful" by some in society. Unfortunately, greed is a symptom of a deficient consciousness when it fails to achieve its goals or to satisfy them.

The main manifestation of greed is the ability to gain without effort, and to gain more with less effort. Whether it is the lottery or gambling, the essence of it is to use a small amount of money to gain more money, to give you the expectation and desire for something in the future, and at the same time, to slowly change your values.

The value of slowly changing your values, and the value of the valuable is based on a certain sense of human thinking and make a cognition, understanding, judgment or choice, that is, people identify things, debate the right and wrong of a thought or orientation.

In a class society, different classes have different value concepts.

Before any kind of thought is absolutely denied, the perspective, background, judgment and meaning of the thought will have a certain degree of objective value.

The value of this idea lies in the degree of its recognition and meaning, that is, the human understanding and perception of this idea, which is the simplest and most real assessment in human thinking.

This is where the simplest and most real assessment of human thinking lies, which also determines whether an idea is great and whether it can be the source of values.

Values have relative stability and permanence. In a particular time, place, and condition, people's values are always relatively stable and long-lasting.

For example, there is always a view and evaluation of something good or bad, and this view will not change if the conditions remain unchanged. However, as people's economic status changes, and as their outlook on life and the world changes, these values will also change.

This means that values are also in the process of development and change. The value is a concept about a value that is formed by a person, which has a lasting and stable characteristic and will always govern the person's daily behavior and activities.

Therefore, the longer the gambling period, the more fatal the impact on people is the change of values, the fatal attraction, the irredeemable subversion of values and the negation of the value of labor.

The reason is that values reflect people's cognitive and demand conditions; values are people's evaluation and view of the objective world and the results of their behavior, and therefore, it reflects people's outlook on life and values in one way or another, reflecting the subjective cognitive world of people.

I can't imagine that I won hundreds of thousands of dollars in three months, while working hard at a normal job, it's hard to save 30,000 to 50,000 after a year of paying off the mortgage, eating and drinking, and raising a car.

I can't get interested in the things I want to buy, I want to go to the gambling table to win back, so I have unknowingly become a real gambler without knowing it.

(2) Fallen stage.

If you win once in a while, you start to fantasize about winning every time, and every time you can experience the thrill of winning money and spending it recklessly. In the process, you will experience good luck, the table on the win turned away and began to squander.

You're not willing to continue after losing, but you won the last time anyway, and you ended up winning back. So you have a lot of table experiences, you lose and win back, you win and don't leave and win back. In the process, you have two changes.

One, your betting chips seem to be getting bigger and

bigger, a strong desire of human nature to grab far more than their own needs of money, material wealth or physical satisfaction, etc. greed, you are slowly raised to have a fat round little devil of a calf, gambling scene, will only get bigger not smaller, win or lose hundreds of thousands of gamblers.

The actual fact is that you will not be able to go back to the casino to gamble with a few hundred dollars, because that has not been able to hook his desire, when you lose 10,000, you will be pressed 100,000, thinking that a pokies to make things right, there is always a %50 chance of winning back the speculative psychology.

The first thing you need to do is to get a good idea of what you want to do.

The actual example is that you can find a lot of people who have been in the business of gambling for a long time. Take a real-life example. A friend likes to go to Las Vegas, USA, to gamble.

He lost more than $100,000 in chips one time, and when he had only the last $10,000 left in his pocket, he became discouraged and thought he would finish gambling, go to a VIP room with a few friends, have a little get-together, and then never gamble again, and made up his mind to lose the last $10,000 and leave.

When he put the money on the wheel of fortune, ready to lose and then pat his buttocks and leave, things happened so coincidentally, it just opened the number he had pressed, and in an instant more than 300,000 came back, causing a commotion, a burst of thunderous applause, and he felt that the god of luck was casting a loving look at him.

The fortune had begun to change, and he was able to turn his defeat into a victory. Since that surprise, he has been taking a chance every time he loses, but there has never been another miracle. The one time he won, he might be able to boast about it for a long time, but the number of times he has lost is too many to count. There is bound to be a time when a gambler is particularly unlucky at the tables, loses a lot of money right from the start, and this is the game that sends you over the edge and marks you as a complete gambler!

(3) Remedial Stage.

After the accumulation of gambling time and time again, you have created a big hole in the bottom of the deep, years of savings has been little left, in order to take care of their image in society, family, they need to remedy. How and with what? After a big defeat, the gambler still tries to use gambling to make up for the hole created before.

The gambler's consciousness is vain and his psyche is desperate, "self-deception" is a bow without a return arrow, he has already lost his family's money and will never have a chance to turn around unless he wins back.

The reason why he "took a discount" is because he still has the mentality of "self-deception" of taking chances and always wants to gamble again when he has money to win back what he lost, and wants to pay back the debt through his "efforts" to save his face.

The result is predictable. After a few repetitions, he lost more and more money and his family gradually lost trust and patience with him. This is the result of self-deception and deception as a result of behavior, self-deception, and this deception also deceive others.

But in Schatt, he clearly defines self-deception and deception as two kinds of behavior, two states of existence. One is the denial of the external world, that is, the denial of truth by lying.

The other is an attitude of denial of the truth as it is known to the self, which is a kind of internal denial. The former is deception, the latter is self-deception (la mauvaise fois). Schart said: "Self-deception is lying to oneself, and lying is an attitude of denial.

The fundamental difference he wants to make is that "the essence of a lie is that the liar knows exactly what he is doing to cover up the loss of gambling money.

We can be sure that the liar is fully aware that he is lying, and that all he wants is to achieve his goal, which is to deny a real world.

So it is easy to see that deception is an act between two consciousnesses, one induced by lying, the other believing in an illusory appearance that does not correspond to the nature of things, and following their own agenda.

Self-deception, on the other hand, is the act of using mental activities such as sensation, perception, thinking, and memory to perceive and recognize the changes in one's physical and mental state and in the environment of people, things, and objects.

That is why Schart said, "Self-deception has a structure of lying on the surface, but the fundamental difference is that in self-deception, I am hiding the truth from myself. There is no duality here between the deceiver and the deceived. Rather, self-deception involves a singularity in one's ability to perceive one's environment and one's self, and in the clarity of that perception.

Schart obviously knows that self-deception is a psychological act, but he wants to give it an ontological basis, because he examines self-deception more in terms of phenomenological analysis.

He said, "All individuals use sensory, perceptual, thinking, memory and other mental activities, their own physical and mental state and the environment, people, things and changes in the comprehensive awareness and understanding, are the special function and activity of the human brain to existence, is the unique human reflection of the objective world.

This is the path of intentionality theory, which follows the path of "the integrated awareness and recognition of one's physical and mental state and the changes of people, things, and objects in the environment by using mental activities such as sensation, perception, thinking, and memory, and the ability to recognize something in the environment and the clarity of cognition.

But when it comes to self-deception, it is impossible to avoid psychology, so Schart has spent a considerable amount of time analyzing Frode's psychoanalysis. But he was dissatisfied with Froude's analysis of the duality of consciousness and the separation of the subconscious from the consciousness.

He thought that man is a self-realized being who has the ability to choose, so his personal use of sensation, perception, thought, memory, and other mental activities, the integrated awareness and knowledge of his physical and mental state and the changes of people, events, and objects in the environment, is not a real, non-false reality.

In his book "Being and Nothingness", Schartre breaks down human things and concepts into their simpler components and examines them separately to find out their intrinsic properties and their connection with each other, always with the free choice of the will.

He considers that the doctrine of treating human beings as mere objects or objects is biased against all objects or phenomena that exist objectively in nature.

Then why do people have self-deception? From Schart's "Being and Nothingness", the human beings' knowledge, experience, facts, laws, cognition, and tested hypotheses about natural and social phenomena can be logically summarized into two main purposes.

By generalizing and deducing, the logical conclusion can be divided into two main purposes: [to conceal the truth] and [to escape from freedom]. For in some cases, both truth and freedom are things that bring suffering.

In fact, these two purposes are causally related, both arising from anxiety about existence. Internal conflicts arise emotionally or psychologically, leading to irrational feelings of anxiety or fear. One could say that self-deception is a rebellion against anxiety.

Schart said, "It is in an emotional anxiety of fear, worry, and tension that one gains freedom, the ability to recognize the environment and the self, and the awareness of the degree of cognitive clarity. It can be said that the creation of internal conflicts due to emotions or psychology, which lead to irrational feelings of anxiety or fear, is the free way of existence of consciousness".

The Danish philosopher Kierkegaard, who inspired

the philosophy of existentialism, has already been mentioned. He introduced the concept of anxiety as a philosophy of holistic, fundamental, and critical inquiry into the real world and human beings. He pointed out that man is indistinguishable from good and evil when his spirit is still in a state of chaos.

His spirit in this state is pure nothingness, but spiritual nothingness cannot remain permanently peaceful, and it inevitably generates anxiety, which is the desire of the spirit to find a way out, to rush out and do something.

As long as people are alive, this desire is inevitable. In modern times, this sense of "anxiety" is even stronger, and we ourselves have had the experience of experiencing extreme emptiness and boredom.

We ourselves have had the experience of experiencing the extreme emptiness and boredom that brings about the "anxiety" of anxiety, and action is the process of making nothingness real. This is what Schart is referring to when he says that man is judged to be free, as explained in the previous section. Anxiety," says Schart, "is the "anxiety" that arises when a person is faced with freedom.

One is often in a state of "anxiety" when one is faced with the dilemma of whether to make or not to make a choice, whether to choose this or that. Kierkegaard even considers 'anxiety' as a connotation of spiritual culture, saying: 'The deeper the 'anxiety', the deeper the culture. Only an impoverished folly would treat it as a disorder.

Both anxiety and melancholy have the same meaning when freedom traverses its historical imperfections and realizes itself in the deepest sense. So he concludes: "The beast has no "anxiety".

Why is man "anxious"? Because he feels his responsibility, and responsibility can only come from his own choice, and choice means commitment.

The "anxiety" about commitment is the source of self-deception. Because commitment is putting yourself in freedom, and freedom means that you must be responsible for the consequences. It is more than just doing something; in many cases, it means a moral judgment of conscience.

Schatt said."The Nazi camp guards knew that his actions were evil, but he could not admit it, and his conscience would torment him, and the universal moral code of mankind would point out that he was a man of evil. So they would find all kinds of excuses to convince themselves that the truth was not the truth, that is, that the crime was not what it was, but what it was not".

They often justify themselves by saying that they obeyed the command, and thus a fact of self-deception is established. So self-deception is an escape, an escape from something, from freedom.

In this situation, one is not "being-for-itself", but "being in itself". Because the so-called "for oneself" is the existence of active choice, while freedom is the existence of passive and unconsciousness.

We can say that "for oneself" is a human existence, while Freedom is characterized by the factuality of objectification. But self-deception is to give up "being-for-itself" and become free, choiceless and irresponsible, making the subject a passive object. It no longer has freedom, and its original characteristics disappear completely from all people in the world. All mass psychology is a structure of self-deception.

(4) The stage of madness: - the stage of self-deception

The gambler has lost almost all of his property by this

time, and some have even entered the debt stage. At this time, the gambler's consciousness has become seriously unbalanced in terms of will, and long-term gambling has made you deny the value of labor, and there is no labor that can keep up with your own gambling values.

The actual fact is that you can't get the best out of your own gambling values.

The so-called "One Road to Hua Shan" describes the gambler's current perilous "situational awareness". Why? Because, to climb Mount Hua, one must start from Yuquan Yuan, follow the clear and cool mountain springs, pass through Wuli Pass, Shimen, Salopeng, Mao Nui Cave, and Yunmen to Qingke Ping.

From Qingke Ping to Wuyun Peak (North Peak), it is one of the toughest walks on Mount Hua, passing through the Xinhui Rock in the middle and going up to the "Thousand Feet" where the cliffs are so high and the sky is so high. This mountain is full of precipitous cliffs and steep paths, making it a thrilling climb.

The same as the Thousand Foot Block, the Hundred Foot Gap, the Laojun Plow Ditch, the Upper Heavenly Staircase, and the Cang Long Ling are all chiseled out on the cliffs, and Harrier turns over all three sides of the air, and it is impossible to go home at this time.

The company is looking forward to relying on a gamble to turn a defeat into a victory. The actual fact is that you can find a lot of people who are not able to get a good deal on the actual money. The actual fact is that you will be able to get a lot more than just a few of the most popular and popular items.

The reason is that gamblers always fantasize about meeting a heavenly game and winning back all of their lives, winning back the money they once lost, winning back the glamorous years and the intoxicating life they once had with people yelling at them to get out of the way and people around them to protect them.

This is the time when a gambler is gambling for a long time if he has money, and when he is not gambling, he is on his way to raise money everywhere.

This is the time when he already does not care about his relatives, friends, brothers and sisters, as long as there is an opportunity to raise money, whether it is cheating, cheating, stealing, its methods can be said to be the most extreme, for the extent of self-deception, no moral guilt to speak of, let alone the value of shame.

The most important thing is to have money to go to the gambling table again. At this point, gamblers are already trading with the devil and have completely degenerated into the living corpse of the gambling ghost in name only.

Why do people have to lose before they leave the casino? The reason is that people have the mentality of "self-deception" and do not give in to defeat, believing that they will win one day. Even the last chip must be bet on, because that is the hope, the hope of winning back everything!

The more you lose, or the more you fail, the more you refuse to admit defeat in your bones, and these gamblers are super stubborn and stubborn and refuse to admit defeat.

Although it is clear that they have suffered countless times, the gambler's mind is still thinking, I want to make a profit! One day, I'm sure I'll win it back, and when I win back the money I lost in my last bet, I'll be able to win it back.

I promise to God, I offer my heart's most sincere assurance that I will never, ever gamble again! (See, gamblers swear again. Remember, gamblers swear many times.) Often, things don't go as planned, and "self-deception" only gets you in deeper and deeper.

2. The principle of nine losses in ten bets

The company's main goal is to provide a good service to its customers. The probability of a bankruptcy is greater than that of a casualty because of the different rules of mending, about two percent.

The actual fact is that you can find a lot of people who have been in the business for a long time, and you can't even get to know them.

The reason is that you are going to lose $5,000 because of the draw. See, the longer you stay in the casino the more times you gamble, the more you tend to have a negative return, and the casino funds are unlimited, your funds are limited, long-term gambling, resulting in the final loss is bound to be you.

Therefore, experts in legal research believe that quitting gambling addiction: the focus is on mental "transformation" Experts in legal research believe that gambling is primarily a psychological and spiritual satisfaction for oneself.

The two are similar in many ways, but they are different in that gambling is primarily mentally devastating, but drug use is not only mentally devastating to the addict, but also physically devastating.

Each time I can not control myself, the result is often a loss of even no money to eat breakfast, time and time again the lesson to make me awake, this kind of internal conflict due to emotional or psychological, and then trigger irrational anxiety or fear feelings of anxiety.

Introducing anxiety Pray Kierkegaard and Heidegger have both described "anxiety," Pray Kierkegaard describing "anxiety" as "anxiety" in the presence of freedom, and Heidegger treating "annoyance" as a grasp of nothingness.

Schart: "Anxiety", which arises from internal emotional or psychological conflicts and leads to irrational feelings of anxiety or fear, arises from nothingness, while "anxiety", which arises in humans or animals in a state of panic and urgency in the face of real or imagined dangers, things they hate, etc., comes with fear.

As long as we prove the artificiality of the motive, the non-primary nature, and the inwardness of consciousness. In these nothingnesses we realize the limitations of our being human.

In these anxieties, in order to escape from them, we hide in the shells that we find ways to be comfortable with (Chapter 2 will talk about this). In this section, however, the reasons and evidence that Shat uses to support or deny certain matters freely precede the nature, or the existence of possible conditions that precede the nature.

05. Why is anxiety a rare thing?

Anxiety reveals freedom (the transcendental consciousness that intervenes in the instrumental complex, and anxiety's recognition of the possibility of that possibility for me (the degree to which it can become a fact). Freedom reveals value (freedom is not in the intuition, freedom is only through the recognition of things or phenomena that

It exists as a value by recognizing that things or phenomena have a positive meaning for people or groups, are valued by people, or are satisfying to people, and become like values that people respect or are interested in pursuing. (Freedom is the basis for the positive meaning and usefulness of various objects to the subject.)

Things or phenomena in the context of ordinary people have positive meanings for people or groups, are valued by people, or satisfy people, and become objects that people respect or are interested in pursuing directly, like warning signs (signs that warn vehicles and pedestrians of dangerous places).

In fact, the perceptual content and tendency to feel anxious about the positive meaning and usefulness of the object to the subject are embedded in people's existing experience, knowledge, interest, and attitude.

Thus, it is no longer limited to the perception of individual properties of things, but is an acquired indirect phenomenon. Anxiety is the grasp of freedom itself on the reflection of freedom. The warning signs have a direct warning effect.

Schart also discusses the escape from anxiety: for example, psychological determinism, a kind of determinism or determinism, also known as Laplace's Creed, is a philosophical position that any event that occurs, including a human decision made of free will, has external conditions that determine the occurrence of that event and not some other event.

Various theories of determinism are woven throughout the history of philosophy. In direct opposition to determinism is non-determinism. Determinism is also often contrasted with free will.

Determinists believe that free will is an illusion, that the human will is not free, but that the mind is treated as if it were the same as a thing. The mind's deterministic free will: is the way in which I am bound to become X. Because I am X myself.

In essence, it is the integrated awareness of the individual's state of mind and body and the changes of people, events, and things in the environment, using mental activities such as sensation, perception, thinking, and memory.

In the process of self-reflection and consciousness, the possibilities that emerge from "nothingness", my possibilities, are regarded as external possibilities, or as the possibilities of others.

So their possibilities are irrelevant to me, and the selected possibilities are my only concrete possibilities.

Such a way of escape not only resolves the anxiety of what I will become in the future, but also attempts to dissipate the threat of the past. That is, I reject my nature [transcendence].

It is the view that historically constitutes a person or environment other than myself. I exist as a religion, occupying our freedom through God, virtue, or morality. The freedom in the self is taken to be the freedom of others.

Since the advent of Sartre's existentialism, many have viewed existentialism as a doctrine of inaction, as the discipline of analytical thinking that explores and reflects on fundamental questions about life, knowledge, and values.

For example, "Is there an objective standard of morality?" and "What is science?" In matters of nature, since it is a mistake to think about the nature of existence, and since it is difficult to think about the nature of existence, does this mean that the rational and active process of reflection on the real world or any object in an "inward dialogue" is abandoned?

In fact, such a view is in essence a misinterpretation of the phrase "existence precedes essence", and focuses too much on existence as a kind of theology (partly

Kierkegaard), or irrationalism (Bergson), which is mainly directed against "rationalism".

The basic view of rationalism is that the objective world is a rational and harmonious whole; that the human mind can understand the objective world and has objective ways to reveal its structure; and that people can use their understanding of the objective world to improve their lives.

Man is by nature a rational animal, and through the operation of reason, man can free himself from ignorance and realize his potential. The irrationalist opposes the basic view of rationalism in various ways.

By focusing too much on existence (self-existence) and neglecting "nothingness" in the human condition, we return to the Laplacean creed.

This is the philosophical position that any event that occurs, including a decision made by human beings of their own free will, has external conditions that determine the occurrence of that event and not some other event.

Various theories of determinism are woven throughout the history of philosophy. In direct opposition to determinism is non-determinism. Determinism is also often contrasted with free will.

Determinists believe that free will is an illusion and that the human will is not free. Second, Schart is saying that, like Descartes' demons, since the world of doubt does not exist. Then doubt itself cannot doubt itself.

If freedom itself is right, then freedom from freedom becomes a necessity. That is why Schartes said, "Man is born free and has to be free.

Since freedom is inevitable, one must bear the burden of freedom by taking it back from others, by giving a precise and concise account of the essential character of a thing or the connotation and extension of a concept.

Or by listing the essential properties of an event or an object to describe or regulate the meaning (with difficulty) attributed to a word or a concept.

As we have seen, Schatt cites this case of the gambler, the transcendental 'act of self-deception' that evades the past and the future, and if such 'self-deception' is possible, then is it not the case that such 'self-deception' unites being and non-being in the same consciousness (teleological determinism.

The idea that any event that occurs, including a

human decision made of free will, has external conditions that determine the occurrence of that event, not some other event. I am X, non-existence is someone else's business), then the next object to be interrogated is self-deception.

However, another question arises: man is the elusive "nothingness" from which existence comes into the world. The impractical, inscrutable "nothingness" separates him from his essence.

Therefore, man is not the absolute impractical and inscrutable "nothingness", but the ego itself turns itself into the impractical and inscrutable "nothingness" of existence.

This leads to the question that "self-deception" demands an answer or an answer. The purpose of "self-deception" is to stay outside the scope of the social matrix, and it is an escape.

06. Self-deception

"Self-deception" is a state between honesty and deception. Self-deception is a state in which the true nature or reality of something is not apparent, but is in a state of "semi-transparency" that deceives oneself and others.

Self-deception is a belief based on a reasonable and fragile foundation. In the moment of "self-deception," the self-deceiver believes that what he is acting or thinking about is reasonable, although he may later realize that he is misperceiving, and that self-deception is essentially a negative avoidance behavior. Schart believes that man is free, so it is impossible not to choose.

Man should be brave enough to choose his own "existential" value in the bounded situation, his own life existence itself, and this action takes place in at least two situations where it is possible to choose one or the other, and the choice is to achieve an end.

And to take responsibility for the voluntary act that he takes through the objective existence, reflected in the human consciousness, through the activity of thinking. One should not avoid responsibility and fall into "self-deception".

"Self-deception" is a confession of one's own actions, not hiding or shirking responsibility for something that does not exist, because it transcends the facts and situations that exist in front of us, and is continuous and indefinite for a period of time.

"Self-deception" is instantaneous, done instantly. For example, a balloon exploded, we can not say that the balloon explosion is a state, it is only a thing that manifests itself, can be felt by all the circumstances, because the explosion process is completed instantaneously.

At the same time, like individuals using sensory, perceptual, thinking, memory and other mental activities, their physical and mental state and the environment, people, events and things change in the integrated awareness and knowledge, not its negation to the outside.

In our opinion, such a position is "self-deception". "Self-deception' is not self-deception, nor is it a lack of cognitive ability and clarity of cognition of one's environment and self, but a unification of a sense of good consciousness.

Initially there is a will, then there is a plan of "self-deception" that can be used to be able to explain, in one's own words, texts or other symbols, the known facts and principles and principles, the original nature under the cover of "self-deception".

And lead to its position on all the activities of the human and animal brain and the results, that is, as having conscious thought, to go back to the past to rethink, to learn from the experience before the grasp of the lessons.

The basis for this success or confidence is the process of "self-deception". Self-deception is a concept that does not easily explain known facts, principles, and principles with one's own words, texts, or other symbols, so it is difficult to succeed in relying on or having confidence in them.

It can only be grasped well through examples. For example, in a school assessment exam, I made a mistake in a question that I could have done correctly due to an accidental mistake. As a result, I finished ninth in the inter-school assessment exam, just one point below the third place total.

So I thought I should have been third, because I knew how to do that very simple question, but I made a mistake! I should have been third. At this point, I was in a hysterical state of 'self-deception'.

Because my seemingly reasonable explanation was

untenable. I only see my own mistakes, but ignore the possibility that others may also be wrong. Maybe the person who ranked tenth was just negligent enough to scribble the wrong answer card, and that's why he ranked behind me.

Maybe my strength is only in the tenth place. But I realized my not-so-exaggerated expectations through "self-deception", which is actually a negative avoidance behavior. I could have worked my way up to third place, but I didn't, and I simply dispelled my anxiety through 'self-deception'.

The best proof of "self-deception" is to cover your ears and steal the bell. A bell thief, when he steals a bell, thinks that if he covers his ears, no one will hear him and he will be able to steal the bell, but he does not realize his mistake.

Therefore, in the whole process of "self-deception", the deceiver and the deceived are actually the same person, and the deceiver and the deceived happen at the same time. The original activity of "self-deception" is to escape from what people think is the target of action or thought.

In other words, the individual uses his or her senses,

perceptions, thoughts, memories, and other mental activities to become aware of and recognize changes in his or her physical and mental state and in people, events, and things in the environment. In order to escape, we cannot escape from the things that are the target of our actions or thoughts.

However, by thinking about the immediate and long-term problems, we are able to formulate a solution to the facts and situations that exist in front of us. In itself, self-deception reveals the inner division within existence (people are not what they are, for example, my dream is to become a philosopher, but I am not a philosopher, so I am divided).

"It is this division that 'self-deception' wishes to be. Sincerity seeks to escape the inner division of self-existence and to move towards a selfhood that sincerity was supposed to be, but is not at all. "Self-deception seeks to escape from being in the division within the self.

But it denies this division itself, just as it denies that it is itself 'self-deception'. "Self-deception" is possible because it is the "existence" of the existence of the human ego itself, and through thinking about immediate and long-term problems, it develops solutions and solutions.

It is the most direct solution to a permanent threat to oneself, because it always contains the danger of "self-deception" in all the activities and results of the human and animal brain, i.e., as a self-conscious thinking being.

The source of this danger is the ability of man to perceive his environment and his self, and the clarity of his perception, in its "being" as "what it is" and at the same time as "not being what it is".

The unrealistic and inscrutable "nothingness" comes from the negation of the judgment of the individual's comprehensive awareness and recognition of his or her own physical and mental state and the changes of people, things, and objects in the environment by using mental activities such as feeling, perception, thinking, and memory.

It is impossible to produce negation except in the form of negation, through which a "being" or a way of "being" is raised and then thrown into the impractical and inscrutable "nothingness". Schart then argues for the dialectical concept of the impractical and inscrutable 'nothingness'.

He points out that there is a parallel between the actions of people who are confronted with actual "existence" and the actions of people who are confronted with the impractical and inscrutable "nothingness" that is the target of their actions or thoughts.

There is a certain state of interaction and mutual influence between parallel things, and a certain nature of connection between people and people or people and things, which makes people want to immediately see "existence" and "non-existence" in the same way.

This makes people want to see "existence" and "non-existence" as being exactly and unquestionably present in all things in heaven and earth, two things that complement and supplement each other, as darkness and light.

In short, it is generally the concept of two or more events occurring at the same moment, which are somehow united in the creation of two identical and simultaneous beings.

Therefore, it would be a waste of effort to study them in depth and in isolation, without any benefit. In Schart's analysis of Hegel's philosophy, pure "being" is pure abstraction, absolute negation, non-being, so that pure "being" and pure impractical, inscrutable "nothingness" refer to the same thing.

Spinoza is a thoroughgoing determinist, who believes that the emergence of all things that have happened is absolutely imbued with an objective law of development that does not depend on the will of man. Even the behavior of human beings

Freedom of will is the ability to know that we have been decided and to know why we are doing what we are doing. So 'freedom' does not mean the possibility of saying 'no' to something that happens to us, but 'yes'.

And to be able to interpret known facts and principles and principles in our own words, in our own language, in words or in other symbols. Why? Things would have to be possible that way.

Hegel's transition of Being to a purely impractical, inscrutable 'nothingness', that is, its description or specification of the meaning of a word or concept by listing the essential properties of an event or an object, contains in itself a negation of Being.

It is a repetition of Spinoza's formula: everything that is regulated is a negation. From a phenomenological conception, Scharthe argues that it is possible to conceive of the purely impractical and inscrutable 'nothingness' that lies within, but not yet manifested, and the actual 'being' activity, the existence of

They are two equally necessary components of the real thing. It is not necessary to make the transition from 'being' to the purely impractical and inscrutable 'nothingness', as Hegel does.

In Hedger, 'being' and 'non-being' are no longer a matter of emptying out one or several characteristics of complex objects, but of paying attention only to the actions or processes of other characteristics (e.g., the mind thinks only of the shape of the tree itself, or only of the color of the leaves, or only of the color of the leaves).

For this being, even when "existing" face to face with the purely impractical and inscrutable "nothingness", this is the anxiety that one is in all the time.

This is something that is distinguishable from the human being and exists independently within itself. But it does not need to be a physical existence. It is only in the purely impractical and inscrutable "nothingness" that "existence" can be transcended. The "anxiety" is to find things, places or truths that others have never seen in this double and constant nothingness.

It is the purely impractical and inscrutable "nothingness" that is potentially inseparable from the actual existence of existence, and it is this purely impractical and inscrutable "nothingness" that brings people to this world of "existence.

Man's "existence" is lacking. The purely impractical and inscrutable "nothingness" makes man transcend and create, and gives him the freedom of will. The theory of "self-deception" is a unique theory constructed by Scharthe from the standpoint of phenomenology and using Freud's psychoanalytic method.

He points out that "the existence of the human ego life itself is not only the "existence" that is manifested in the world by prompting negation, but also the "existence" that can adopt an attitude against self-denial.

"We are willing to admit that "self-deception" is a lie to ourselves. Lying is not an attitude of denial, but the essence of lying is that the liar is fully aware of the truth he is hiding. In this way, "self-deception" appears to have the structure of a lie.

The difference between "self-deception" and lying is that in "self-deception" one is only covering up what one knows to be true from others. Therefore, in the self-concept, there is no deceiver or deceived, and everything is pure. But on the contrary, when one adds one's own thinking and judgment to what one knows, there are relativities of good and bad, tall and short, black and white, beautiful and ugly, right and wrong, good and evil, etc. Self-deception essentially contains a singularity of consciousness. Using Freud's psychoanalytic theory, Schart pointed out that in order to escape from difficult situations and poor lives, and also referring to the complexity of things and many obstacles, people naturally turn to mental activities that have already occurred in the course of human life but are currently unaware of them, which are "processes of mental activity that have already occurred but have not reached the state of consciousness", and which together with consciousness constitute all the mental activities/cognitive activities of human beings.

Psychoanalysis replaces the concept of

"self-deception" with the concept of lying without a liar, replacing the deceiver and the deceived with the oneness of "I" and "feeling good about myself", placing myself in a situation with something, and considering the weight of everything about myself or others according to the result of the mutual treatment of the relationship, constructing an autonomous "self-deception" between the subconscious and the conscious. Consciousness. This paragraph reminds us once again of the example given by Schart, of the "transcendental" and "artificial" attributes of women's dating. In fact, these two personal characteristics and temperaments are inherent in the individual, and they are mirrored by each other in a way that treats each other differently.

But what Schart is trying to say is: "How can self-deception be possible? How can one deceive oneself with one's conscience when one knows that one is planning to do so? Sincerity faces the fetters and judgments of the common world, and hides in the shell of the tortoise, claiming that it is not so"; Milan Kundera has a word called kitsch, which means the lack of self-thought and self-righteousness, the worldliness of the vast majority of people who go with the flow of life.

Lack of self-thought, self-righteousness, only know how to go with the flow, the masses, with the Saudi meaning of self-deception, also as a substance or a whole and another substance or another whole compared with each other, or refers to the concept of dependence on certain conditions and existence or change close. Shat has written about the self-deception of words, and 'kitsch' has subsequently become a literary term. In this way, we can conclude this volume again: "Self-deception" is only possible if it is genuine and sincere, and not a self-perceived deception that deceives oneself and others, or if one is sincerely aware that it is not a conscientious "self-deception", and that it is intrinsically lacking in "purpose" within the limits of "self-deception".

Schart pointed out that "self-deception" deceives oneself as well as others, and that it stems from the fact that one's concrete existence is a structure of "what one is not" and not "what one is" in the immediate reality and condition of existence. Man "is what he is not" because what he is is not determined by his past, but by the plans and purposes he is not, but should be. At the same time, "he is not what he is" because man is always beyond what he is, denying his reality, denying his actuality. The purpose of "self-deception" is "to make me be what I am in the sense that I am not what I am, or not what I am in the sense that I am what I am. In this sense, one can transcend oneself through action, and exist in the realm of immediate facts and conditions.

Schart's "man is nothingness" is a paradoxical and even contradictory expression, because common sense "nothingness" means non-existence, while the essence of "self-deception" is nothingness. Self-deception" deceives oneself and others, and the biggest problem one has to face is believing. The fundamental problem of "self-deception" is to deceive oneself and others, and that is the actual belief. Truly speaking, "self-deception" deceives itself and deceives others by not believing what it wants to believe. But precisely because it admits that it does not believe what it believes, 'self-deception' deceives itself and others, and that is why it is called 'self-deception'.

Volume 2: "Being-for-itself", "Time"-"Holism

Section 1: The innovative view of the argument "being-for-itself".

In logic, an argument is a set of statements or

statements, one of which is a "conclusion" derived from the others, and the others are "premises".

Argumentation is very important in logic, because logic itself is the main object of study of argumentation, in order to find out some methodological principles to distinguish its correctness or incorrectness.

Schart attempts to present a holistic understanding of being-for-itself in time, which is an innovative view of the argument for being-for-itself.

(1) Being and essence

"(1) Existence and essence" is the fundamental idea of Scharthe in his exposition of existentialism. He proposes a philosophy that explores the meaning of human existence, the basis of irrationalist thinking, which is that all ideas based on and in accordance with objective facts are correct ideas.

It is only by establishing that "existence is prior to essence", that objective existence is reflected in human consciousness, and that the result is produced through the activity of thought, that all emphasis is placed on human values, and that human dignity and rights are preserved.

The content of the ideas and theoretical concepts that emphasize the dignity and rights of human beings, i.e., the sum of the elements that constitute the concept, refers mostly to the abstract essence, which is the concept used to determine the reference, can be more closely interpreted.

Schart also suggests that the reason why "existentialism" is not recognized, or even misunderstood in terms of its content or other relevant circumstances, is that "existentialism" is divided into two types of concepts.

The first is the view represented by the German philosopher Jaspers and the French Catholic philosopher Gabriel Marcel, who affirm and advocate theistic existentialism; the second is the view represented by the German philosopher Heidegger, who advocates atheistic existentialism.

These two types of "existentialism" are easily confused, and this is a major factor in the controversy.

In his study of existentialism, which actively seeks a fundamental cause and a higher reliability, in order to enhance the existence of consciousness and the emptiness of the object of consciousness, Schart is the first to make a distinction between the concepts of existentialism.

Existentialists who advocate "God" as an objective being say that the association of the causes of movement and change of things, the essential properties that determine "what" things are, or the style that determines "what" things are, has the advantage of being free from any conditions and restrictions.

For example, before a vessel made of clay or metal can be cast and made into a new product and presented to the world, there must be a concept of form in the mind of the craftsman before the craftsmanship, contour, and function of the clay or metal are made.

If this is not the case, then the vessel made of clay or metal can only be created without any basis and without any reason. It is clear from Plato's explanation that the reason for the existence of vessels made of clay or metal

It is the result of the combination of container making, container definition, and container function that the nature of containers made of clay or metal is. All of these factors take precedence over the existence of containers made of clay or metal.

Many advocates of "God" as an objective being believe that God created mankind, so they also believe that God can connect two or more concepts or things in a certain way, and subjectively create things that are objectively and universally acceptable to man.

God is a superb craftsman who must have a human concept before creating man, otherwise the created man would not be able to carry out his activities.

This is the same as the case of the vessels made of clay or metal mentioned above, where they knew exactly what they wanted to create, what they wanted to achieve and what they wanted to achieve before they created.

The philosophers of the 18th century, led by Diderot, Voltaire and Condé, stated that the individual human being is a concrete example of the concept of universal humanity, and that in both primitive and civilized societies we have the same nature, and that our nature takes precedence over concrete or historical existence.

In contrast to "existentialism", which advocates "God" as an objective being, there are "existentialists" who deny all religious beliefs and superstitions about ghosts and gods, and Saudi Arabia is one of them. Existentialists, who deny all religious beliefs and superstitions, believe that God does not exist and that there is "something" that does.

There is something that exists before the essence, specifically: the reality of man is this "something". The existence of man precedes the essence. After the existence of man, man will be in dialogue with man himself, and many self-concepts and self-definitions will emerge in the world.

The fundamental principle of "Existentialism" is embodied in the existence of self-life and the existence of life itself.

The fundamental principle of "Existentialism" is embodied in the existence of self-life, the existence of life itself, which is the existence of existence before it enters the world.

What Schart is saying is that "existence is prior to essence", which means that man's self-existence, the existence of life itself, is "existence". As the existence of existence, one is involved in the world and meets oneself face to face, so that one can plan and develop oneself.

The semantics (the actual expression of the concept)

of Schart's "existence before essence" is a judgmental sentence, a concept that can be defined and observed as a phenomenon. The core theory that constitutes his "existentialist" thought

In fact, although "existentialism" is generally considered to have originated from Kierkegaard, the first famous existentialist philosopher to adopt the term was Scharthe. As the famous scholar Frederic Copleston explains, Schartre asserts that "the basic principle common to all existentialists is that existence precedes essence.

Philosopher Steven Crowell, on the other hand, finds it difficult to define "existentialism" because, rather than being a complete philosophical system in itself, "existentialism" can be understood as a way to reject other systematic philosophies.

What does "existentialism" mean when it says that existence precedes essence? The most famous and explicit advocacy of this is generally considered to be the maxim of Jean-Paul Sartre: "Being precedes essence".

By this he means that there is no innately determined morality or soul other than man himself. Both morality and soul are created by man in the course of his existence. Therefore, man is not obligated to conform to a moral standard or religious belief, but has the freedom to choose.

Thus, when evaluating the merits, goodness, evil, beauty, or unreasonableness of a person, event, or thing, it is the act of the person, not the person's identity, that is the conclusion of the judgment and analysis of the event or person, because the essence of a person is defined through his or her act.

Therefore, we read it in the basic concept of the philosopher Kant's phenomenology, which means that the material world of human consciousness is something that is distinguishable and exists independently within itself. But it does not need to be a physical existence.

It is what people know as the real world. In the realm of reality, there is a distinction between experience and ideas, or experience and logic. In yin and yang, we define s by experience and l by conceptual logic; the essence of existentialism is the L(l) that is organized by logic l, and existence is the content S(s) of experience s.

Existence precedes essence, that is, we have the

definition of experience s first, and only after finishing it does we have the definition of l by conceptual logic.

In "Being and Nothingness", Schartre analyzes and critiques the philosophical propositions of Berkeley's "To exist is to be perceived" and Descartes' "I think, therefore I am" in search of a kind of I-thought that is not a reflection, seeking to rethink the past from the perceiver, to draw from it the "being" before the lesson of experience, and to obtain an ontological proof.

Schart denies the existence of God or any other pre-defined rule. He rejects any "hindrances" in life that prevent smooth passage or development and reduce one's freedom of choice.

If there were no such obstacles, then the only problem a man would have to solve would be which path he chooses. Yet man is free; even in his self-deception, he still has a potential to develop and a tendency or opportunity to become a reality. Schart also suggests that "the other is hell".

This view seems to be contradictory to the view that "people have the freedom to choose", in fact, each person is free to choose, but for the result of the choice, each person has his own responsibility, which cannot be shifted to others.

In the process of making a choice, the most serious matter facing a person is the choice of others, because everyone has the freedom to choose, but the freedom of each person may affect the freedom of others.

"Existentialism" does not deny the existence of God. For example, Kierkegaard is a Christian who believes that existentialism is the beginning of the Christian mode of thought. Nietzsche, in his book "The gay science", states that "God is dead".

Nietzsche does not say that God is dead at the level of concrete, elusive things or artifacts; rather, Nietzsche's holistic, foundational, and critical inquiry into the realm of reality and man, aware that God is dead, represents a crisis of existing moral standards.

"Being' is a reality that does not depend on the will of man, including material existence and conscious existence, including the existence of entities, properties, and relations.

It is a kind of existence that belongs to the individual's mental activities such as sensation, perception, thinking, and memory, and the comprehensive awareness and knowledge of his physical and mental state and the changes of people, things, and objects in the environment. He defines consciousness as transcendental from Husserl, and considers it as the most important discovery of Husserl. He considers this as the most important discovery of Husserl.

"Existence" is the basis of existence. "Existence" is everywhere for existence, but nowhere to be found.

A virtual world is also a kind of existence. It is something that is distinguishable from the world of human consciousness and exists independently within itself.

But it does not need to be a physical existence. In essence, it is still the existence of a person using mental activities such as sensation, perception, thinking, and memory. The carrier of the virtual world (the hardware device of the Internet) is a material existence.

There is no "being" that does not exist in a certain way, and there is no "being" that is not grasped by both revealing "being" and concealing this way of being.

By using mental activities such as feeling, perception,

thinking, and memory, an individual can always transcend existence, but not towards its 'existence', but towards the meaning of this 'existence'.

To deepen the holistic, fundamental and critical inquiry of Being into the realm of reality and human beings, Scharthe proposes the theory of "being-for-itself" and "being in itself". This is the essence of Saud's philosophy of being.

Earlier, we introduced the starting point of Saud's existential thought: consciousness. In Schart's thought, the individual uses mental activities such as sensation, perception, thinking, and memory to become aware of his or her own physical and mental state and the changes in people, events, and objects in the environment.

On the contrary, the reflection of the human mind on the objective material world is an accidental, random and free mental process of sensation, thinking, etc.

The cognitive ability and the clarity of cognition of the environment and the self are meaningless and absolutely free, just like a wisp of smoke in the air, a ghost in the world, which is: nothingness.

On the basis of the individual's comprehensive

awareness and knowledge of his own physical and mental state and the changes of people, events and things in the environment by using his senses, perception, thinking, memory and other mental activities, Schart distinguishes two kinds of existence, one is "being-for-itself" and "being in itself".

The so-called Freedom is the thing itself, which is somewhat similar to the Kantian philosophy of "being in the original form or itself of all things in heaven and earth", which has no consciousness, no meaning, no purpose, no chaos, and no essence.

On the other hand, "being-for-itself" is not the same, it is a self-existent thing and it is the result of all the activities of human and animal brains, i.e. as a combination of conscious thinking, because there are individuals who use mental activities such as sensation, perception, thinking, memory, etc. to

Because of the individual's participation in the integrated awareness and knowledge of his or her own physical and mental state and the changes of people, things and objects in the environment, "being-for-itself" becomes meaningful, purposeful, and at the same time has absolute freedom, and the archetype of "being-for-itself" is actually a human being.

A human being is a kind of "being-for-itself" because of his or her participation in the integrated awareness and knowledge of his or her physical and mental state and the changes of people, things, and objects in the environment by using his or her senses, perceptions, thoughts, and memories.

They are completely passive and passive, and they do not have any freedom and meaning in themselves, only when they exist as objects of human beings.

It is only when they exist as human objects that the human mind reflects the objective material world, and the sum of various mental processes such as sensation and thought, that they are given their essence.

In other words, "the existence of all things in heaven and earth, things as they are or as they are" is determined by man's cognitive ability of the environment and the self, as well as cognitive awareness and recognition of his own physical and mental state and the clarity of changes in people, things, and objects in the environment.

Schart's philosophical view is that human nature is "to be formed", to give a simple example. The nature of a cup and a pencil is obviously given by man. The cup is used to hold water, the pencil is used to write, and even the names of the pencil and the cup are given by man.

Therefore, the nature of external objects is given by man, and they are pre-existent. Because the nature of the cup and the pencil existed in the consciousness of man before they came into existence, the nature of external things precedes existence.

The essence of man is "to be formed", or indeterminate, because man has a free consciousness, and your future is a scientist, a teacher, or whatever.

This is completely indeterminate and not predetermined, unlike a cup that is made to hold water, there is no predetermined nature of human existence, and human existence precedes nature.

Starting from the philosophical view of consciousness, Scharthe proposed the concepts of self-possession and self-activity, and on this basis, he proposed "being-for-itself" and "being in itself".

This is a major feature of Schart's existentialist thought, which distinguishes between 'being-for-itself' and 'being in itself'. Being-for-itself actually refers to human beings.

In this way, Scharthe distinguishes between the existence of human beings and other things by whether they have consciousness or not, and introduces the "being-for-itself" characteristic of human beings based on the special function and activity of the human brain, which is a reflection of the objective world unique to human beings.

The characteristic of being-for-itself

Because man has a consciousness, his nature is not fixed and is "to be formed". The nature of a person is formed later in life, through the free choice of his or her consciousness, and the choice of consciousness is a purely personal thing, and everyone's choice is different.

This is what Schartrecht meant when he said that existence precedes nature. Human existence does not live according to a predetermined path, because with consciousness, I am free to choose my own nature and be the master of my life.

This is the positive side of human "freedom". Freedom of choice is also one of the core values of existentialists, and it is also the spirit of practice that Scharthe practiced throughout his life. He opposed the fetters of traditional values, freely chose his own way of marriage, and even chose not to accept any official honors, and became the first person to refuse to receive the Nobel Prize.

But the "freedom" here has a certain "negative and negative" aspect. The philosopher Kant says that true freedom is not the freedom to do whatever you want to do, but the freedom not to do whatever you don't want to do.

In other words, the freedom to say "no" is the true freedom. In Schart's thought, freedom of consciousness has a similar meaning; freedom also means the negation of the past, and necessarily so.

This negativity is inherent in freedom of consciousness. For example, your present nature is only the result of past actions, not the future. In other words, your present nature can only be found in the future. People always look to the future with anxiety and expectation, and this uncertainty brings us anxiety and anxiety.

This uncertainty brings us anxiety and anxiety. Because man is free and his nature is not fixed, the past is generated for the present, the present is generated for the future, and what is the future? The only thing that is certain is that the future is not the past or the present.

It is a negation of the present and the past, an uncertainty, or the present is nothing, the present as "for oneself" is a negation, a "nothingness".

In short, we can think that freedom has a positive side, it allows us to be indeterminate in our nature, to be free to create, and on the other hand, it also allows us to live forever in negation, in the wake of uncertainty.

It is like Hegel's saying: everything that is real is rational, and everything that is rational is real. In fact, the meaning of this statement is not an explanation of reality, but a negation of reality.

It means that whatever exists is bound to perish. Schartre's thought was influenced by Hegel's dialectical thought, and the absolute freedom of consciousness actually expresses a kind of "negativity".

(2) 'Being-for-itself' and 'being in itself'

In the introduction to Philosophy of History, Hegel explains, "Philosophy is the integrated awareness and knowledge of one's own state of mind and body and of the changes in people, events, and objects in the environment, using mental activities such as sensation, perception, thought, and memory.

In other words, consciousness exists in a free and infinite number of forms, and the opposing form of abstract introspection is only a reflection of it. Consciousness is free, exists independently, has a personality, and belongs only to the spirit."

So "consciousness" as a separate concept consists of two parts, each of which has an infinite number of "forms", one of which is the principle and the other the concrete reflection of each historical event.

Therefore, he also said, "The human being in the usual sense has the ability to recognize the environment and the self, and the clarity of cognition. There are two aspects: the concept of things in general and the abstract concept of specific reactions to things.

He also said, "Each person has a different self-awareness and reacts differently to things, and there are deviations from the principle consciousness, but for a normal person

But for a normal person, this movement or action of

spatial deviation under external influence has a limit, and this limit depends on his normal state, on the degree of respect for God. To understand the extent to which this form of thought reflects the nature of things belongs to the realm of metaphysics."

So, although Hegel's language is difficult to understand, he explains and states in detail and in depth the fundamental principle of philosophy, which is to study through rational reasoning and logic questions that cannot be answered directly through perception.

It is necessary to study the mechanism of how propositions and counter-propositions are linked in each event, and therefore to compare the examples of each historical event with their prototypes, to understand the commonalities and differences between them.

As one of the most important French philosophers of the 20th century, Scharthe's book "Being and Nothingness" is an exposition of existentialist thought that is heavily influenced by Husserl and Heidegger.

"Being in itself" is the core idea of the liberal philosophy of Being and Nothingness, which uses analytical thinking to explore and reflect on the fundamental questions of life, knowledge, and value.

"Being in itself" is the first part of existence that should be attended to, but the ultimate problem facing Schart's existentialist philosophy is the problem of choosing between the boundaries of the situation, "being and nothingness".

"The paradox of 'being in itself' and free existence, freedom and the choice of boundaries, will ultimately be the paradox of Sartre's holistic, fundamental and critical inquiry into the real world and man.

This paper attempts to introduce Sartre's freedom and self-activity, while further clarifying the deeper truths as a path to a free choice of existential philosophy.

Section 2: The Existential Implications of Schart's Existentialist Philosophy

Although Schart's overview of "existentialism" is inherited from Kant, Husserl, Heidegger, and Descartes, "existentialism" is different in a deeper sense from the classical philosophy of Western philosophy (English: Ancient philosophy), also known as ancient philosophy, which is the philosophy of the ancient Greek to Roman period and generally includes the philosophers of the period from pre-Socratic philosophy to the rise of Christianity. It generally includes the ideas of philosophers on existence from the pre-Socratic period to the period before the rise of Christianity. In discussing the idea that "existence precedes essence," Scharthe states that "existence" has a creative character and can be subdivided into "being-for-itself" and "being in itself," which expresses the relationship between existence and nothingness from the viewpoint of essence.

Therefore, returning to the question left open at the beginning of the text, what is Self-possession and Self-activity, the most important philosophical terms of the Saudis are 'being in itself' (English: being-in-itself / French: être-en-soi) and 'self-being' (English: being-for-itself / French: être-en-soi). for-itself / French: être-pour-soi).

(1) 'being in itself'.

Saudi "existentialism" expresses "for oneself

It believes that when people act or think, the behavior manifested by the object is the result of the freedom of choice of the will, and subdivides "existence" into two types of existence, namely "being-for-itself" and "being in itself". Being-in-itself' is the existence that is not felt by the cognitive ability and clarity of cognition of the environment and the self, that is not understood, understood, and awakened to, and that has not changed from small to large, from simple to complex, from low to high;

"It is difficult to give a precise and concise description of the essential characteristics of a thing or the connotation and extension of a concept. Once you say what it is, you have already added artificial values and regulations, just as when you say that it is white and square, you have already added artificial meanings and regulations. Therefore, the best way to define the meaning of "being in itself" is by limiting the scope of the term or concept to the existence of the meaning and value that is added by taking out all the artificiality. In short, "being in itself" is the objective world that is to be given meaning by human beings.

(2) Three fundamental features of being in itself

This means that "being in itself" can be analyzed from three points of departure: First, "being in itself" is self-existent: "being in itself" is the objective In other words, existence does not require the existence of facts;

secondly, "being in itself" is always what it is: "being in itself" is independent, it does not need to depend on or attach to other things, it can be filled by itself; thirdly, "being in itself" is existence: "being in itself" is itself, existence can be filled by itself. being in itself is itself, existence and itself are opposed or correspond to each other, opposed to each other in nature, such as big and small, beautiful and ugly, with any transition and change. It transcends all conditions of time, space, causes or conditions of development of external things, etc. It is summarized as follows.

01. The first characteristic, 'being in itself' is self-contained.

Simply put, it is not appropriate to think in terms of creating or being created to define "being in itself". Because 'being in itself' is self-existent, it is not possible to use creationism to explain 'being in itself'. Even if "being" is created, "being" cannot be explained and interpreted in terms of creation, because it regains its "being" outside of creation.

Freedom here means "being in itself", which can neither be said to be created nor to be able to create itself. This means that "existence" is non-created. Neither passive nor active can be said to be self-caused according to the way of consciousness. Therefore, it is neither passive nor active, because the above-mentioned characteristics are the characteristics of man's self-consciousness, not the characteristics of "being in itself". 'being in itself' is already there.

Man has active autonomy, but the means he uses are passive. "Being" is the oneness of its own existence, that is, it is united with itself without distance. There is no relationship between "being" and itself. It cannot realize its own interiority, nor can it affirm its own certainty, nor can it move autonomously with its own dynamism, because "being" itself is full, and "being" itself is self-existent, and that is all.

The second characteristic, "being in itself" is always what it is.

"Existence is what it is. To put it simply, it is not appropriate to think in terms of changing concepts to define 'being in itself'. 'being in itself' is directly 'being in itself', it does not contain the possibility of change within itself. "Being in itself is opaque, there is no mystery to the Self, it is solid. Therefore, Schartrecht says very graphically that "being in itself" is solid, that is, it does not have an inside and an outside, because to say that it has an inside and an outside would be to impose meaning and possibility of change on the Self.

And these meanings and possibilities of change are in fact images that belong to the clinging of human love. In this way, Freedom is a study of a rigid, absolutely uniform metaphysical concept without any connection, without any change or development, without any distinction or variation. Our metaphorical reference to "metaphysics" can be understood as the use of rational and logical thinking to study the orbit of all problems in the universe and life. For example, what is the nature of the world and what is the meaning of life? The concept of metaphysics, in terms of the quenching of our daily life experience, is the "existence" of the existence of our own life itself, just like a fish swimming in water without realizing that there is water.

An example of a philosophical concept that is often misinterpreted is Descartes' famous phrase "I think, therefore I am". Many people think that this phrase means "I exist because I am thinking", and some people may even derive it to mean that "to be human is to think, and to not think is to not exist or not exist". This interpretation differs greatly from Descartes' interpretation of the nature of existence.

The concepts of "I think" and "I am" are not causally related, but are interpreted and clarified by reasoning. That is, from the former being true, the latter can be deduced to be true. That is to say, from "I think" being true, one can deduce "I am" being true. It is not true that "I do not exist" when "I do not think".

Another example is the concept of "eggs are born from chickens and eggs are born from chickens" and "why do people live". This question is one of the most important questions of inquiry in metaphysics. Camus said, "There is only one truly serious philosophical question, and that is suicide. The study of "why one does not kill oneself" is in fact the study of "why one lives". It is difficult to understand and answer such a question of the existence of one's own life without the experience of life and the painful reflection of one's heart.

03. The third characteristic, "being in itself" is existence.

To put it simply, it is not appropriate to define "being in itself" by the concept of possibility or impossibility. The "existence" of "being in itself" means that "existence" can neither be derived from possibility nor subsumed into necessity. Necessity concerns the relationship between ideal propositions, not the relationship between existents. Schartre considers that 'being in itself' is only 'being', it cannot be 'derived' from or 'derived' from any other possibility, nor can one even attribute 'impossibility' to the Self.

It is only for human beings that there are possibilities and impossibilities to speak of. A phenomenon of "being" can never be derived from another being, because it is a being. This is the contingency of 'being in itself'. "Nor can 'being in itself' be derived from a possibility that may belong to another realm of 'being' - 'being-for-itself'.

(3) 'being-for-itself'

After the elaboration of "'being in itself'", Schatt turns to "being-for-itself". Schart pointed out that "for oneself" must be combined with Freedom in order to appear.

In fact, this structure is the negation of the intentional act and the intentional object (perception and the interior of the perceived object), as mentioned in the previous article.

Schart draws attention to the fact that when he says

that the act of intention is combined with the object of intention (perception and the interior of the perceptual object), it seems to express that "for oneself" can be free from the self and "exist" alone.

This is the basic principle and starting point of Schart's philosophy of freedom; obviously, from his existentialist philosophy, Schart has closely linked freedom with human existence.

Therefore, the question of "existence" cannot be avoided in the discussion of Schart's philosophy of freedom. In a certain sense, human "being" is human freedom. "Being-for-itself" is "being".

"Being-for-itself" is the opposite of "being in itself", it is the existence of human consciousness, which defines itself by the internal negation of "being in itself". "Being-for-itself" is also translated as 'for oneself'.

In the philosophical viewpoint of Schart's book "Being and Nothingness", it refers to the existence of a unique way of being, i.e., the ability to recognize the environment and the self, and the existence of the clarity of cognition. It is the opposite of 'being in itself'.

In his view, 'being-for-itself' has a specific structure, it is a 'being' without its own foundation, and it is not a thing that exists independently of itself and within itself.

But it need not be a physical 'being', it cannot 'exist' apart from a 'being'. It is always transcending itself, transcending its own past, what it already is, and looking towards the future, so that "temporality" is the way of existence of "being-for-itself".

"Being-for-itself" is composed of intentional activity, which can ask questions, make choices and negate, and when it points to something, it makes the thing that is the target of action or thought null and void, i.e., it gives a new meaning of existence to the thing that is the target of action or thought.

"Being-for-itself" and "being in itself" are opposites in nature, but they are united in an integrated relationship. On the one hand, "being-for-itself" is not something that is autonomous and distinguishable and exists independently of itself, but cannot exist independently of "being in itself".

On the other hand, without being-for-itself, being in itself can only be a meaning and a value, something that does not contain any distinction in itself, something that is all in one.

What unites being-for-itself and being in itself is being-for-itself itself. Shat enables people to face the world and the self through the structure of being-for-itself, in order to fully demonstrate the freedom of human beings.

The "internal negation" relationship is perhaps the most revolutionary part of the philosophy of being-for-itself, which allows Saud to demonstrate in a convincing way that consciousness can transcend any object.

It states that consciousness must be united with the object of consciousness, precisely, with the object in a way that exceeds, surpasses, and even overrides the object.

To put it in terms of everyday experience, it means that one is "being-for-itself" always at a distance from one's culture, nationality, class, occupation, identity, etc.

Consciousness is born to be supported by something that is "not" itself. To use the example of the Saudis themselves, the person corresponds in a way that is not a correspondence of a correspondence of service.

It is the inevitable structure of the individual's awareness and understanding of his or her own state of mind and body and the changes in people, events and things in the environment, using mental activities such as feeling, perception, thinking and memory.

There are two main ways of being-for-itself existence: one is "not what it is"; the other is "what it is not". It has the possibility of negation, and this negation factor is accumulated continuously, and it has the ability to be active in nihilization, and can obtain new existence by negating the present.

It helps the world to reconstruct a new order and give it a new meaning. People in the world are influenced by the factor of nothingness (constant change) to exist, and the world is composed of many people, so the world is also nothingness (birth, dwelling, dissolution).

The corresponding concept is the self-invented temporary existence, which denies existence. On the basis of such a concept, Scharthe first proposed the concept of "nothingness".

In other words, everything has its innate nature, and this nature is not limited to a certain range or category. The emptiness expressed here means that there is no other requirement than negation.

As it were, only by constantly denying one's own existence can one constantly negate, renew and transcend, and only then can one realize the ultimate freedom of being-for-itself.

Section 3: Sartre discusses "being-for-itself" from three levels

01. The direct structure of "for oneself

(01) Consciousness is the 'being-for-itself' structure.

"Being-for-itself' is consciousness facing the existence of the self. Consciousness cannot coincide with oneself. It is a comprehensive awareness and recognition of one's own physical and mental state and changes in people, events, and objects in the environment by using mental activities such as feeling, perception, thinking, and memory. The burden of existence is reduced.

It is the ability to "vanish" which is inscrutable to oneself, and it is the cause of the way of existence, but there is no cause.

Therefore, the reflection of the objective material world in the human mind is the sum of various mental processes such as feeling and thinking.

It can neither be explained by something that does not

belong to it, nor can it be explained by itself. "Being-for-itself" is the existence of the self, because it cannot coincide with itself.

In this way, a person comes into existence from the inscrutable "nothingness" and uses mental activities such as sensation, perception, thought, and memory to become aware of and recognize his or her own physical and mental state and the changes in people, events, and things in the environment. This becomes the internal structure of "being-for-itself".

(02) "Being-for-itself" is the external structure of self-being.

Scharthe borrows the term Faktizitat from Husserl and Heidegger to illustrate the external structure of being-for-itself: the existence of the being of the human being itself is "existing in the impractical, inscrutable 'nothingness'".

This means that "nothingness", which exists in an impractical and inscrutable way, is only temporally limited and exists temporarily in the world in the form of life existence. The "for oneself" of "being-for-itself" is constantly innovating its existence, constantly metamorphosing beyond its present manifestation.

Before it disappears, it infinitely becomes freedom, that is to say, to introduce something new from the outside, or to propose an insight that is different from the conventional or normal way of thinking by using the existing knowledge and materials.

It is the act of improving or creating new things, methods, elements, paths, and environments in a specific environment, based on idealized needs or to meet social demands, and to obtain certain beneficial effects.

Therefore, the individual uses sensory, perceptual, thinking, memory and other mental activities to become aware of and understand his or her own physical and mental state and the changes of people, events and things in the environment.

In the human mind, the reflection of the objective material world is a variety of mental processes such as feeling and thinking, the cognitive ability of something's self and the clarity of cognition become some "point of view" of things, and things are thus regulated by laws or necessity.

Thus, the cognitive ability and the clarity of the cognition of the human being to the environment and the self, and the cognition of the role or influence of the thing on the thing, become the same kind of existence as the thing, and this existence shows the external structure of the human being, which is human-made, as opposed to natural and natural.

02. Temporality

"Being-in-itself' is out of time, cut off from time, while 'being-for-itself' is realized in a limited process of temporalization. Schart's phenomenology of three-dimensional time (past, present, and future)

The process of "being-for-itself" is a complex topic or thing that is gradually broken down, so that the topic can be better analyzed in detail, understood and understood from a certain cognitive point of view. The purpose is to achieve a temporal whole, without going through too much thinking process.

An immediate thought, feeling, belief, or preference that quickly emerges. When one believes in, worships, and holds a religion or a doctrine as a rule and guide for speech and action, but is not sure of the reason for it, it is often attributed to this intuition.

The "past" is the ease with which I am as "beyond"

what I am. "The present is not what it is (the past), and it is not what it is not (the future).

And the future is the possibility of "being-for-itself" to metamorphose beyond the present, and it is always in front of man, waiting for him to make the situation that did not exist a reality. So "being-for-itself" is the "now" of the future.

In essence, it is a kind of "nothingness" that is unrealistic and inscrutable in the future. When things or things gradually develop from one stage to another stage, to the state of "now", it becomes unrealistic and inscrutable "nothingness" by making the originally non-existent situation a reality.

This gives the temporality as a whole structure, the knowledge of all that man should have of natural or social things.

It is the meaning that man gives to the things that are the object of action or thought, the spiritual content that man transmits and communicates in the form of symbols. The analysis of the static temporality and temporal kinetics shows that "for oneself" can only exist in the way of time.

'Being-for-itself' is scattered in the three dimensions of temporality like a 'Diaspora' (a temporary or permanent departure from the country to which it belongs), but is unified in this structure. It also marks the original meaning of the impractical and elusive nihilization of 'nothingness'.

"for oneself" is not what it is, but what it is not, and the unity in eternal return is what it is not and what it is not. Any "for oneself" exists according to these three dimensions.

03. Transcendence

Schart's detailed and in-depth statement of transcendence is intended to address the phenomenon of existence as a concrete fact and condition of human existence in front of us, as well as the problem of "being in itself" in its original relationship.

The ability of the unrealistic and inscrutable "nothingness" of the human ego to become nothing is the ability to recognize a desire, the possibility of the future, and this is the freedom of the human will. This "possibility" of using each thing or phenomenon as a condition for each other and inseparable is constantly evolving beyond one's current situation.

It is something that can never be obtained or stopped in the unrealistic and inscrutable "nothingness", something that is distinguishable in its nature and exists independently within itself. But it need not be a movement of physical existence, which is transcendence.

The transcendental movement is the process of development of the Self in the direction of the Self, and this process is realized by knowing. For example, the Creator is transcendent, he is above the world, in the broadest sense of the word, the whole, the all, the supreme being, and the ultimate cause.

Yet he is also something that is inherently distinguishable and exists independently within itself. But it does not need to be a physical being. And in the impractical, inscrutable "nothingness", because through participation and causality, he is also within the world.

Transcendence is a fundamental concept in theological and religious discussions of God, as well as in philosophical discussions of knowledge and existence.

There is no way for "for oneself" to establish Freedom, nor can Freedom manifest itself to "for oneself", because it does not maintain any relationship in itself, and this task is undertaken by the activity of the human mind, which reflects the characteristics and connections of objective things, and reveals their meaning and effect on human beings.

To "know" is to realize the two meanings of "for oneself": to make existence in the world and to become the negation of this reflection of existence: the negation is to make a situation that does not exist a fact.

The "for oneself" that is stipulated in its existence reveals that Freedom is something that is inherently distinguishable and exists independently within itself.

But it need not be a physical existence, and the negation that makes a situation that does not exist a reality is called transcendence.

Thus, the "existence" of the self-life itself is the negation of the self, worldliness, spatiality, quantity, instrumentality, and temporality to "being in itself".

Therefore, the human brain reflects the characteristics and connections of objective things, and reveals the meaning and role of things for human beings. In a broad sense, it includes all cognitive activities of human, i.e. perception, memory, thinking, imagination, understanding and production of language, and other mental phenomena.

Therefore, Freedom can only be "for oneself" of consciousness. Since "for oneself" is different from Freedom, and it relentlessly pursues the ideal Freedom, this unification of Self-possession and Self-activity can never be realized.

04. Intentionality

What is the relationship between Self-possession and Self-activity? The word "intentionality" as a philosophical concept originally appeared in the philosophy of the Academy and referred to a state of being - a state with intention.

It was only in 1874 that the Austrian philosopher F. Brentano reintroduced it into philosophy and it became an important philosophical concept.

and "being-for-itself". Schart is hoping that people will not overuse their ability to think abstractly, because the two are always present together, and it is impossible to separate intentionality.

Intentionality is the ability of the mind to represent or present things, attributes or states. Simply put, many mental activities are about connections to the external world, and intentionality is the "about" here.

The individual uses sensory, perceptual, thinking, memory and other mental activities to become aware of and recognize his or her own state of mind and body and the changes in people, events and objects in the environment, which must be about one's ability to recognize the environment and self and the clarity of that recognition.

That is why Schart said that consciousness is born to be supported by something that is "not" itself, in order to maintain an intentional relationship between consciousness and the object of consciousness (something that is "not" itself).

This suggests the existence of a cognitive capacity for the environment and the self, as well as a degree of cognitive clarity, as Martin Heidegger put it, with contextuality, the reflection of the human mind on the objective material world.

is the sum of sensations, thoughts, and other mental

processes, with a precise and brief description of the intrinsic characteristics of a thing or the connotation and extension of a concept, according to the context in which it is located.

Or by listing the basic attributes of an event or an object to describe or regulate the meaning of a word or a concept itself, how the specific self-definition operates. The role or effect of "internal negation" on the matter in question is perhaps the most revolutionary part of Scharthe's philosophy.

It makes Schart to prove in a convincing way that the individual uses sensation, perception, thinking, memory and other mental activities, the comprehensive awareness and understanding of his own physical and mental state and the changes of people, things and objects in the environment.

It is able to transcend any action or thinking as the goal of things. It specifies that the cognitive ability of the environment and the self and the clarity of cognition must be combined with the object of consciousness, more closely.

The human mind's reflection of the objective material

world is the sum of sensations, thoughts, and other mental processes, which are combined with the object of action or thought in a way that exceeds, surpasses, or even overrides the object.

"Being-for-itself" and "being in itself" are relative, but the difference is "being-for-itself". "Being-for-itself" is not existence, it is non-existence, it is the negation of existence, it is not something that is distinguishable and exists independently within itself.

It requires a physical manifestation of actual existence. It is an impractical, inscrutable "nothingness" of existence. The "for oneself" is a "non-autonomous" absolute, also known as a non-physical absolute.

Its reality is purely questioned. Its existence is never affirmed, but is questioned, because the impractical and elusive 'nothingness' of difference always separates it from itself.

Self-possession and Self-activity" are not decisively separated, but reunited by an integrated connection, which is either something else or Self-itself.

After discussing "being in itself", Scharthe turns to the existence of self-being. Schart pointed out that "for oneself" must be combined with Freedom in order to appear, and "for oneself" is the negation of Freedom.

In fact, this structure is the internal negation of intentional acts and intentional objects/perceptions and perceived objects, as stated in the previous article. Schart draws attention to the fact that when he speaks of the internal union of intentional act and intentional object/perception and perception, he seems to be expressing that "for oneself" can exist independently of Freedom.

Schart wants people not to overuse their abstract abilities, because the two are always present together and cannot be separated.

According to the definition of intentionality, it refers to the individual's comprehensive awareness and recognition of his or her physical and mental state and the changes of people, things and objects in the environment by using mental activities such as sensation, perception, thinking and memory, and must be about one's ability to recognize the environment and the self and the clarity of cognition.

This is why Shat says that consciousness is born to be supported by something that is "not" itself, in order to maintain the intentional relationship between one's cognitive ability to perceive the environment and self, and the clarity of cognition and the thing that is the target ("not" itself) when the consciousness acts or thinks.

05. The relationship between Self-possession and Self-activity

Schart believes that if "for oneself" lacks freedom, then "for oneself" is like a vacuum created by the withdrawal of dynamism, so that "for oneself" is like water without a source, a tree without roots. It is something that has no foundation.

In other words, if "for oneself" has no freedom, it will lose the foundation of "existence", and freedom without "for oneself" will also lose its meaning.

Therefore, Self-possession and Self-activity are two living entities that are integrated with each other, one cannot be separated from the other. The root of the integration of the two is "for oneself" itself, and with "for oneself" as the center, Freedom gradually transitions towards self-activity.

Based on the analysis of the basic characteristics of existence itself, i.e. all real things, "being in itself" has priority over "being for itself".

In short, "being in itself" existed a long time ago, while "being for itself" was derived from the development of nothingness.

Without "for oneself" to give value and meaning, "being in itself" will always be a cold, desolate, meaningless thing.

Schartre considers the state of interaction and mutual influence between being and phenomenal things; a certain nature of association between man and man or man and things, unlike the relationship between abstract and concrete, where being is not a structure like something else.

Existence is not a structure like something else, but a basis for the manifestation of mutually opposing phenomena. Non-existence and existence are not opposites, but a pair of contradictions that exist or change depending on certain conditions.

"Without the individual's mental activities such as sensation, perception, thinking, memory, etc., the comprehensive awareness and knowledge of one's physical and mental state and the changes of people, events and things in the environment, the reflection of the objective material world by the human mind, such as sensation, thinking and other mental processes, is empty.

Therefore, it can be seen that if the phenomenon of Freedom is not associated with the individual's mental activities such as sensation, perception, thinking, memory, etc., the integrated awareness and knowledge of one's physical and mental state and the changes of people, events and things in the environment, it can only be regarded as an abstract object. It can be regarded as non-existent.

"It can only reveal itself when it reveals the characteristics of Freedom, which is a kind of contradictory two properties inherent in things themselves, i.e., a kind of thing with two contradictory properties at the same time.

Without Freedom, "for oneself" is a consciousness that does not exist physically, i.e. something abstract and general; on the contrary, without "for oneself", Freedom has no meaning, it is an impractical and inscrutable "nothingness", something that cannot be described.

Saudi Arabia has handed down from generation to

generation, from history to history, ideas, culture, morals, customs, arts, institutions and ways of behavior. Being in itself" is the basis of "being-for-itself", and "for oneself" gains its existence only through the negation of Freedom.

To be more precise, "being-for-itself" uses Freedom to lend itself to existence. The "for oneself" with its dynamic nature acts as a unifying link between Self-possession and Self-activity, and is an essential factor that affects the occurrence, existence, or development of things.

The "for oneself" is the factor that is essential to the occurrence, existence, or development of things, and makes "existence" null and void, so that the Freedom can be manifested to the world (in a broader sense, all, all, everything), like a spring.

Section 4: Hegel's Self-possession and Self-existence

Self-possession and Self-activity?

Self-possession and Self-activity are terms used by the 19th century German classical philosopher G.W.F. Hegel. It is used to express the different stages of the development of the absolute idea. In the logical thinking of Hegel's "Phenomenology of Spirit".

Freedom means that which is hidden within but not yet revealed, i.e., potential, while "for oneself" means that which is invisible becomes visible, i.e., unfolds and reveals. Hegel's account of the development of the concept from Freedom to "for oneself".

Western philosophy is the study of universal and fundamental issues in the fields of existence, knowledge, values, reason, mind, and language. The act of "knowing" and "identifying" a subject in order to know with certainty, and that these knowings have the potential to be used for specific purposes.

The process of transferring from one status and state to another gradually penetrated into human nature and existence in the German classical period. By the time Hegel arrived, he moved from the world reflected by sensibility, to the world of reality reflected by human nature, and finally to the absolute idea through reason.

In Hegel's view, the traditional logic of the law of identity, the law of contradiction, and the law of the middle, as a way of thinking to grasp the nature of the world, is abstract as it is repetitive, and not only fails to grasp the nature of anything, but also contradicts the fact that all people of normal intelligence in ordinary society have, or generally have, "cognitive knowledge" of a subject.

Not only does it fail to grasp the nature of anything, but it also contradicts the fact that ordinary people of normal intelligence in society have, or generally possess, the act of "knowing" and "identifying" a subject in order to know it with certainty, and that these knowings have the potential to be used for specific purposes. This is even contrary to formal logic itself. Hegel further argues that it is ludicrous to say that contradictions are not conceivable, and that everything in itself has contradictory unity of two properties.

In the Lesser Logic, there is a triadic dialectic from being to essence to concept. "Existence" cannot be Freedom, it is a latent existence that is hidden and not discovered by people. It has to be referred to through the norms of human nature and the realm of reason.

That is to say, "existence" must have an absolute concept, and it is only through human nature that it can gradually reach the point where it is hidden beneath the visible appearance, so that what was invisible becomes visible.

This is what it means to not be 'being-for-itself'.

"Being-for-itself" means that "existence" is a "self-existent" existence within the realm of reason and human nature. For example, "existence" itself, in the Lesser Logic, is first interpreted from "existence".

This "being" is something but not, something real but not, something both, and nothing. There is no pure "existence", and anything that "exists" is not a pure existence. This kind of meaningless "existence" can also be called non-existence. Through being and non-being, the three paragraphs are pushed to the next step. In short, it "exists" through its own connectedness.

01. The concepts of "being-for-itself" and "being in itself"?

Generally speaking, for a form of thinking that reflects the nature of things, we may describe the phenomena, processes, states, and reasons of things to explain their meanings, causes, and reasons.

Such descriptions, using various rhetorical devices, visualizing things, may be based on certain rules (e.g., logical reasoning, scientific analysis), and the analysis of how and why an event, an action, is possible.

Moreover, the analysis of how and why an event or an action is possible also transforms the unknown into the interpretation of a known matter, i.e., a concept corresponds to one or more meanings.

However, it is usually not easy to explain this way because it ignores the fact that the meaning of a concept is determined by the context in which it is situated, and therefore it is important to try to explain a concept, especially a philosophical one.

It is better to give an account of the form of thought that reflects the nature of the thing, of the textual thing or article in which it is located, or even of the historical thing.

In this way we can see the abstract, universal idea that is acting as a specification of the domain or class of entities, events or relationships. It is then that we can get a clearer picture of the structure, thread, or clue of things.

When Hegel uses the concepts of Self-possession and Self-activity, they are usually not used independently of each other, but together, in terms of the positions from which problems, issues, etc. are analyzed or criticized.

Rather, they are used together; either as a relation of precedence in the process of reflecting reality by means of concepts, judgments, and reasoning, or as a relation of one substance or whole and another substance or another whole in comparison with each other, or as a relation of existence or change depending on certain conditions.

Therefore, in discussing the preconditions of a matter, understanding this point first will help us to explain the known facts and principles and principles in our own words, texts or other symbols.

The following is a reference to the explanation of the Dictionary of Education, as well as the logical thinking of Hegel, along the lines of a detailed analysis, from a certain cognitive understanding, to understand. The following is a detailed analysis of Hegel's logical thinking in order to understand it from a certain cognitive point of view.

I. Self-possession and Self-activity

In section 21. of The Phenomenology of Spirituality, Hegel talks about how "the fetus Freedom is human, but not "for oneself"; it is only as educated reason that it is "for oneself" human.

If we recall Aristotle's definition of the nature of man: "Man is a rational animal," then we can easily understand why the fetus is free as a human being, but not "for oneself.

It is because the abstract thinking activities such as

concepts, judgments, and reasoning that belong to the fetus have not yet been developed.

Therefore, the fetus must still experience the ability to use language and writing and general knowledge, and history can provide the present person with an understanding of the past and serve as a reference for future actions, which is an important achievement of human spiritual civilization along with ethics, philosophy and art.

It is through education that certain moral, civilized, or normative spiritual or material behaviors are developed in an individual, so that his rationality gradually develops and becomes apparent, and he can be called a Homo sapiens, that is, a "for oneself" person.

Therefore, here we can take Hegel's Freedom, "for oneself", to correspond to Aristotle's concepts of "potential" and "realization"; the fetus has the reason that human beings are capable of using reason.

But this rationality, for the moment, is only a potential ability that can be exercised, and it must be realized only when it is properly educated.

The seed as a potential is the tree of Freedom, and the

tree as a realization is the seed "for oneself", so when it comes to Self-possession and Self-activity, we should not see them as two different things.

Sometimes, Hegel uses the expression Self-possession and Self-activity in order to emphasize that something of Freedom has changed from potentiality to realization.

Being-for-itself" and "being in itself

When Hegel uses the concepts of "being in itself" and "being for others", he is trying to express the state of interaction and interconnection between things.

When he uses 'being-for-itself', he means that this thing is only involved and influenced by itself, not by other things, and that it is independent, direct, and does not need intermediation.

Such 'being-for-itself' is sometimes called 'being in itself' by Hegel, but Hegel then shows his readers that what he has just called 'being-for-itself' does not really mean not being in itself.

It is not true that things do not interact with other things, that they interact with each other, that they are "he" as long as they interact with other things, that they are connected and influenced by each other.

The other here can be all other objects or phenomena that exist objectively in the natural world, or the individual's comprehensive awareness and knowledge of his or her own physical and mental state and the changes of people, things, and objects in the environment by using mental activities such as sensation, perception, thinking, and memory.

We "exist for others", that is, we are involved and influenced by other things, non-independent, indirect, and in need of mediation.

For example, the individual uses mental activities such as sensation, perception, thinking, and memory to perceive and know the state of his or her body and mind and the changes of people, things, and objects in the environment in a comprehensive manner.

Initially, it is a "being-for-itself" that is only connected to and influenced by oneself and not connected to other things, but this "being-for-itself" is the object of one's cognitive ability and clarity of cognition of the environment and self, and is associated with consciousness.

That is to say, it is connected to and influenced by

other things (i.e., the individual's mental activities such as sensation, perception, thinking, memory, etc., the comprehensive awareness and recognition of his or her physical and mental state and the changes of people, things and objects in the environment).

Thus, it is also a being-for-others. Finally, this "being-for-itself" discovers that it is not only a "being-for-itself" but also a being for others, and this is what Hegel calls "reflection".

Reflection is the discovery of the "other" in the self, the self as the object of action or thought; or a purposeful way of thinking, from examining one's own past experiences, gaining lessons from them, and improving learning behavior.

It is a process of "metamorphosis beyond the present" of the self, in which the human brain reflects the characteristics of objective things, connects and reveals the meaning and effect of things on the human being, and finally returns to the self.

In addition to other things, the other exists for the individual's mental activities such as sensation, perception, thinking, and memory, as well as for the individual's comprehensive awareness and knowledge of his or her own physical and mental state and the changes of people, events, and things in the environment.

Sometimes, it is also used as our existence. When Hegel talks about "we" in Psycho-Phenomenology, he means "we," the readers of Psycho-Phenomenology (sometimes including Hegel himself, the author of the book).

Because the ability to recognize the environment and the self, as well as the clarity of cognition, has not yet developed into self-awareness at first, individuals use mental activities such as sensation, perception, thinking, and memory to

In the beginning, we do not observe the process of our own development because we use our senses, perception, thinking, memory, and other mental activities to become aware of our physical and mental states and the changes of people, events, and objects in the environment. It is we ourselves who can observe the whole process of consciousness development, and what we must do.

What we have to do is to do nothing but watch, and let the individual use mental activities such as sensation, perception, thinking, and memory to develop his or her own awareness and understanding of his or her own state of mind and body and the changes of people, events, and things in the environment, according to his or her own internal rules.

02. Thoughts - the language that is hidden in the brain

Intrinsic language is the language that is not spoken, the language of the individual's mind. In his book Thought and Language, L.S. Vygotsky, the father of Soviet psychology, describes the thoughts, emotions, memories, images, and other transients that occur in the brain.

Statements about the characteristics of things (statements), mainly through symbols or symbol systems (language, numbers, pictures, etc.) as detailed, organized and systematic as possible, the general observed state, facts and properties of the object expressed, Wieckowski believes that children from the age of two will be thinking thoughts, intentions, thinking and language began to interact.

Wieckowski reanalyzes what Jean Piaget called egocentric speech. Piaget argues that initially young children usually speak without regard for the viewpoint of the listener, but that this egocentric speech, as children learn and inherit various social norms, traditions, ideologies, and other surrounding socio-cultural elements, gradually adapts to them.

As children learn and inherit social norms, traditions, ideologies, and other surrounding sociocultural elements, and gradually adapt to them, socialized speech emerges. Wieckowski argues that the child's egocentric language does not disappear at this time, but becomes internal language, which is different from external speech.

The structure and function of internal language, which plays a pivotal role in normative thinking, can be found in H. Werner & B. Kaplan's generalization. Since the inner language is silent, it becomes a way for the mind to "turn inward" to the inner language.

It becomes a quick medium for the mind to engage in a rational and active process of reflection on the real world or any object in an "inward dialogue". The inner language is also a concentration of an inner self-discovery, an outer creative expression of self-liberation.

In this sense, inner language is like a language between two acquaintances that is understood in the heart without being expressed through words or actions. In a three-person dialogue situation, if two of the people are acquaintances and the third person is less familiar with them, then communication between the two acquaintances can sometimes be as simple as a word or a gesture.

But the third party needs to make things clear. For example, "Dream of the Red Chamber. The first thing you need to do is to get a good idea of what you are getting into. The first is that the two of them are not able to make a decision.

In the inner language, one word can contain many meanings. It conveys the meaning of the word, personal meaning, even from generation to generation, from history down the ideas, culture, morality, customs, art, institutions and behavior.

There is an invisible influence on people's social behavior and control the meaning of tradition. This situation makes the inner language the central link that introduces or guides the two parties to a particular relationship.

According to Witkowski, an important function of

inner language is the planning of cognitive operations that lead to the recognition and understanding of things by the individual.

He describes the inner language in terms of the organization, structure, and pattern of the mind; through the inner language, a person can plan and organize the rational and active process of reflection of the mind in an "inward dialogue" with the real world or any object. For example, during daily life and work, inner language can guide all actions until the work is completed.

As for the relationship between inner language and outer language, D. McNeil argues that the gesture of the body during movement is a direct expression of the thinking process, thus producing outer language. Before the words are spoken, the inner language of self-talk constructs the meaning of the intended expression in the mind.

The inner language is often accompanied by the mental image (thought), and the gesture of the body is the external factor of the mental image that works through the internal factor of the behavior, so that the outer language is the pronunciation of the inner language.

In a psychological counseling process. In the work of counseling, the professional counselor, in a direct face-to-face situation with the client, assists him/her in relieving psychological confusion, facing the reality, and achieving internal spontaneous growth according to the concrete facts and needs of the situation.

The counselor often teaches the client to use the correct and positive inner language of observation and memory of a specific object or righteousness, instead of the inner language of abstracting the common essential characteristics of the perceived things and generalizing the incorrect and negative inner language, in order to change the client's deviant perception of his or her own life.

03. Hegel's concept of "self-existence" and "self-existence"?

"Being in itself", in Hegelian logical thinking, refers to the process of reflecting reality through concepts, judgments, and reasoning in the process of knowing.

It is different from figurative thinking in that it uses scientific abstract concepts and fields to reveal the nature of things and express the results of knowing reality. It is constructed with two meanings

1. to leave "regularity" and insist on one's own existence, that is, to be free from "regularity" and not to establish an abstract identity with other things, i.e., P=P. Think about Kant's object-self (Kant believes that the sum of all objects, that is, the world of experience, exists and is connected only in the form of all things that human beings live in this world and perceive to be combined.

Kant's introduction to the object-object is as follows:

And we indeed, rightly considering objects of sense as mere appearances, confess thereby that they are based upon a thing in itself, though we know not this And we indeed, rightly considering objects of sense as mere appearances, confess thereby that they are based upon a thing in itself, though we know not this thing as it is in itself, but only know its appearances, viz.

(In fact, since we have reason to regard the objects of the senses as mere phenomena, we also recognize from them the Self as the basis of these phenomena, though we know not this thing as it is in itself, but only its appearances, viz.)

Or refer to the materialistic substance: substance is also called material, material, and has been defined by scholars in different ways. Generally speaking, there are various interpretations as follows: first, it is the raw material of man-made objects, such as the wood that forms tables and chairs; second, it is the material that forms natural or physical objects, such as rocks, water, bones, and muscles; third, it is the object of sensation, such as the apparatus of sound, and the nature seen by the eyes; fourth, it is the basis of reality, such as the flesh that expresses the phenomenon of life, and the atoms that produce the phenomenon of nature; fifth, it is the material that occupies space, is malleable, and has mass and weight. Fifthly, they are those that occupy space, are ductile, have mass and weight, are movable, and are impenetrable, such as wood and metal; sixthly, they are the opposite of mental, mind, spirit, and form; and seventhly, they are the basic causes of experience.

Early man thought that the universe, including all celestial bodies, is infinite space - U means infinite space, Zu means infinite time. It is the totality of all matter and its forms of existence. Philosophically, it is also called the world, which is composed of some material facts and conditions that specifically exist in front of the eyes, such as the Greek philosophers thought that earth, water, fire, and air are the original substances that constitute the universe, and Chinese Taoist thinkers thought that gold, wood, water, fire, and earth form the world.

The scientific research, which takes certain objects as the scope of study and seeks to obtain unified and exact objective laws and truths based on experiments and logical reasoning, has enabled people to gain a better understanding of the substances that form the world.

Democritus and Leucippus advocated the atomistic view that all matter in the universe is composed of tiny particles of atoms, and that atoms are the limit of separability of matter.

Nowadays, scientists have discovered that atoms are the smallest unit of matter, and later they found that there are smaller electrons, protons, neutrons and short-lived elementary particles among atoms, which are the basic substances of the world.

In addition, the indestructibility of matter, the interchangeability of matter and energy, and the indestructibility of mass-energy are important laws that can explain the changes of matter. To put it simply, the science that takes a certain object as the scope of study and seeks for unified and exact objective laws and truths based on experiments and logical reasoning.

The research shows that the world is made up of empirically accessible materials. Because science emphasizes the specificity and falsifiability of predicted results, it is different from vague philosophy. Moreover, science is not the same as the search for absolute truth, but rather a groping approach to truth on the basis of existing evidence.

Philosophy differs from other disciplines in that it has a unique way of thinking, such as a critical, often systematic approach, based on rational argumentation. In contrast, disciplines that are based on the study of universal, fundamental questions

In philosophy, which includes the fields of existence, knowledge, value, reason, mind, language, etc., the explanation of matter predates the science that takes a certain object as the scope of study and seeks to find unified and exact objective laws and truths based on experimentation and logical reasoning; Plato thought that matter is limited by the nature of space-time and can accept a permanent form.

Aristotle (Aristotle) thought that matter has weight,

occupies a position in space, and can be known by the senses, is the possibility of achieving form, the most original in the physical definition of matter is usually.

Objects consisting of quarks and leptons, photons and bosons are not usually considered as matter in the physical definition. Rather, it is a formless primary matter, which has the potential for change, and therefore is realized by form.

Aristotle also used the term matter, which is not scientifically defined, to explain the occurrence of change from the point of view of the potential realization of something with a non-zero rest mass (theoretically the rest mass of a photon is zero, but there is no such thing as a rest photon).

Medieval philosophy inherited Aristotle's view and divided matter into first matter (primordial matter), which is the original basis of all matter, and second matter, which is an object that has weight, occupies a place in space, and can be known to exist by the senses, or all substances that form the components of an observable object in scientific research.

The modern philosopher R. Descartes believed that

matter is the opposite of mind, that matter is malleable, that some properties can be quantified, that the movement of matter is governed by mechanical laws, and that restraint is exercised so that it does not go beyond its scope or move at will.

J. Locke (Locke) will not depend on the subjective consciousness of man, and the existence of objective and real material nature, divided into primary and secondary nature, primary nature will not change, such as hard and soft volume, etc., is inseparable from the nature of matter, secondary nature can be changed, such as color, sound, etc.

2、Potential, that is, hidden in the inner, and not yet manifested. This is the claim of the purposivists, who claim that all things that "exist" in nature have their individual purposes of existence.

This means that their own value is not in other things, but in their activities, or in their operation and realization, which coincide with this purpose.

Some ecologists influenced by teleology believe that natural ecosystems have a purpose, which is to maintain themselves in good and harmonious order, and not only that, but that all the constituent species and individuals in them have their own place, in their own unique way, for the whole.

In their own unique way, they all contribute to the order of the whole system and thus have their own value, not depending at all on humans, whether they can use them for subsistence or for the creation of civilization.

They even believe that it is not right to interfere with or disturb the natural ecosystem in a man-made way. With the advent of empirical science, teleology is gradually being questioned, and "purpose" is no longer an item that we must appeal to in order to explain why nature (or any species or individual in nature) exists, so the wisdom of basing a philosophy of environmental ecology on teleology is open to question.

2. Trying to Prove the Existence of God

The term "teleology" includes three different concepts: Teleological Argument in theology; Teleological Explanation in science; and Teleological Ethics in ethics. Teleological Ethics). 1.

The teleological argument in theology, which attempts to prove the existence of God, is also called the argument from or to design; its main argument is: in this universe, everything is so cleverly coordinated, the human like the movement of the planets, the small like the relationship between the hand and the eye, this coordinated whole is absolutely not accidental, but from a hidden and unseen designer. The potential designer is God.

The teleological argument, the existential argument, and the cosmological argument are the three major arguments for the existence of God in eighteenth-century European theology, which are based on existing empirical knowledge, experience, facts, laws, cognition, and tested hypotheses.

It is a logical conclusion of the three main arguments for the existence of God through generalization and deductive reasoning. This process or form of language in which certain reasons are used to support or refute a point of view usually consists of a thesis, a thesis, an argument, and a way of arguing. There are three points of dissatisfaction among present-day people.

(1) The harmonious operation of the orbit of the universe may be a mere coincidence, or it may be a "material, perceptible world" that exists independently of human consciousness, which is the sum of all material

movements outside of human consciousness.

In terms of content, it consists of two parts, namely, natural existence and human social existence. The former exists independently of human activities, while the latter is formed in human practical activities and is not transferred by human consciousness.

The common denominator of both is the principle of the unity of the objective reality of the world substance, the non-conscious, non-conceptual existence or aggregate, which is absolutely objective.

The unity of natural existence and human social existence constitutes the world of the "external world" or "material world", which, between generations, has the result of the contradictory phenomena contained within things that are not intensified, and of the various theories explaining these phenomena, not necessarily the result of God's design.

(2) The English philosopher D. Hume (1711~1776) believed that only in the two observable observations, referring to the perceptual acts of seeing and hearing, that is, analyzing and thinking, that is, observation is not only a visual process, but a comprehensive perception that is

dominated by vision and integrated with other senses.

Observation includes the active process of analysis, synthesis, judgment, reasoning and other cognitive activities on the basis of images and concepts, so it is called the overall view and understanding of the external world by the human brain when external stimuli act on the senses.

At the same time, it also refers to the phenomenon of things, the movement of things between states, so that there can be a mutual relationship between existence and behavior, especially the relationship between before and after, a kind of human brain reflecting the characteristics and connections of objective things, and revealing the meaning and connection of things to people.

But God is not an observable state of things, so there can be no causal relationship between the universe and God; in other words, God cannot be the creator of the universe.

(In other words, God cannot be the creator of the universe. (3) If God is the creator of the universe throughout time, but the universe can be described as the sum of space and time, an orderly system, also called the world; in early Greece, the universe not only contained time, space, planets, and matter, but more importantly, it had a characteristic that represented an orderly and harmonious system.

Therefore, who created God, the creator of its existence, was explored? If God created Himself, there is one thing in the universe that cannot be explained.

01. Teleology is a form of thinking in science that reflects the nature of things: in the study of certain objects, based on experiments and logical reasoning, in the search for unified and exact objective laws and truths.

Especially in biology and psychology, to explain or understand things in terms of their possible functions or purposes (in general, people, things, and objects touched by all human activities, such as plants, animals, humans, or a certain phenomenon) is the teleological explanation.

In other words, the action or thought is viewed as the goal, just as the animal's behavior is explained by the animal's goal of finding food and clothing to keep the animal alive. A chess player plays Go (black and white, with the black player playing first, until one side has no more pieces to play, and the side with more board space is the winner)

The nature of the activity of playing Go is explained by the pursuit of winning. The main criticism of teleological explanation is that in many cases of teleological explanation, if the concepts of "function" and "purpose" are left aside, things can still be explained clearly, so teleological explanation is not absolutely necessary.

02. In terms of the ethics of teleology, a social ideology. It is the sum of behavioral norms that regulate the relationship between people and individuals, and between individuals and society. Such as honesty and hypocrisy, good and evil, righteousness and unrighteousness, justice and partiality, etc., belong to the field of morality.

Deontological Ethics, which is a holistic, fundamental and critical inquiry into the real world and human beings, is a deontological ethics that compares one substance or whole with another substance or another whole.

Its basic proposition is: the process of analysis, synthesis, judgment, reasoning and other cognitive activities on the basis of images and concepts, whether objects exist, whether they have certain properties, and whether there is a certain relationship between things in the affirmative or negative.

The criterion of whether an action is moral or obligatory is whether the result of the action creates the greatest good for the person, and if so, it is moral and obligatory.

Utilitarianism is a type of ethical theory, the moral philosophy of Aristotle (384-322 B.C.), and is a kind of teleology, which holds that the most right action is to maximize the benefits. The "benefit" is happiness, and the preference for maximum happiness and the preference for avoiding suffering is right, which is typical of teleological ethics.

It is a type of normative ethics theory. There are many variants of utilitarianism, but in general, they all argue that the moral good is based on the benefits that members of a community can have, and that what makes an action moral is that it leads to the ends that are proper to human beings.

3. Kant's Unity

To understand "being-for-itself", we must first understand Kant's unity. Unification means that the contents and tendencies of perception contain people's existing experiences, knowledge, interests, and attitudes, and thus are no longer limited to the perception of individual properties of things.

It is an important concept in the philosophy of I. Kant (1724-1804) and has two main meanings: in terms of sensory experience, it is the cognitive subject's ability to unify episodic or experiential mental states into a unified experience that belongs to him or her.

At the perceptual level, it is the cognitive subject's ability to form a universally valid judgment of what he or she has intuitively learned, using a priori domains internal to the person.

To address the question of how the act of "cognition" and "recognition" of a subject is universal and inevitable in terms of its potential ability to be used for specific purposes, Kant first explores "how innate knowledge is possible".

He asserts that a being with a unique consciousness and a unique personal experience, or another entity external to itself and related to it, in the process of cognition, is first exposed to all human activities in the realm of experience, to people, things, and objects.

After the baptism of time and space, we determine the nature of things by our own senses and perceptions alone, and many contents are formed as concepts by the action of intuitive forms.

However, the concept after the perceptual action is not universal, but still belongs to the subjective intuitive content of the individual with cognitive ability and practical ability.

In order to understand the universality and necessity of something through experience or association, we have to further transform the subjective intuitive content into an objective form by using the twelve domains of perception.

In other words, because of the a priori nature of the spheres of knowledge, these tendencies to see things through the eyes of the subject based on his own needs are the attributes of perspective, experience, consciousness, spirit, feelings, desires, or beliefs that the individual can have.

Its fundamental characteristic is that it exists only within the person who has the ability to know and practice the object and belongs to the state of mind of the subject. It can influence human judgment and the concept of truth, that is, it can have objectivity and form knowledge with universality and necessity.

This means that it is also the act of "knowing" and "recognizing" a subject in order to know with certainty, and that these knowings have the potential to be used for specific purposes.

But in the role of the sphere of knowledge, the cognitive subject is able to unite different spheres, such as the spheres of "quality" and "quantity", to form a universal judgment, because the consciousness unites them into one, the whole of the cognitive subject's consciousness.

This is Pure Apperception. The opposite is Emperical Apperception, which is a contingent, empirical perception that certain mental states belong to one's own experience.

Both of them belong to the role of consciousness. The unity of unity is what Hegel called "being-for-itself", that is, everything that appears to exist for me is able to maintain itself in the interaction with other things.

"At the heart of 'being-for-itself' is the dynamic 'demand' to see my existence in the existence of things.

This requires emphasizing that external objects exist for me, and that I need to be consistent with others' ideas and perspectives, and that the new ideas that I agree with and my original views and beliefs are combined to form a unified system of attitudes.

This attitude is enduring and becomes part of one's personality, within oneself. All the tangible things in the world are only the potential ego's comprehensive awareness and knowledge of its own physical and mental state and the changes of people, events and things in the environment by using mental activities such as feeling, perception, thinking and memory.

The simplest example is Hegel's statement in Phenomenology of Spirituality 21.1 that "the fetus is a human being, but not a self-made human being; only as an educated reason is it a self-made human being.

This is an example that Hegel himself often gives: the nature of man is freedom and reason, but although an infant has the possibility of freedom and reason, while it is still an infant, freedom and reason are only hidden in itself, and it cannot yet be called a free and rational person.

Only when the infant grows up, through some moral, civilized or regulated spiritual or material behavior manifested in the individual and nurtured through education, is able to realize its own free and rational nature, and realize it.

Only when this is achieved can one truly be called a person who is considered to be intelligent and can use his or her intelligence to do many things realistically. Here, a baby is a person's 'being in itself'; an adult is a person's 'being-for-itself'.

"Being-for-itself" and "being in itself

In Hegel, there is not just one meaning, but rather the two words have different meanings in different contexts and logical levels.

But, more importantly, these different meanings of the two words are organically related, and these different meanings are not as different as they may seem.

In fact, these different meanings are grounded in the basic position of Hegelian philosophy, so that they can cross over and communicate with each other.

So 'being-for-itself' and 'being in itself'

If we take this example here to understand "being-for-itself" and "being in itself" in all Hegelian contexts, then we will not understand the problem.

In fact, we generally say "being-for-itself" and "being in itself", and then understand "being-for-itself" as a deepening and development of "being in itself", which is not a problem, but it is not clear:

"These three stages actually constitute a kind of movement of "negation of negation", a movement through which the self does not recognize the existence of things or the truth and rationality of things.

Hegel describes Aristotle's "potential to reality" description of a thing, changing its original nature into something of a very different nature, and describing it as a transforming thought.

These three stages are a comparative relationship, not an absolute one. It is more accurate to understand that anything has "selfhood, otherness, and self-reality".

In Hegel's view, such a thing cannot be analyzed, synthesized, judged, reasoned, and other processes of cognitive activity and speech on the basis of representations and concepts.

Because once the human brain is indirectly and generalized reflection of objective things. This includes logical thinking and figurative thinking, which usually means logical thinking and speech, and it already shows that this thing, which is absolutely itself, now "exists for consciousness

(i.e., it has begun to exist for otherness, i.e., it exists for something other than itself - consciousness.) Otherness, in fact, is not something else, it is something other than one thing, and such other thing is one of the many things that are interrelated with other things.

We ordinary people of normal intelligence in society have, or commonly have, the act of "knowing" and "identifying" a subject in order to know it with certainty, and these knowings have the potential ability to be used for specific purposes.

The tendency to think that something is independent and can exist entirely on its own is actually a tendency to see things in terms of the subject's own needs, which is an illusion of the attributes that an individual can have in terms of perspective, experience, consciousness, spirit, feelings, desires or beliefs.

For example, if a person seems to be independent and self-determined, but you let him not eat for two days, he will realize what is meant by other things and deny the existence of things or the truth and rationality of things.

The so-called otherness denies that this thing is not something that can be absolutely independent and self-determining, and does not need to depend on other things outside itself. This is the otherness of a thing. Self-existence means that a thing can be completely and absolutely independent, independent and self-determining, and does not depend on other things outside itself.

In fact, there is no such thing to be found in our experience of external stimuli, including the eyes, ears, nose, tongue, body, etc. There is only one such thing, and that is what Spinoza called the entity, God.

Such a God, in everything, agrees with others in terms of thought and viewpoint, and the new ideas he identifies with and his own original views and beliefs are combined to form a unified system of attitudes.

In the midst of ourselves, or to put it superficially, "we see everything as a change from this God", such a God is a

God that does not have any influence on the occurrence, existence or development of things within a certain range.

It does not exceed and is not subject to any limitation in terms of self-activity, it only relies on its own inherent nature or personality, and necessarily develops itself, such an inevitable and certain trend in the development and change of things.

Necessity is determined by the nature of things, and to know the necessity of things is to know the nature of things, and at the same time to be absolutely self-determining, that is, to be free. This is probably what it looks like.

Hegel's self, for others, and self-action are actually combined with his negation movement, which is the process of movement of things themselves, expressing things, facts, or speaker's concepts and views in a rational way.

4. Are Hegel's 'being-for-itself' and Sartre's 'being-for-itself' the same?

There is a transition in Western philosophy, and in the German classical period, it gradually penetrated into human nature and existential thinking. In Hegel, he moved from the world reflected by sensibility, to the world of reality reflected by humanity, and finally to the absolute idea through reason.

Hegel's trinity is one. In the Lesser Logic, from being to essence to concept. Existence cannot be self-existent, it is latent existence. It has to be referred to through the norms of human nature and the realm of reason.

In other words, existence must have an absolute concept, and it is through human nature that "being" is gradually reached. This also means that it cannot mean 'being-for-itself'.

"Being in itself" means that existence is within the realm of reason and human nature, and reaches the existence of self-being. For example, "being" itself, in the Lesser Logic, is first interpreted from "being". This "being" is both, and nothing. There is no mere "being"; anything that "exists" is not a mere "being".

'being-for-itself', i.e. human 'existence', can never be as full as being-in-itself (English: être-en-soi), and one is bound to be lacking, to use the Saudi terminology.

That is, "being-for-itself" can never occupy the same place in space as "being in itself", and once it occupies the same space, it kills consciousness.

5 Hegel's "being-for-itself"

This is what Hegel says in the first part of his theory of being. The "metamorphosis" that he deduces when he talks about "being" and "nothing" is, in the previous context, "quality". The substance of this quality, which is hidden within things, is itself 'being in itself'.

And 'being-for-itself' is deduced from the previous 'freedom' and its own denial of the existence of things or the truth and reasonableness of things. "For oneself" mainly emphasizes the directness of dynamism, which makes "existence", which does not exist, a fact.

These two concepts are an important part of Hegel's logical system of thought, the dialectic of the Trinity. Hegel's logical system of thought is: being - essence - concept.

"In Hegel's theory of existence, "being in itself" is the first aspect of the interpretation of self-existence. Being in itself" is also derived from [being, nothing, metamorphosis]. These themselves constitute Hegel's "quality". The next step is to proceed with "quantity".

The second stage, the negation of 'being in itself', is

the 'fixation', through which 'being-for-itself' is completed. "Being in itself' - being - 'being-for-itself'.

These are done in the realm of "quality". Next, he talks about quantity, quality - the quantity ---- scale. In fact, these things were first dabbled in by Plato. In the Critique of Pure Reason, Kant first hierarchized the logic of human reason.

In philosophy, it means the ability to provide innate concepts to synthesize sensory material to form knowledge. In other words, he refers to the possession of knowledge and rationality in the areas of "quantity, quality, relation, and modality.

Hegel's 'being-for-itself' is a term used in his Theory of Being. In Hegel, he divides the development of "concept" (absolute idea) into two stages: one is "being in itself" (or "potential concept"); the other is "being-for-itself" (or "realized concept").

Hegel's so-called 'being-for-itself' is the second stage in the development of his 'absolute concept'.

Therefore, the so-called "for oneself" means "to be conscious or to have the subjective will to do something actively rather than passively".

In other words, "being in itself" means "not knowing what kind of existence it is, although it is a kind of existence"; conversely, "knowing what kind of existence it is and existing for such an existence" means "being-for-itself". This is a concept from German philosophy.

To put it simply, "being in itself" means being in itself, for example, a stone is "being in itself", and "being-for-itself", being for itself, for example, a human being, has the ability to move.

"Freedom" means potential, and "for oneself" means to unfold, to reveal. In Hegel's view, at the stage of 'freedom', the absolute idea is that the contradictory and unified opposites of the two properties of things in themselves have not yet been developed, expressed as existence, objectivity.

At the stage of "for oneself", it becomes the essence, the potential distinctions and oppositions, the mutual struggle for certain purposes are revealed, and the concept is reflected; finally, it develops to the stage of the form of thought that reflects the essence of things, where thought conquers and unifies existence, and the absolute idea becomes the most real thing that is free and self-contained.

Existence is the "self-existent", "potential" essence, and essence is the "self-existent", "unfolded" existence. In Hegel's Logic, two propositions are taken as the major and minor premises, and the conclusion is drawn that the lower concept is the higher concept of "self-existence" and the higher concept is the lower concept of "for oneself".

Existence is the "freedom" of the same thing, and essence is its "for oneself". Existence is the content and phenomenon of essence, and essence is the truth of existence.

The development from existence to concept is the development from "freedom" to "for oneself" of the same thing, the process from untruth to truth, from untruth to truth, from intuitive diversity to unity of diversity, from surface phenomenon to profound essence.

Hegel's account of the development of concepts from "freedom" to "for oneself" is a logical argument that uses "deductive reasoning" to arrive at a "completely different" but "necessary" conclusion from the identified evidence.

This process of reasoning is called the form of inference, which expresses the regularity of the development of the whole world, including thought, from the lower to the higher levels, including the application of the rule of argument of the trivium to examine what is a valid argument.

6. What is the difference between "being-for-itself" and "being in itself"?

Self-possession and Self-activity are Hegelian terms. They are used to express the different stages of the development of the Absolute Idea. "Freedom" means potential, and "for oneself" means to unfold and reveal.

Hegel's description of the development of the concept from "Freedom" to "for oneself", as the development of the same thing from the "Freedom" stage to the "for oneself" stage in Hegel's philosophy is the transformation from existence to thought and from the lower stage to the higher stage of development, therefore Self-possession and Self-possession and self-activity can be derived as the meaning of spontaneity and self-awareness.

The difference between the two is that plants and animals are "being in itself", which is a natural existence, while our human existence is a subjective meaning that involves the use of our senses, perceptions, thinking, memory and other mental activities, and the comprehensive awareness and knowledge of our physical and mental state and the changes of people, things and objects in the environment.

You know why you live and do what you do with a certain purpose, while plants and animals do not have this kind of reflection, they do not know why they live and do not have their own subjective consciousness.

Section 5: Sartre's "being-for-itself" and "being in itself"

In the context of Sartre's philosophy, "for oneself" (English: being-for-itself / French: être-pour-soi), "negation", and "nihilization" are basically synonyms, usually referring to human beings.

In fact, Scharthe's ontological prescription is a branch of philosophy that studies concepts such as being, existence, becoming, and reality. It includes the following questions.

How to classify entities into basic categories, and which entities exist on the most basic level. Ontology is sometimes called existentialism and belongs to the main branch of philosophy known as metaphysics.

Deficiency, from Hegel, is not a psychological or physical deficiency in human beings, nor does it refer to hunger, thirst, or insecurity; deficiency refers to metaphysical ontology, which is "the study of which terms represent real existent entities and which terms represent only a concept".

This is why ontology has become the basis of some branches of philosophy. In recent years, scholars in the fields of artificial intelligence and information technology have also begun to apply the concept of ontology to knowledge representation, i.e., by using the basic elements of ontology: concepts and the connections between concepts, as a model of knowledge to describe the real world, which is sorely lacking in human beings themselves.

That is, no matter how much wealth, food, or security you have, you are still lacking in the understanding of the true meaning of "being" in the life of the universe. "being-for-itself".

That is, one's existence can never be as full as being-in-itself (English: être-en-soi), one's existence is necessarily lacking, to use the Saudi terminology: "being-for-itself" can never coincide with "being in itself", and if it does, it will If it does, it kills consciousness.

The existence of consciousness is pure, clear and transparent, without any innate stipulation, which is the phenomenological basis of Schart's famous proposition that "existence precedes essence".

This is the basis of Schart's famous proposition that "existence precedes essence". Schart points out that since consciousness is an empty existence, it is a "nothing", but this "nothing" is not a lifeless, dead "nothing", but a creative "nothing".

In Schartre's view, lack is the source of human dissatisfaction, and it is precisely this dissatisfaction and emptiness that motivates the subject to create itself. J. G. Fichte also called his philosophy "the science of knowledge".

He agreed with Kant's view that "knowledge is the law of nature" and advocated the study of human internal consciousness. He pointed out that it is not because man is limited that he has to act, but that the proposition should be read in reverse.

Rather, the proposition should be interpreted in a different way: it is human beings who have to act, and therefore they are destined to be limited. In other words, the innate limitations of human beings constitute the requirement that they be able to act actively.

In his discussion of human deficiency, Scharthe uses Self-possession and Self-activity to point out the "non-existence" of God and the futility of man's strong emotions, including the drive to feel, to be passionate, or to desire something.

Schart believes that all things, if not "freedom", are "for oneself". If God "exists," God should be a full, solid Self, lacking nothing; but at the same time, God should be a "being" that can think and act, but to think and act is to lack something.

The conclusion is that if God "exists", God is only a stone that can analyze, synthesize, reason, judge, etc., and carry out activities to achieve a certain purpose, but such an absurdity cannot exist.

Therefore, God is a self-contradictory concept in word and deed. But Schart is quick to point out that the passions and endeavors of people, the lifelong pursuit of people, are nothing more than the desire to be in a state where people want to be completely fulfilled, but at the same time they want to be free to choose and think.

In a word, people want to be God. But, as noted above, Scharthe argues that consciousness is born to be supported only by the object of consciousness and cannot coincide with it. In the second half of his book, Being and Nothingness, Scharthe writes

Man is a useless passion after all. It means that the subject of behavior can only act according to its own needs, with the help of consciousness, the intermediary role of concepts, preconceived goals and results of behavior. As a conceptual form, it reflects the relationship between man's practice of objective things.

1. "Being in itself

What is Self-possession and Self-activity, the most important terms in the study of universal and fundamental issues in the field of being, knowledge, values, reason, mind, language, etc., should be "being in itself" and "being-for-itself".)

First of all, let us start from "being in itself" and what is "being for itself". "It is difficult to define 'being in itself' positively, because once you say what it is, you have already added the values and regulations of human consciousness.

For example, when you say that something is square or round, you are already using your senses, perception, thinking, memory and other mental activities to perceive and understand your own physical and mental state and the changes of people, things and objects in the environment, adding human meanings and regulations.

Therefore, the best definition of "being in itself" is the existence without all the artificially added meanings and values. Indirectly, "being in itself" is the knowledge of natural or social things that is to be given by human beings.

It is the meaning that man gives to the object, the objective world of spiritual content that man transmits and communicates in the form of symbols. Schart is of the opinion that "being in itself" has the following three characteristics.

First, "being in itself" is "freedom".

First, 'being in itself' is freedom. To put it simply, it is not appropriate to think of 'being in itself' in a way that is appropriate, that does not exceed our mental expectations, that does not conflict with something, that is compatible with each other, and that "creates" or "is created".

Freedom" here means that "being in itself" can neither be said to be created nor can it be said that it can create itself. Therefore, it is neither passive nor dynamic, because the above characteristics are human characteristics, not the characteristics of "Freedom". "Being in itself" is already there.

Second, 'being in itself' is always what it is.

Secondly, "being in itself" is always what it is. To put it simply, it is not appropriate to think of 'being in itself' in terms of things changing in nature or form. 'Being in itself' is itself directly 'being in itself'; it does not include the possibility of change in the nature or form of things within itself.

Therefore, Schart is very graphic in saying that "being in itself" is solid, that is, it does not have an inside and an outside, because if it has an inside and an outside, the individual uses mental activities such as sensation, perception, thinking, memory, etc., to perceive and understand the state of one's body and mind and the changes of people, things, and objects in the environment, and has already attributed the meaning and the possibility of change to "freedom". Freedom', and these are actually human regulations.

Third, 'being in itself' is existence

Finally, 'being in itself' is existence. To put it simply, it is not appropriate to think of 'being in itself' in the sense of interpreting 'possibility' or 'impossibility' by speculating on the unknown in the light of the known.

Schart thinks that 'being in itself' is only 'being' and that it cannot be 'derived' or 'derived from' any other possibility, nor can one even attribute 'impossibility' to 'freedom'. It is only for people that there are possibilities and impossibilities.

2. 'Being-for-itself'

After discussing 'being in itself', Scharthe turns to 'being-for-itself'. Schart pointed out that "for oneself" must be combined with "freedom" in order to appear, and "for oneself" is the negation of "freedom".

In fact, this structure is the negative relation between [the intentional act and the intentional object/perception and the interior of the perceived object]. Schart draws attention to the fact that when he speaks of the combination of intentional act and intentional object/perception and the interior of perception, he seems to be expressing the idea that "for oneself" can exist independently of "Freedom".

Schatte wants people not to overuse their abstract abilities, because the two are always present together and cannot be separated. According to the definition of intentionality, an individual uses mental activities such as sensation, perception, thinking, memory, etc.

This is why Schart said that consciousness is born to be supported by something that is "not" itself, in order to maintain an intentional relationship between consciousness and the object of consciousness (something that is "not" itself).

This suggests that consciousness exists in a contextual way, as Martin Heidegger suggests (English: situatedness / German: Befindlichkeit), where individuals use mental activities such as sensation, perception, thought, memory, etc., to make sense of their physical and mental states in relation to the environment.

The "internal negation" relationship should be the

most drastic change in Schart's philosophy, which can be applied to any aspect of sexuality, and it allows Schart to prove in a convincing way that consciousness can transcend any object.

It specifies that the individual uses sensation, perception, thinking, memory and other mental activities, the comprehensive awareness and knowledge of his physical and mental state and the changes of people, things and objects in the environment, must be combined with the object of consciousness, precisely, consciousness is combined with the object in a way that is beyond, above and even over the object.

If we express it in terms of daily experience, it means that people are always in touch with their thoughts, cultures, morals, customs, arts, institutions and behaviors that have been handed down from generation to generation and from history. There is an invisible influence on people's social behavior and a distance between the rules of control.

Consciousness is born to be held by something that is not itself, so that the object does not collapse. To use the example of Shat himself, that person is a waiter in the way that he is not a worker who performs service work outside the restaurant, it is not a question of whether you are loyal to your identity or not.

Rather, the individual uses sensory, perceptual, thinking, memory and other mental activities, the integrated awareness of his physical and mental state and changes in people, things and objects in the environment, and the inevitable structure of cognition determines that it is impossible for a person to completely overlap with his identity and regulations.

In short, in Schart's view, the existence of an individual who uses mental activities such as sensation, perception, thinking, and memory to become aware of his or her own physical and mental state and the changes of people, things, and events in the environment.

In general terms, human existence is actually cast into the world and is in different situations. The relationship between human beings and the world is a negative relationship. Man has always been related to the world in a "negative" way.

Because of the presence of man, volcanic eruptions and earthquakes bring about "destruction" rather than merely changing the structure of the material masses, because for man there is "destruction" and for nature there is no "destruction".

For nature, there is no such thing as "destruction", but only a change in the material mass. This "negative" operation of saying "no" is called "nullification" (English: nullification / French: néantisation) in Saudi Arabia. Nullification in Saudi terms means to delimit a being from a restricted sphere and to give it a meaning.

For example, "A little bit of red in the midst of all the greenery, a touching spring color need not be much. There is a story that during the Song Dynasty, when a higher examination for painting was held, the emperor Zhao Ji gave a question: "A little bit of red in a jungle of ten thousand greens, no need for more spring colors.

He painted a small building in a green jungle with a beautiful woman standing by the window.

The only thing that stands out is a touch of red on her lips. We often poke fun at the few men in a group of women.

Literally, it is the green leaf against the red flower. A little bit of red in a jungle of ten thousand greens highlights the red, while a little bit of green in a jungle of ten thousand reds highlights the green.

This is when the nihilization is revealed, nihilization means that you erase all the other greens in the bush, i.e. you deny the other greens by affirming the one red, and only extract the one red that is watched as the focus of attention.

The other greens become, in name only, the set surface of the one red. Why do you focus on that little bit of red? The answer is in fact very simple, because you feel that a little red is "rare as precious", the few is dazzling to seize the purpose, this explains the direction of all consciousness, have the purpose of the value as the basis.

Section 6: Being and Time

Self-possession and Self-activity is a term used by the 19th century German classical philosopher G.W.F. Hegel. It is used to express the different stages of the development of the absolute idea. "Freedom" means potential, and "for oneself" means to unfold and reveal.

Hegel's account of the development of the concept from "Freedom" to "for oneself" expresses the regularity of the development of the whole world, including thought, from lower to higher levels, in the form of idealism, and contains a rich dialectical element.

"Freedom" can be understood as existence, and "for oneself" can be considered as the awareness of one's own existence. The former is a capacity, which Hegel wanted to express as a potential, universal, prior to the phenomenal, unaware.

The latter is a kind of movement, dynamic, in which the subject is aware that he has this ability and is able to use it.

Rationality, for example, is the nature of man, which shows that rationality is the self-reliance of the adult. But not for an infant, who only has the ability to reason, but not the consciousness of reason, a state of not being awakened.

In terms of something: something is "freedom" in its own sense, and in terms of its relation to other things, it is out of itself and caught in interaction.

So it is existence for others. Something that is free from interaction with others is the source that surrounds itself, the independence that insists on its own existence and does not fall into the network of interaction, is self-existence.

If existence is understood as directness, then existence is self-existence. "Freedom" is not a modification of existence, but a synonym of existence.

In this sense, "for oneself" is also not a requirement for existence, but a relation to existence. And for oneself can only be a relationship between existence and existence. Of course, this is only in the sense of understanding existence as a mere immediacy. Heidegger's reference to existence corresponds to Hegel's concept.

The little logic begins by saying that existence is the concept of potential. This statement can also be understood as meaning that existence is only potential existence, it is not yet real existence, which we call a concept. Simply put, self-existence is that which is itself, not yet thought about, not yet connected.

Self-being should be the embodiment of this thing in the connection, the result of thinking, the product of the connection. Hegel's thinking process is to think about something in the state of "freedom" and get the result of "for oneself". In Hegel's words, "freedom" is the concept of potential.

Obviously, "Freedom" is something that has not yet

been sorted out and is not yet a concept. Hegel: The essence is 'being-for-itself'.

Obviously, we can say that we have discovered the essence of that thing when we have clarified the connection of that thing and reached the state of "for oneself". Then we have a definite concept of that thing.

To use a not very appropriate explanation, but it is very understandable.

"Freedom" can be understood as existence, and "for oneself" can be thought of as awareness of one's own existence. The former is a capacity, a potential that Hegel wants to express, universal, prior to the phenomenal, unaware.

The latter is a kind of movement, dynamic, in which the subject is aware that he has this ability and is able to use it. Rationality, for example, is the nature of man, which shows that rationality is the self-reliance of the adult.

But not for an infant, who has only the ability to reason, but not the consciousness to reason, a state without awakening. "Freedom" = potential; "for oneself" = realization

1. "Being in itself": that is, why I am me, because I exist as me, so I am me. In fact, this logic is very overbearing and unreasonable, indeed, it has a kind of what you see is what you get feeling. On this level, I am only concerned with me, and nothing else is hot and relevant. 2.

2. 'being-for-itself': why I am me, because I am not you, not him, not Zhang San, not Li Si, not anything other than me, so I deny all other things once, then I can only be this me.

And through what to deny I am not other things? Spiritual phenomenology tells us in the chapter on perception - traits. I'm not you because I'm 5'8" and you're 5'7", I'm not him because I have the quality of being white and you have the quality of being black. And so on.
『being-for-itself』=associating with oneself only

3. Being for him: Why I am I can exist because something else exists. Why is the past, because there is a present. Compared to the present, yesterday is the past. Without the present, the past cannot exist. To exist for him = to be associated with something else.

Of course, this is only one of Hegel's explanations of

self-existence as other. Self-existence can also be derived from: objective existence, where things are not aware of their existence (similar to natural emergence without external influence) Self-existence can be derived from: objective existence, where things are aware of their existence.

Objective existence, and the thing itself is aware of its own existence (similar to being aware of itself and taking the initiative to do so; feeling itself; being aware of itself. Self-awareness means internal self-discovery and external innovation of self-liberation. It is the essential law of all human practices, which is expressed in the inevitable maintenance and development of human self-existence)

Hegel, who reflected the world from sensibility to the real world reflected by human nature, and finally reached the absolute idea through reason.

Hegel's trinity. In the Lesser Logic, from being to essence to concept. Existence cannot be "freedom", it is latent existence. It has to be referred to by the norms of human nature and the sphere of reason.

That is to say, existence has to have an absolute concept to gradually reach "being" through human nature. That is, it cannot be "being-for-itself". "Being in itself" means that existence is within the realm of reason and human nature, and that it is a self-existence.

For example, "being" itself, in the Lesser Logic, is firstly interpreted from being. This existence is both, and nothing. There is no such thing as a pure existence, and anything that exists is not a mere existence. This meaningless existence can also be called non-existence. By having and not having, the three paragraphs are pushed to the next step. In short, it exists through its own connectedness.

Georg Wilhelm Friedrich Hegel was a German philosopher. Many people believe that Hegel's thought marked the culmination of the 19th century German idealist philosophical movement, and had a profound influence on later philosophical schools, such as existentialism and Marx's historical materialism.

What is more, since Hegel's political thought combines the essence of both liberalism and conservatism, his philosophy undoubtedly provides a new way out for those who feel that liberalism is facing challenges because of its inability to recognize the needs of individuals and to realize the basic values of human beings.

1. Temporality

The concept of "temporality" has been briefly mentioned before, the characteristic that things are effective, meaningful or useful only within a certain period of time. It is not easy to understand time in the eyes of philosophers. It has the meaning of intrinsic time, of a clock that is in the same place as the event in relativity theory, and can be measured in meta-time.

It is also the only "framework for interpreting a particular event and providing resources for the proper interpretation of that event", and is therefore a relative concept that can only be interpreted with respect to a focused event within a contextual framework, not independently of the time in the framework.

The framework refers to its constraint, but also to a shelf - to its support. It is a basic conceptual structure used to solve or deal with complex problems. The materialized time refers to the property that things are most valuable and productive within a certain period of time, and not valuable enough afterwards. For example, 15 minutes is used to measure the consumption of a task.

Metatime, on the other hand, is a process in which time becomes material time and is intrinsic. Subjective time - also known as internal time - is the duration and sequence of experiences or mental acts that are not communal and cannot be measured.

In phenomenology, consciousness is always about the consciousness of something, and every act of consciousness is directed toward something. We can find this in Augustine, Bergson, Brentano, Husserl, Heidegger, etc.

For example, when we listen to a piece of music, we first ask ourselves whether we are hearing a succession of sounds or a continuation of sounds.

A continuous sound means that we hear a section of sound and it is one after another. Continuous sound means that I hear something continuous, but I'm not sure if it's a sound. If you go with the former, you may be a solipsist in phenomenology.

In the preface to his Critique of Pure Reason (1781), the German philosopher Immanuel Kant (1724-1804) suggests that he thought that anyone who does not know the limits and nature of the cognitive faculty

He thinks that anyone who does not know the limits and nature of the cognitive faculty, that is, who does not yet think on the basis of observing the limits and nature of the cognitive faculty, and who can reasonably explain the causes of change, the connections between things, or understand the laws of their development. It is solipsism to jump to a definite conclusion.

Inspired by the English Empiricist philosopher D. Hume's (1711-1776) Scepticism, Kant believed that all human acts of "knowing" and "recognizing" a subject, in order to know it with certainty, and that these knowings have the potential capacity to be understood.

These knowledges have the potential to use content for specific purposes and must be based on experience in order to be valid. Traditional metaphysics relies only on rational thought, using pure concepts for the construction of knowledge, without examining how reason can access these concepts.

In other words, to affirm all kinds of truths in metaphysics without critiquing the power of reason makes metaphysics, which is originally called "the queen of all sciences," full of all kinds of unscientific and confusing arguments, that is, it becomes solipsism.

Therefore, Kant's view that all Rationalists are solipsists is what Kant calls "the awakening from the long dream of solipsism" in Prolegomena to any Future Metaphysics.

In terms of philosophical attitudes, solipsism is contrasted with skepticism, the former being absolutely certain of the connection between subject and object, while the latter is generally skeptical of knowledge, events, opinions, or beliefs, or of claims that are taken for granted. Doubt about the possibility of such a connection.

As far as the development of philosophy is concerned, there were two camps in the philosophical world at that time: European rationalism and English empiricism, with R. Descartes (1596~1650), B. Spinoza (1632~1677), G.W. Leibniz (1646~1716) and others representing the former, and J. Locke (1646~1716) representing the latter. J. Locke (1632~1704), G. Berkeley (1685~1753) and Hume.

Rationalism starts from universal skepticism but goes to arbitrariness, while empiricism first affirms experience as the basis of knowledge, but finally moves to total skepticism. Skepticism has its own significance in the development of philosophy.

The basic spirit of skepticism is to deny the recognition of knowledge, but because of this basic attitude, other ideas related to metaphysics, theory of knowledge, and ethics emerge.

Kant argues that we must avoid falling into solipsism, but at the same time not condone skepticism, because solipsism creates illusory metaphysics, while skepticism destroys all metaphysics. He stands for criticalism, which advocates a critical approach to the problem of knowledge

Presentationalism is a "reduction to the thing itself," which is essentially a sophisticated rethinking of established concepts. Just as the sky is blue is a solipsism, it should be me who thinks the sky is blue.

When we listen to a piece of music, what we hear is more of a continuation of the sound. So if you pass through a continuous time, you are actually passing through a continuous time. Continuous time is the result of the movement of time. The temporality of sound is a sense of internal time.

We use the piano melody to illustrate that on a piano keyboard, two adjacent keys (both white and black) form a chromatic; two keys separated by a single key form a whole tone.

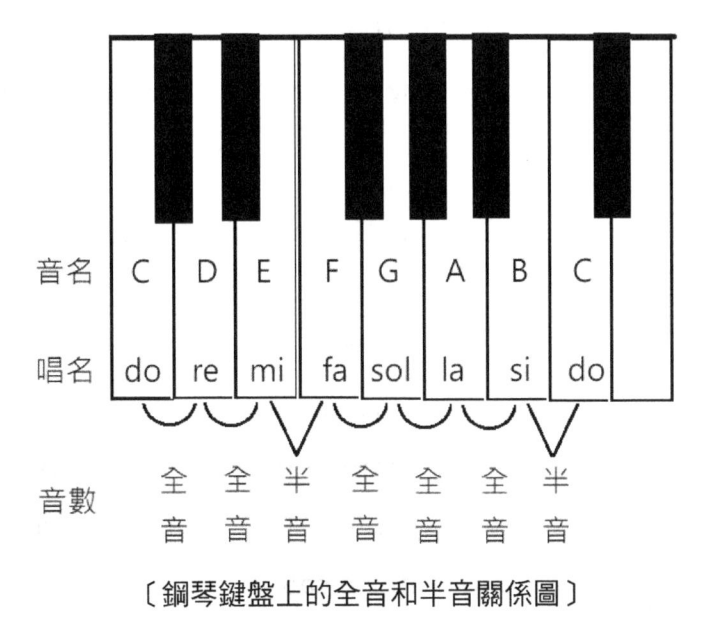

[鋼琴鍵盤上的全音和半音關係圖]

First of all, I played the whole tone of the CD key in P1.S, and the semitone between BC in P2.S. Why in our mind, there is only a do.re-si.do melody. If the CD (whole tone) disappears and only the BC (semitone) remains, then we can only hear the BC (semitone).

But in fact we hear the melody do.re-si.do. So the CD (whole tone) has obviously stopped and not played again, so where did the sound of the CD (whole tone) go, obviously it has not completely disappeared, still lingering in the cognitive ability of the environment and self and the clarity of cognition.

At this time the state of consciousness Husserl marked it as do.re-si.do', then this CD (whole tone)' and the original CD (whole tone) you heard there is no change, in fact there is.

It is lingering. But has it passed? No, in fact it's in your head in progress. The lag is still in the realm of perception along with the present moment of consciousness. The present consciousness receives the BC (semitone) while remaining open to the new present, because we know that there will be another tone next. In such an open anticipation, we call it "pre-taking".

Thus we find that the perception of consciousness is not just the present act of consciousness, but a threefold field of experiential liminality that combines [the present], [pre-taking] and [lingering]. That is, the perception of temporal consciousness is not simply a mechanical superposition of past, present, and future.

The present includes the present and the not-present: the present of [lag] and [foreshortening]. The [lag] and the [foreshortening] not only enter into the [present], but they also become the preconditions and conditions for the [present] to be established.

Therefore, Husserl's view of time essentially emphasizes that time consciousness is longitudinal, and longitudinal intentionality indicates the instantaneous interconnection of thoughts, emotions, memories, images, etc., that emerge in the brain, i.e., time consciousness.

The intention that emanates from itself, that points to itself, functions to make time consciousness fluctuate as a "unified" thought. When any subtle thought arises, there are four stages that are constituted: birth, dwelling, dissimilation, and extinction.

Husserl's time consciousness eventually follows the same path as Kant's, stating that time consciousness is the lowest level, the form that constitutes the cognitive ability and clarity of cognition of the environment and the self, and constitutes itself as a comprehensive awareness and recognition of one's physical and mental state and the changes of people, events and things in the environment by using mental activities such as sensation, perception, thinking and memory. The "occurrence" is a "primordial occurrence".

The "temporality" of Heidegger's "Being and Time

Heidegger: That which Husserl named time

consciousness is time in the sense of "primordial". For example, the recent mechanical view of time represented by Newton, in which time and space are objective existences. But the fatal flaw of this view of time is that we talk about the present, and the present is always in the past tense. As soon as we talk about the present, we have already passed away.

The past and the future do not exist. The present is a completely imaginary present. Secondly, the present is the center; the main part (in terms of the relationship between things) of existence. The future comes from the present, the present from the past.

In fact, this means that everything is already in the past, and everything exists only for a short time (birth, dwelling, dissimilation, and death) without being created. In other words, this view of time is a divine view of time. Only God can manipulate the past, present, and future to a certain extent, quantity, or degree, and this is Hegel's view of temporality.

Temporality is a way in which time blossoms, and temporality is what makes the "here and now" of senior existence, the proper perspective for understanding the meaning of existence, which Hedegger tells us here in advance is time.

Without a point of view, we cannot understand the object; but with a point of view, it limits the field of our understanding. It is the field of view that makes possible, and thus it is also the final field of view that makes possible the understanding of "this being" with respect to other beings, with respect to the existence of the cast, it is the initial understanding and preparation.

Heidegger says that man is the most unique of all kinds of beings, that man is the "blossoming being". It means that man is not just concerned with himself, not just clinging to himself; he can leave himself as "himself", and he also thinks that time also has the following four characteristics: 1. blossoming, 2. horizon, 3. temporalization of himself, 4. primordiality.

02. Time blossoms.

The future, the past, the present. (Corresponding to Husserl's foreshortening, lingering, and impression) The three modalities are intertwined and established with each other. The future: signifies the origin's movement towards itself. Heidegger sees this presence as the blossoming of time and speaks of three conditions of "this presence".

(a). The future = the future, the time that has not yet arrived. For example, past mistakes are future wisdom and success; a lazy teenager will be an old man in a ragged coat; success does not come in the future, but is accumulated from the moment of decision to do it; that is, the coming day is longer than the past year.

(b). Past = any moment or period of time before the present moment we are in; it can be a moment, but mostly refers to a period of time. The terms that we use everyday to represent past time already exist in ...: for example, yesterday, the day before yesterday, before, etc., are all branches of the past. , (b).

(c). Now = this time, referring to the time of speaking, sometimes including a long or short period of time before or after speaking, implying existence in ...: the present. (Distinguished from 'past' or 'future'): How is his current situation? | Now he is a factory manager

In essence, it is because of the relationship between these three.

(a). The future is close to us, so that we can have the ability to be in the future.

(b). We are thrown into the situation of being, thinking about the past. Thrown=past=was. But to be thrown is to be thrown into the future. So the past comes from the future. The future makes the past appear.

(c). I return to myself from the future and present the future in my own situation, I release the present from the future (energy) of the past (essence). <Being in time P387>

Time reveals itself as time by blossoming (revealing) itself, and we know what it is before it blossoms.

2. Sightedness, originally refers to the area within the field of vision, in the field of architecture or art, means the geographical area or regional characteristics that can be seen from a particular location. By this human in the process of understanding

The common characteristics of the things we perceive, from perceptual knowledge to rational knowledge, are extracted from the essential attributes. We are also justified in applying the concept of horizon, which Husserl apparently tries to capture, meaning all the limited intentionality, to the transition to the vital continuity of the whole.

A field of vision is not a rigid boundary, but a kind of

integrated awareness and recognition that flows with your personal use of sensation, perception, thought, memory, and other mental activities, of your state of mind and body, and of changes in people, events, and objects in your environment.

If the apparatus is in the state of being in the hands, then it is the absence of these two modes of departure and their corresponding visualization patterns, the future and the past, because as the apparatus in the hands, there is no history, it only exists temporarily as a functional role.

3. Temporality, temporality is the characteristic that things are valid, meaningful or useful only within a certain period of time, and it is the reason why time is established, when a certain action or thought is taken as a target, or as a trust or basis, often with an "unchanging" starting point, distinguishing the order of change, with physical existence, or explicitly rather than abstractly, in general terms, and thus with past, present and future.

The French philosopher Henri Bergson (1859-1941) introduced the concept of durée to illustrate temporality. He believed that the individual uses mental activities such as sensation, perception, thinking, and memory to become aware of and understand his or her own physical and mental state and the changes of people, events, and objects in the environment.

In other words, we are in a continuous process of awareness and integration of external things. It is only when one looks back, in the continuous flow of continuity, that one returns and notices the continuous moments of the past.

In this way, one can recall the "past" by going back, rethinking the past, and learning from it. M. Heidegger (1889~1976) continued the position of E. Husserl (1859~1938) in analyzing or criticizing the problems and issues based on the study of "inner time consciousness", and from the perspective of existentialism, human beings have been able to analyze natural and social phenomena according to the existing empirical knowledge, experience, facts, laws, and cognition.

The description of a logical inferential summary of one or many problems, events, studies, etc., in a single, or inductive exposition, by means of generalization and deductive reasoning, etc., as well as the hypotheses that have been tested and confirmed.

He also presents the problems and solutions to them

with his own understanding and narrative temporality. He believes that the meaning of "care" (care; Sorgen) is based on temporality.

It is because of temporality that "this being" (Dasein) can constantly show the instrumental world or "suspend" other "this being". If someone wants to go from one place to another, it shows the world from this place to the other place, possibly on foot or by car.

At this point all kinds of possibilities are revealed, based on the fact that the general concept of "past" includes knowledge and skill. It is the world of tools that has been shown to the mind after experiencing or observing a particular event or event and applying it to subsequent operations.

When one makes a judgment and assertion about something "in the present", it immediately reveals a world that points to the "future". Therefore, Heidegger thought that any "this being", when it shows its existence, is based on the "past"; in the "present" it shows a world that points to the "future".

The "past," "present," and "future" are in fact integral, and this integral is "temporality," an objective form of material existence, and the distinction between the continuous system of past, present, and future is based on temporality.

A. Schütz (1880~1953), on the other hand, has a systematic and organized law or argument about temporality. It is generalized from practical verification, or derived from conceptual derivation, and applied to sociology.

He continues Bergson's, Husserl's, and Heidegger's view by distinguishing between the motives for the generation of each social action: the motive of cause and the motive of purpose. The former is the reason why an action arises, and the latter is the purpose that an action aims to achieve. Usually, when an action arises, it has both of these motives, but the distinction is not made in detail.

For example, when a person uses an umbrella in the rain, he may do so because he "used to" catch a cold from the rain, or because he does not like the feeling of wet clothes, so he uses the umbrella to achieve the purpose of not getting wet.

The former is the cause motive and the latter is the

purpose motive. In other words, the cause motive is based on the "past" experience, and the purpose motive is directed to the "future", so an action is a combination of "past", "present" and "future".

From the above discussion, we can deduce that temporality is fundamental in the allocation of school hours. The need for motivation and the review of prior knowledge is to evoke past experiences and to help achieve learning outcomes.

This, in turn, is based on the continuity of knowledge, i.e., time. On the other hand, when students want to learn a new curriculum, they should connect with their past experiences instead of showing a gap.

This shows that students' learning activities are not limited to the time spent in the classroom, but that the past learning experiences and the knowledge content to be presented should take into account the temporality or consciousness of the learner as a whole.

It is always blossoming in different modes, and it blossoms as the unity of the blossoming itself. Only its own temporalization can allow the appropriate perspective to blossom for the understanding of meaningfulness, and

Heidegger tells us here in advance that it is time.

Without a point of view, we cannot understand the object; but with a point of view, it limits the expansion of the field of our understanding and activates itself.

The originality of this being, as the meaning of the existence of this being, temporality is the precondition for understanding the type of existence of any being that is not this being, i.e., in the practice of the blossoming field of vision - the figure.

The original meaning of "existence" is the actual existence. However, from a linguistic or grammatical point of view, especially with the infinitive "to be" (to be), there are different ways of describing "existence".

The main ones are: (1) existential, as in "these things will be ......" or "there is ......"; (2) predicative, as in "this is red" or "this is the table"; and (3) unpretentious. ; (3) veridical, such as "...... is true"; (4) constitutive, such as "this house is made of bricks and mortar"; and (5) representational, based on perceptions formed in the mind with a sympathetic The mind is formed with a sympathetic attitude and a kind heart to observe things.

(5) Presentational: The image of a sympathetic

attitude and kindness in observing things based on perceptions formed in the mind. It is easy to show emotions and values the harmony of interpersonal relationships.

In addition, the expression of existence sometimes refers to the primordial present, such as "he is tired", and sometimes it is timeless, such as "the square of two is four". In short, the meaning of "existence" can be described in terms of the basis, root, nature and state of existence of things.

Western philosophy, which has its roots in Greece, has always been concerned with understanding the question of "existence" as its highest mission, the basis of existence of all things, and the core issue of "existence". In the pre-Socratics period, however, the focus was on the cosmological problem of understanding the changes in the creation of the universe.

For example, Thales (624~546 B.C.) advocated water as the "beginning" of the universe, and later, thinkers of various schools of thought also proposed the infinite, air, or deity to represent the beginning of the universe.

It was not until Socrates opposed this purely

intellectual inquiry and shifted the center of philosophy from knowledge to action, from cosmic problems to life problems, that he began to concern himself with the problem of human existence.

Plato followed, taking up Socrates' view of the "universal concept", that things have a widely shared quality, not limited by time and space, and that there is a "good self" independent of the phenomenal world, called "Idea", which is an eternal "being".

In the Middle Ages, Christianity emerged as a study of universal, fundamental issues, including the fields of existence, knowledge, values, reason, mind, and language, emphasizing that God created by the "freedom of the will" and that the world is therefore a product of God's will.

This view of the existence of an intelligent being, which can be called "life-centered," is very different from the ancient Greek discipline of analytical thinking and reflection on the fundamental questions of life, knowledge, and values.

The Renaissance and the Reformation have led to a rise in humanism and to a modern period of holistic, fundamental, and critical inquiry into the realm of reality and the human person.

René Descartes (1596~1650), the pioneer of modern philosophy, broke away from the traditional will of God and re-examined the entirety of knowledge, using the method of doubt as the starting point of the search for knowledge.

Only the main "I" can be guided by certain values and purposes of life, and the ability to discover and solve problems under the premise of following certain social laws and natural laws.

The strict method prescribed by the "I" includes both objectively existing and observable things (such as people, trees, houses, abstract ones such as prices and freedom) and imaginary things (such as deified figures) to anything that can be perceived or imagined externally. In this way, the rationalism of Dualism of Substance is formed.

As for the objective existence reflected in the result of thinking activities in human consciousness, or the viewpoint formed, and how the subject of the conceptual system can reach the object "existence" with spatial extension, the concept of "God" is the starting point of the "existence" of objective things existing outside the subject.

We think that God, the infinite being, is the basis of the existence of "I", and that it is possible to reach a certain level of cognition between finite beings, mind and matter, and human beings through mental contact or direct communication.

Since Descartes established this way of analyzing, synthesizing, reasoning, and judging the subject-object opposition (the two contradictory and unified properties of things), the philosophers who followed him have been able to analyze the subject-object opposition.

The philosophers who followed him sought, through different propositions, to use the mental activities of sensation, perception, thought, and memory of a unique individual to perceive and understand the state of one's body and mind and the changes of people, events, and things in the environment.

The essence of existence is grasped by a being with a unique personal experience, or by the cognition of another entity external to and related to itself.

By F.W. Nietzsche, this capacity, role, personal view, and status of man in the process of practice, i.e., the status and character of his autonomous, active, dynamic, free, and purposeful activity, reached its peak.

The foundation reaches its peak; in its holistic, foundational, and critical inquiry into the real world and man, a being with a unique consciousness, or with a unique personal experience, or another entity external to itself and related to it, as the unbedingter Wille zur Macht; will to power power)

That is, a will that issues commands according to itself and, through itself, constantly increases its power, even to the point of persecuting or oppressing other existents, making them compromise to external pressure and give up their resistance to its will.

The objective existence of subjectivity is reflected in the individual's use of sensation, perception, thinking, memory, and other mental activities, the comprehensive awareness and understanding of his or her own physical and mental state and the changes of people, events, and things in the environment, the result of thinking activities, or the formation of views and conceptual systems that have developed to such an extent that people doubt the possibility of mastering the truth of existence.

Martin Heidegger (1889~1976) criticized the philosophers for their analysis, synthesis, reasoning, judgment, and other thinking activities about existence, saying that these statements have deviated from the original meaning of existence.

This is because if one determines "existence" by the laws prescribed by the subject, one has to rely only on oneself to determine truth and "existence".

To the extreme, man will expand his own power to predict and dominate nature, and even to rule others, resulting in the separation and even opposition between man and nature, and between man and man, two things that originally belong to each other naturally or harmoniously.

Heidegger therefore wanted to follow the phenomenological approach advocated by E. Husserl (1859~1938), which does not presuppose any position on which to study and analyze or criticize problems and issues.

Rather, it goes directly back to "existence" itself to see how it is apparently represented and to describe what it shows in a factual way. It refers to statements about the properties of things, mainly by means of symbols or systems of symbols (words, numbers, pictures, etc.) that express what is generally observed in as detailed, organized, and systematic a manner as possible.

The so-called observation refers to the act of seeing, listening, and other perceptions, which means analyzing and thinking, i.e., observation is not only a visual process, but a comprehensive perception that is dominated by vision and integrates other senses.

And observation includes active analysis, synthesis, judgment, reasoning and other cognitive activities on the basis of the image, concept.

Therefore, it is called the higher form of perception. It is the process of examining the phenomena and movements of people, things, and objects that are touched by all human activities; it is the process of examining or investigating

Generally speaking, the description of the characteristics of something is described in the direct perceptive experience (perceptive experience) to present, but the description should still be from the surface of things.

But the description should still go from the surface state and phenomenon to the essence of the thing, its cause, and even the basis of its existence.

Therefore, Heidegger looked into the original meaning of existence from Etymology and found the following meaning of "existence" as physis and logos in the ancient Greek texts.

Physis is the study of the nature and properties of matter and energy in natural science. Since matter and energy are the fundamental elements of all scientific research, physics is one of the most fundamental disciplines of natural science.

Physics is an experimental science in which physicists observe and analyze various phenomena of nature based on matter and energy to find patterns in them, not in the external objective world of nature as commonly referred to, but as a continuous emergence from itself.

(The logos, on the other hand, is an enduring and continuous one, which refers to the relationship between people or things that occurs between the two sides of the internal contradiction and the relationship between things, and which enables the existent to stand out.

Thus, it can be said that "existence", in order to reveal itself, breaks through its concealment by something else, and makes the original system, rules, qualifications, rights, etc. lose their effect of unnecessary binding, restriction, etc., and allows people to expand their field of activity freely.

Through this, the various existents are combined into a system of intrinsically related parts that act or think as a goal.

Each component must have some kind of interactions, interconnectedness, or complementary functions, or common interests, or coordinated actions, etc.

Simply put, it is an organized objective existence in the natural world, all objects or facts that can be observed and observed.

It is usually used for more specific things. That is, the existence of a person is not an objective form of material existence.

The existence of a person is not an objective form of material existence, a spatial existence expressed by length, width, and height, alongside the existence of other existents, but the existence of "Sein-in-der-Welt" (Sein-in-der-Welt), which is a common characteristic shared by individual objects.

In other words, man and the things in the world are originally one, and man continues to preserve himself by continuously revealing the world and using the things in the world, so it is a dynamic expression of the concept of existence of "birth, dwelling, and death".

Everything in the world is changing all the time, and the only thing that remains unchanged is change, and things change from time to time, all along the trend of "formation and maturity, stable development, decay and mutation, destruction and disappearance".

Whether it is the change of specific materials such as mountains and rivers, or the change of thoughts, feelings, or social organizations, all of them cannot escape from this trend of change, only the length of time varies. Things are changing every moment, quantitative changes will eventually occur qualitative changes, the core of things qualitative changes, it will be completely alienated disappeared.

This trend of change in things should be an important

part of our worldview. In other words, no matter what we face, we are unconsciously incorporating this viewpoint and applying it.

It is worth reflecting that we can all accept the position on which this study analyzes or criticizes problems and issues, but our cognitive ability and clarity of awareness of our environment and self do not allow us to put this perspective into practice.

Heidegger's existentialism does not place the subject of cognition in the activity of apparent thought exploration, but uses the active and creative mode of action generated by the power of thought in the process of thinking, and regards the subject as an ever-changing "being" (person) in space and time.

In this way, it is recognized that all human beings' logical inferential summaries of natural and social phenomena obtained by the subject's thinking, in accordance with existing empirical knowledge, experience, facts, laws, cognition and tested hypotheses, through generalization and deductive reasoning, have an element of uncertainty of existence.

The search for "reality" becomes an approximate

process of approaching the truth of "existence". Otherwise, it is very easy to treat "existence" as "existent" and to distinguish people or things in the outside world through social division of labor.

In other words, something is treated as a material (labor) object, something that can be controlled, decomposed, manipulated, changed, transformed, exchanged, consumed, produced.... The real meaning of existence is hidden.

The above abstract, universal idea of "existence" serves as a developmental context that specifies the domain or class of entities, events or relationships.

The study of this kind of existence is called "existentialism" or "Ontology" and can be divided into three parts: the first part explores the basis of being a being and the essence of existence; the second part explores the properties of the essence of existence; and the third part discusses the domain of being and the various departments it represents, such as the inquiry into entities and accident. The third part discusses the domain of the existent, and the various departments represented by it, such as the inquiry into entities and accident.

The third part discusses the various aspects of being, such as the question of entity and accident. By exploring "being", we can understand the existence of the self, the meaning and purpose of being itself, and the relationship between human beings and other beings.

Since the emergence of Heidegger's existentialism, which reevaluated traditional "existentialism," the development of existentialism has had a great impact on the indirect or invisible way of acting or changing the behavior, thought, or nature of [people or things], making Existentialism one of the major philosophical trends of the twentieth century.

The greatest influence on education is that the people, things, and objects touched by all human activities should be treated as a specific and unique being, so that education should be aimed at the whole person.

Learning itself is not an observer but an intervening process, and the motivation, ability, or characteristic of acting as a subject in an activity according to one's own will is a topic covered by many disciplines such as philosophy, political science, ethics, and jurisprudence, with different connotations given to it by different fields.

The "subject of action" includes: biological individuals, groups, organizations, etc.; "acting according to one's own will" includes: free expression of will, making independent decisions, pushing forward the course of action on one's own, making practical decisions, using things around oneself, etc.

By using mental activities such as feeling, perception, thinking, and memory, the individual perceives and understands his or her own state of mind and body and the changes of people, events, and things in the environment.

The act of obtaining true "cognition" and "recognition" of a subject, in order to be sure of the knowledge, and the potential ability to use this knowledge for specific purposes.

It means to be able to understand and grasp (a situation) through observation or experience, through experience or association, and then to be able to explain known facts and principles and principles in one's own words, texts or other symbols, in order to enrich one's own existence.

In conclusion, from the previous discussion, we trace

from Husserl the cognitive ability of man to his environment and self, as well as the clarity of cognitive temporality, and through analytical research seek to understand the temporality that inscribes Heidegger, as an abstract, universal idea, and at the same time acts as a specification of things in which

A necessary essence, without which a thing cannot be established, a distinguishable domain or class of events or relations, and a certain kind of thing that exists independently within itself. We also see the lessons that Hedegger and Husserl can learn from each other, so that we can make up for our shortcomings by taking what we have.

(1) Transcendence itself has three drawbacks.

In the introduction at the beginning of this paper, the question is posed: What is the state of interaction and interconnection between the reality of human concrete existence and the primordial things of phenomenal existence [self-existence]? That is, what is the state of a certain nature of association between man and man or man and things?

Schartre rejects actualism [the view that ontological

reality is independent of human senses, beliefs, concepts and ideas. Unlike some idealists who believe that reality originates from or is influenced by consciousness. The reality of the present world is not the same as the reality of ancient Greek or medieval philosophy, and sometimes it is even the opposite] and idealism [the view that spirit is the driving force behind the formation of the universe and exists before matter, and that all objective things do not depend on human existence, but have form, color, sound, and taste, and are concrete and sensible. It has form, color, sound, and taste, and is concrete and perceptible. It is all constituted by the action of the spirit. The view of things or issues from a certain standpoint or perspective.

As we have already said, "transcendental freedom" cannot be applied to the individual's comprehensive awareness and understanding of his or her own physical and mental state and the changes of people, things and objects in the environment by using mental activities such as sensation, perception, thinking and memory, and at the same time, one's cognitive ability and clarity of cognition of the environment and the self cannot be achieved by looking at things from the perspective of the subject's own needs. tendency.

It is a component of the viewpoint, experience, consciousness, spirit, feelings, desires or beliefs that an individual can have to think or judge the reasonableness of something from a different point of view or perspective.

It is not something that can be distinguished from other people, things, and objects that are touched by all human activities and exist independently within themselves. But it is not influenced by subjective thought or consciousness, but exists independently to "construct" something beyond.

Therefore, the human mind's reflection of the objective material world is the sum of sensations, thoughts, and other mental processes and the original relationship of existence, and cannot be the external relationship that unifies two objects that actually exist in isolation or have actual content as a whole.

These problems are also described in the introduction, which describes the main idea or content of the book, and in the theoretical descriptions of the first and second volumes, where a single or many problems, events, and studies are described, or summarized.

The question of the "existence" of these issues, or a

realistic and understandable approach to them, has been described and repeatedly emphasized. The state of interaction and interconnection between the Self and the Self-being is not a constituent part of the Self-being, but of the Self-being.

Man's ability to make himself nothing is in fact a lack of cognition, a hope of knowledge, a possibility of the future, and this is the freedom to which Sartre refers.

Therefore, Sartre calls this movement of things or phenomena that are mutually conditional and inseparable from each other, that may constantly exceed itself, and that in the impractical and inscrutable, can never be obtained, or cease to be what it is, transcendence.

To put it in terms that do not involve concrete concepts, but only a framework with definite abstract concepts, but without definite figurative concepts or in common language is: "Man is always still unaware of certain limitations and is in a state of confused perception, plunging ahead of himself, whether in space or time. For example, who I am going to be, or where I am going to go.

It is said that "man accepts life without a choice, then

spends it in a condition of no choice, and finally returns it with an irresistible struggle.

And the meaning of the true reality is the visible transcendence. Transcendence in this context. In fact, it is divided into human transcendence and symbolic transcendence. Human transcendence is actually the knowledge of human nature.

Analytical psychology is an in-depth psychology developed by Jung, also known as Jungian psychology or archetypal psychology.

It is the school that Jung began to form after he left Freud's psychoanalytic school in October 1913. In early translations, before the Psychoanalytic Association officially translated it as psychoanalysis, it was often confused and turned into analytical psychology, but later, in newer editions, psychoanalysis or analytical psychology was used to distinguish it from psychoanalysis.

Jung believed that the human mind consists of the conscious ego and the unconscious, and in a seminar, Professor W. L. Song pointed out that Freud's subunconscious - "subconscious" - must be distinguished from Jung's unconscious - "unconscious".

There are two major parts. The conscious ego has a memory and continuity. But Jung believed that the ego is only a small part of the whole mind, and that the unconscious is more influential, and that within the unconscious ego there is the highest spiritual self. The dream is the window that the unconscious opens to the ego.

The integration of one's unconscious nature occurs naturally and the development of one's personality is called individuation.

Jung believed that there are strengths and weaknesses in mental functions, for example, when the thinking function is at the top, the emotions are relatively weak. Jung's view of the personality types is that the dominant and inferior functions do not represent which is good or bad, but rather how high or low the psychological functions stand in the self.

Jung believed that although each of the eight personality types tends to manifest only one, in fact each of the eight personality types has some potential to be experienced and developed, and that if the dominant function dominates the ego for too long, it may lead to unconscious backlash and even relationship problems in real life.

The reason why Jungianism is recognized as a discipline of depth psychology is that Jungian and post-Jungian developments agree that it is necessary for people to recognize different inner personalities in order to avoid the rejection or suppression of certain personalities, which may lead to one of them becoming the dark side of oneself - the shadow of the psyche - and lead to the outbreak of psychosomatic and psychotic disorders.

If a person's conscious self and unconscious contradict each other and cannot be integrated, psychotic disorders, psychosomatic disorders, such as phobia, fetishism, or depression, may arise.

The unconscious must be analyzed in order to understand and recognize the unknown characteristics of the unconscious, such as by analyzing dreams or reactions to artwork or poetry.

Jung believed that the unconscious is divided into the Personal unconscious and the Collective unconscious. The personal unconscious consists of individual complexes, while the collective unconscious consists of the unconscious treasures accumulated by the human race as a whole over the centuries.

The evidence for this is the common symbols of humanity that Junger has collected around the world. He examined the religions, myths, legends, fairy tales, and fables of different peoples to find the archetypes common to all humans.

Thus, in individualization, it is possible to find symbols that are beyond the individual's experience. The content may be seen as fundamental human issues such as life, death, meaning, happiness, and fear, and may also have a spiritual dimension.

We can cite Jung's transcendence, which may place more emphasis on overcoming a hill. But in fact, Schart's transcendence may be more metaphysical, in the sense that man is everywhere transcending in order to become what he is not.

Transcendence itself is a characteristic of consciousness-perception. It is the comprehensive awareness and rational perception of one's own state of mind and body and the changes of people, events and things in the environment by using mental activities such as sensation, perception, thinking and memory, which gives one the transcendence of existence.

The second is the transcendence of symbols. This is actually related to consciousness. We cannot admit that the symbols themselves are self-existent, nor can we admit the existence of a mechanical world of regular symbols.

It refers to a symbol with a specific meaning, which can trigger a conceptual association with a specific object, as in Sothir's Noumenon. In his "Course in General Linguistics", Sothor popularized the concepts of "energy" and "reference", defining "energy" as an expressive type of symbol, a concept that corresponds to reference.

What is referred to is necessarily related. But it is clear that Sothir is a linguist, and that he and philosophers have been able to make logical, supportive, or relevant statements about natural and social phenomena in accordance with existing empirical knowledge, experience, facts, laws, and cognition, as well as through generalization and deductive reasoning.

The disagreement between him and the philosophers on the generalized conclusions of the complex psychological process of seeking or establishing rules and evidence to logically support or determine a belief, decision, or action began in Husserl's time.

The transcendence of symbols is, above all, a kind of "suspension" of lack, which awaits a peek into the cognitive capacity of the person to recognize the environment and the self, as well as the clarity of cognition. This is how words operate, as transcendent beings.

This is also stated in Schart's Word (a word is the smallest unit of language that can be used independently, while a phrase is a whole composed of several words). The term "theory" of the transcendental nature of symbols has many different meanings due to the difference in the scope of application.

For example, the speculation of things, the hypothesis of events, the principle of things, the empirical general principle of things, or the rigorous law of things, etc.

Regardless of the meaning of the intention to achieve a certain purpose, theories refer to the ability of people to analyze, synthesize, reason, judge, or imagine the formation of images, perceptions, and concepts based on some similar events or phenomena, which are not formed through visual, auditory, or other senses, but by the conceptualization of perceptual activities.

Thus, a theory may be a tentative explanation of something seemingly related but confusing, attempting to explain a particular phenomenon, past or present, on a set of plausible grounds.

A theory, too, may be a set of predictions for the process of human cognition, from perceptual to rational knowledge, abstracting and generalizing the common intrinsic features of what is perceived, an expressive framework for the ego's cognitive consciousness, explaining the more likely evolution of certain things in the future, or those things that are expected to happen.

In fact, this use of theory as a tool for reasoned explanation and prediction is the paradigmatic use of theory. Whether it is an attempt to explain something in the past, in the present, or to predict the future, as an attempt to deduce conclusions from known or assumed premises, or to reverse the reasoning from known answers, a theory must be based on a set of knowledge or beliefs.

Among the "sciences," it is especially obvious to take a

certain object as the scope of study and to seek unified and exact objective laws and truths based on experiments and logical reasoning, so much so that some people claim that the "correct" meaning of a theory is a "scientific" theory.

A theory is a logical way of thinking about something that is happening in the world, based on experimental and logical reasoning, in order to find a unified and definite objective law and truth.

It is an explanation that derives its conclusion from known or assumed premises, or from the results of a known answer to its justification; and it becomes a theory only when it is shown to be the only account of facts (empirically observed and experimentally proven), and not merely a hypothetical belief.

Science is a systematic body of knowledge that accumulates, organizes, and tests explanations and predictions about the universe. With a certain object as the scope of study, based on experiments and logical reasoning, the objective laws and truths are unified and confirmed, and the specificity and falsifiability of the prediction results are emphasized.

It is different from the vague philosophy. Theory provides a unified and non-contradictory description of facts, and explains them in terms of necessary laws and cause-and-effect relationships, so that it can not only explain what is happening now, but also predict future events.

However, there is a distinction between explanatory theory and practical theory. Theories are basically [descriptive] and [explanatory], like I. Newton's theorem of gravity, in which a certain object is taken as the scope of study, and experiments and logical reasoning are used to find unified and exact objective laws and truths.

But there are also practical theories, such as educational theories, which basically have a prescriptive or recommendatory function, i.e., they guide educational practice. An explanatory theory tells us what something is, while a practical theory tells us what to do.

It is worth noting that theory is often meant to be opposed to practice: theory refers to pure knowledge and observation, while practice refers to any action other than knowledge. Thus there is often an inconsistency between knowledge and object, i.e., the actor has a cognitive disagreement about the content of the person or thing to which all human activity is directed, or other relevant circumstances.

The actor has a cognitive bias in all human activities, the content of the person or thing to which the actor refers, or other relevant circumstances, and the meaning is inconsistent with the inner will. It is thought that "theory is theory and practice is practice". However, theory and practice are interrelated; practice cannot be based on theory, and theory cannot be connoted without practice.

On the one hand, theories form principles and standards to judge the success or achievement of goals in practice; on the other hand, if a particular generalization from practical verification, or deduction from concepts, does not fit closely with the facts in practice, it is because it is a "bad" theory.

That is because it is a "bad" or inappropriate theory. When a systematic, organized law or argument is further developed as a result of improvement, it does not mean that the system of information organized in "concepts" that exists in people's brains is completely wrong.

It only means that it is not sufficient to explain more new facts. The transcendental theory of symbols, also known as doctrine or doctrinal theory, refers to the theory that human beings, in accordance with existing empirical knowledge, experience, facts, laws, cognition, and verified hypotheses about natural and social phenomena, are able to explain them by means of theories and theories.

The theory of transcendental symbols, also known as doctrine or doctrinal theory, refers to the logical conclusion of natural and social phenomena according to existing empirical knowledge, experience, facts, laws, cognition, and tested hypotheses, through generalization and deductive reasoning. What is not explicitly stated in this statement, but indirectly revealed, is in fact transcendence.

The world of imagination produced by literature is also transcendent, and the atmosphere produced by literary works is also transcendent in the sense of sadness, sublimity, and humor. Secondly, there are some identifiers that people have in social situations.

For example, when we think of a teacher, we think of a refined, gentle and easy-going temperament, which is a kind of transcendence. In an epistemological sense, what is sold in a consumerist society is not material goods, but a transcendent material good that is sold in Japan.

Therefore, in the boundary between the visible and invisible, our consciousness is searching for this transcendence again.

Is transcendence good or bad? It is not possible to analyze "gentleman" as a symbol of transcendence from a methodological point of view, from a disciplined and systematic analysis of written or oral discourse.

For example, Tao Yuanming "not to bend for five buckets of rice" means: I can't stoop to bribe these small people for the sake of a meal. So Tao Yuanming hung up his seal and went away, leaving a perfect image to the world.

This is the best example of transcendence. From ancient times to modern times, transcendence is an exquisite survival instinct. This reading of transcendence itself has three conceptual drawbacks.

1. Transcendence is prone to criticism.

Transcendence (transcendence) conveyed in philosophy is derived from the basic concept of the literal meaning of the word (Latin), i.e., to cross over. However, it has different meanings in different historical or cultural stages.

It is easy to resist and fight for the other side to agree with one's demands because of the different positions based on which each side studies and analyzes or criticizes problems and issues, and causes large-scale attacks or denunciations.

For example: Religious philosophy, the Trinity (Latin: Trinitas, also translated as the Holy Trinity), Christian theological term, is the theological theory of the Christian religion God YHWH (Protestantism often translated as God or Jehovah, Catholicism often translated as God Yahweh), established in the First Nicene Council of the Nicene Creed, is the basic creed of the three major Christian denominations.

The doctrine of the Trinity asserts that the three distinct persons of the Father, the Son, and the Holy Spirit (the Holy Spirit in the Catholic Church, and the Holy Spirit in the Orthodox Church and the Protestant Church) are one essence, one nature, one nature, and one God, and that they express their relationship by means of homoousios (Greek: homoeopathy).

This includes the use of the "name" in the singular in scripture, where the proponents of the Trinity refer to the Father, Son, and Holy Spirit as sharing the name of God.

Some scriptures are considered to imply the idea of the Trinity, which the Fathers and the Apostles argued and preached.

Historically, some Christian denominations and individuals have rejected this doctrine. The mainline Christian churches accept the Trinity as one of the traditional doctrines, while the opposing side supports "Monophysitism", "Formalism", "Trinity", etc. and rejects the Trinity.

This naturally includes various philosophical ideas and systems, as well as the manifestation and verification of knowledge about existence, and does not involve whether or not this is an ontology, describing the transcendental nature of the fundamental structure of existence.

2. Transcendence becomes metaphysics.

It is not particularly theoretical, or not well investigated. It describes the relationship between two things, such as God and the world, animals and plants, the cognizer and the cognized, etc., where one exceeds the other, or is external to the other. It also implies that there is a discontinuity, or a break, between the two things.

Poetry is an artistic work that expresses beauty and emotion in a refined and rhythmic language, but can one achieve the goal or the result of interpreting poetry directly through its original words? Of course, this also gives some room for the humanities to survive.

3. transcendence is not easy to interpret.

This means that people who are ignorant and clumsy, and who do not respond deftly, will be very tired. For example, the film, released in France on December 30, 2015, depicts how the brazen Kikli brags about his fraudulent behavior and says he has not given up his criminal career.

"Kikli is an irresistible temptation. It's a job for him, not a scam," said director Pascal Elbe in a recent interview in Paris.

The French film "Je Compte sur Vous" (Thank You for Calling) brings the story of the con man to the big screen, featuring Gilbert Chikli, a Frenchman of Israeli descent, who poses as a company CEO or French intelligence officer and uses the phone to scam his way in. He pretends to be the CEO of a company or a French intelligence officer and uses the telephone to commit fraud.

With a simple background check and a phone call, Chikli was able to coax or threaten people to transfer thousands to millions of euros immediately to bank accounts in Dubai, Russia, China, and Hong Kong.

The victims put down the phone and still believed they had just spoken to their boss or a real French intelligence agent, one victim said the man had a magnetic voice and confidence, and two other victims handed over cash to Kikli in a Paris bar.

When the trick or charm didn't work, Kikli turned to threats, and another victim was threatened with deportation if he didn't immediately transfer nearly $2 million to the United Kingdom.

"He would adapt to his interlocutors, and it was crazy," Alber said. He improvises, he's quick, he'll say what you want to hear," said Erb.

In the March 6, 2021 film "The Queen of Fraud," the female protagonist of the film specializes in targeting elderly people as fraudulent targets, conspiring with doctors to falsify their medical records so that they can be found incapacitated by the court, then putting them in a care facility and stealing their assets in silence.

The smart, beautiful blonde woman tells the old man she is his legal guardian upon arrival, and that's where the scam begins. She arranges for the elderly to be admitted to a care facility and then takes away their cell phones so they can't ask for help.

It is her as the manifestation of knowledge about existence and the verification of fraudulent means that transcend humanity, ready to silently steal the elderly's property in their name. The long-photograph set to kill! "The Queen of Frauds" is based on a real-life fraud case

The Oscar-nominated Rosa Montpelier plays the Queen of Fraud, who conspires with a doctor to falsify medical records so that the court will find the target incapacitated and commits the crime without a word.

The woman, who lives alone and has money, is not as innocent as she seems and puts the queen of fraud at risk of death. It's like a care fraud that happens around you and me, adapted for the big screen.

In Baudelaire's essay, there is an example of how a donkey can understand the jokes that others play on it. In Nietzsche, too, Nietzsche hysterically criticizes the idol, the twilight of the idol, the optical role in the will to power.

In fact, these are all criticisms of transcendence. So philosophers since Nietzsche have basically been criticizing the transcendence of human nature. Sartre, in existentialism, in fact, gives a larger summary of transcendence.

He is what he is, and he is not what he is. In fact, he is saying that transcendence itself is immoral. This is even more evident in his essay "Words".

1. Recognition as a Type of Self-activity and Self-reliance Relationship Recognition as a Type of Self-activity and Self-reliance Relationship What kind of awareness is awareness. Nassatt borrows from Husserl's intuition.

Consciousness is in the face - in the presence. Then he talks about the character of awareness as belonging to and "for oneself". "For oneself" is such a kind of being ----", "for oneself" is a pure negation.

Perception, intuition, presence, negation, "for oneself" negation is the realization of {1. external negation: the cup is not green (external unconstructed), 2. internal negation: I am not beautiful (belonging to "for oneself" and revealing Freedom at the same time) } negation as a rule that establishes "being in itself 』.

"What is real is the process of realization. We call this negation, which reveals the inner and realized Freedom in the stipulation of being in itself, transcendence."

(2). Negation - Self-action - Revealing Freedom - Transcendence

01. As the stipulation of negation

"for oneself" is given to Freedom in the presence of Freedom, and Freedom is given in "for oneself". Freedom is "this", and "for oneself" makes it exist. "Recognition cannot be reduced to existence.

The "this" is diminished in the undifferentiated whole of existence. The partial negation itself dissolves in the fundamental negation, in order to give way to a new structure of negation, a new "this".

02. Quality and quantity, potentiality, instrumentality

Sartre examines "this" separately, and among the different stipulations in these examinations is the revelation of "this" by "for oneself".

Quality: The existence of "this" when it is examined outside of its external relations.

Quality is "this", but existence is not quality, and quality is revealed as all existence in the limit of "being".

But quality is only the absolute contingent presence of existence. Quality appears only in the revelation, pointing to the pattern of existence that we deny. Quality is 'being in itself'. We used to think of quality as a law of quantitative change.

But in the Sartrean context quality is not a superposition of quantities. For example, lemons are lemons in themselves, not the yellow color of lemons, and the yellow color of lemons is not the competent way to grasp lemons.

The yellow color of the lemon is not the competent way to grasp the lemon, because we have already emphasized that from knowledge cannot be reduced to existence. Lemon is permeated with its various qualities, and the various qualities are permeated with each other.

For example, in the case of dim sum, while we eat the color of the exterior of the dim sum, we also eat the flavor of the dim sum. I put my fingers into the cold jam, and at the same time the jam revealed the sweetness of the jam.

Quality is "this". If we have to prove what it is by

epistemology or dualism, then quality is the combination of flux.

Quality = quality, which is well understood by transcendence. Secondly, the most important thing in French surreal poetry is the sense of flux, and here we can see the thought process of the French literati from generation to generation.

Volume: purely external, such as the three men. The three maintain the unchanging and concrete unity of the team. The quantity is Freedom.

By the inner negation of "for oneself" each "isolated self" obtains a negative ideal connection. Even though quantity is Freedom, it can only come into existence by "for oneself".

Potentiality: 1. The constancy of "this" in Self-activity and Self-reliance, waiting for "for oneself" to turn it into the essence. 2. The essence of "this" is manifested in the form of a call to the essence.

The "for oneself" is the negation of the essence, and the "pure essence/existence" is never to be found. (Green is never green.) The negation of "for oneself" can make green even greener, and the essence as never given and always entangled with meaning enters existence from the future essence.

1 Constancy for a "this" is an example of potential beauty, sublimity, and desolation in accordance with its nature, as I have already mentioned at the beginning of transcendence. This is the best way to understand the potential of self.

Instrumentality: The transcendental nature of "for oneself" determines that "for oneself" cannot be known in a contemplative way. To know "for oneself", one must detach and entangle. The relationship between the object/self waiting for the self to use it, the instrumentality.

A critique of natural scientists: a thing is not first a thing and then a tool; nor is it first a tool and then a thing. It is the thing-instrument. It is impossible for a scientist to seek a purely extrinsic relationship.

(This brings us to Hume's normative and factual knowledge.) Sartre's notion of instrumentality lies in the fact that, since the splitting of objecthood from person

Since it is epistemically impossible, the objecthood of the world is established at the level of instrumentality.

Again, actuality is not the actuality of things, but the method of objects.

(3). The Phenomenology of Three-Dimensional Time

1. The time of the world

This section begins with what scientists see as time, and how such time is generated: the time of the world is similar to the instrumental: mundane time is quantitative and instrumental.

Past: created through a homogeneous moment and reconnected by a purely external relation.

Now: an instant that cannot grasp world time, a mere slip/nothingness. The present does not exist, the movement specifies universal time as the pure present.

Future: The future of world time is the abstract framework of a hierarchical series of small futures, a series of small spatial containers arranged.

The cohesion of Sartre's world time is a pure illusion, an objective reflection of the "for oneself"'s out-of-body (blooming) planning of the "self" itself, an objective reflection of the cohesion of human reality in motion.

However, if time is examined according to time itself,

there is no reason for this cohesion to exist, and it immediately disappears to that which is examined separately,

It would immediately disappear into the absolute diversity of transient moments that are examined separately, and lose any temporal nature and are simply restored to the non-temporal totality of "this".

Therefore, the time of the world is pure Freedom's nothingness, and to have an existence seems to be only by "for oneself" in which to exceed it in order to use its activity itself.

Our initial understanding of objective time is real: it is precisely because I am beyond the possibilities of my common present existence that I find objective time to be the nothingness of the world associated with the possibility of separating me from myself. ...... Time is revealed as the flash of nothingness above the surface of a strictly non-temporal existence.

Phenomenology (English: phenomenology, from the Greek phainómenon, meaning "that which appears", and lógos, meaning "study") is one of the most important philosophical schools of the 20th century, formally founded by the German philosopher Georg Husserl.

Husserl was influenced by Franz Brentano and Bernard Bolzano, who believed that every perceptual image formed on the basis of perception is a form of perceptual knowing. It is a perceptual image of something formed in the mind on the basis of perception, and consciousness is always a consciousness related to something.

At the same time, it also advocates "truth itself"-that is, the rationality of thinking or judging something from different points of view or perspectives, universally, without the influence of subjective thought or consciousness.

It is the study of the nature of consciousness, or the description of the fundamentals and laws of a priori, absolute knowledge, which is called "phenomenology".

Phenomenology is the philosophical study of the structure of experience and the structure of consciousness. As a study of universal, fundamental questions, including the fields of being, knowledge, values, reason, mind, and language, it differs from other disciplines in that philosophy has a unique way of thinking.

Phenomenology was founded in the early twentieth century by Edmond Husserl and later developed by a group of his followers at the University of Attingen and the University of Munich in Germany.

Since then, phenomenology has spread to France, the United States, and elsewhere, and has gone far beyond the "interpretive framework of Husserl's earlier work, which revolves around a particular event and provides resources for its proper interpretation," and is therefore a relative concept that can only be interpreted in relation to a focused event within a contextual framework, and not independently of it.

Phenomenology should not be seen as a unified movement, but more as a collection of different authors with common family resemblances, but also significant differences.

Therefore: a single, final definition of phenomenology is dangerous and may even be as contradictory as the lack of thematic focus. In fact, it is neither a doctrine nor a philosophical school.

Rather, it is a style of thought, a method, an open,

ever-new experience that leads to different conclusions and leaves those who want to define the meaning of phenomenology at a loss.

According to Husserl's conception, phenomenology is first and foremost a structure of integrated awareness and knowledge of the individual's mental and physical state and the changes of people, events and objects in the environment, using sensory, perceptual, thinking and memory activities.

It is also the systematic reflection and study of phenomena that appear in various consciousnesses and behaviors. Phenomenology can be clearly distinguished from Cartesian analysis, which treats the world as various objects, collections of objects, and objects in continuous action and reaction.

02. Overview

In its most basic form, phenomenology attempts to create the conditions for what is often seen as a subjective, objective study of topics, including: the individual's use of sensory, perceptual, thinking, memory, and other mental activities

It includes the individual's ability to use mental

activities such as sensation, perception, reflection, and memory to perceive and recognize changes in one's physical and mental state and in people, events, and objects in the environment, as well as the experiential content of judgment, perception, and emotion, and one's cognitive ability and clarity of awareness of the environment and self.

Although the exploration of phenomenology is based on experimental and logical reasoning in order to find the unified and exact objective laws and truths, it is not intended to be a departure from clinical psychology.

However, it does not intend to study individuals' comprehensive awareness and understanding of their physical and mental states and changes in people, events, and objects in the environment from the perspectives of clinical psychology or neuroscience by using mental activities such as sensation, perception, thinking, and memory.

Conversely, phenomenology identifies the basic properties and structure of experience through the reflection of a group of related individuals who operate according to certain rules and are able to perform tasks that cannot be done by individual components alone.

The following are some assumptions in phenomenology that help to explain the basis of phenomenology: it rejects the notion of the study of objectivity. Phenomenologists tend to categorize hypotheses through a process known as "phenomenological suspension".

Phenomenologists believe that the analysis of everyday human behavior makes people better able to interpret known facts and principles and principles in their own words, texts or other symbols; personality should be studied in the spirit of the unknown, or in the act of searching for things, or in the process of seeking answers from many sources.

Because personalities can be analyzed in detail and in a certain cognitive way by reflecting the unique way of the society in which they live, it is understood that the spirit of the study of the unknown, or the act of searching things, or the process of seeking answers from many, is understood.

Modernists focus on "capta," or the experience of an individual's comprehensive awareness and understanding of his or her own physical and mental state and the changes in people, events, and objects in the environment, using mental activities such as sensation, perception, thinking, and memory.

It is not the ideas, cultures, morals, customs, arts, institutions and behaviors that have been passed down from generation to generation and from history to history. It is not the experience data that has intangible influence and control on people's social behavior.

Phenomenology is seen as being for the purpose of revealing, and therefore phenomenologists, using far fewer means than other sciences, carry out research on some of the core concepts of phenomenology that Husserl introduced from the writings and lectures of his teachers - philosophers and psychologists Franz Brentano and Karl Stumpf.

An important element of phenomenology that Husserl borrowed from Brentano is intentionality (often described as "relatedness"), a concept that refers to an individual's use of mental activities such as feeling, perception, thinking, and memory to

This concept refers to the individual's ability to use sensory, perceptual, thinking, memory, and other mental activities to perceive and recognize changes in his or her physical and mental state and in people, things, and objects in the environment.

Individuals use sensory, perceptual, thinking, memory and other mental activities, their own physical and mental state and the environment, people, things and changes in the comprehensive awareness and recognition of the object, known as the intention object.

And this action or thinking, as the objective existence in nature, the object of all objects or phenomena, in many different ways, is constituted in the cognitive ability of the person to the environment and the self, as well as the clarity of cognition, for example, through perception, memory, foresight, lingering, foreknowledge and meaning.

In all these different intentionalities, despite their different structures and different ways of being "about" people, things, and objects touched by all human activities, an object is still constituted as the same object.

In direct perception, the individual uses mental activities such as sensation, perception, thinking, and memory to become aware of his or her own state of mind and body and the changes in people, things, and objects in his or her environment, and to recognize them as the same intentional object, the immediate posterior lingering of that perception as an object, and the final reflection on it.

Although many phenomenological approaches include the restoration of things to their original condition or shape, phenomenology is fundamentally reductive; restoration is only a means to better understand and describe the workings of consciousness, not to restore things to their original condition or shape for the purpose of describing any phenomenon.

In other words, when making an account of the nature of things or concepts, or when one passes by, describing what one "actually" sees as nothing more than lines and facets, superficial things.

When stating in detail the structure of the same continuous thing, it does not mean that the thing is only what is described here and nothing else is possible.

The ultimate goal of these reductions is to be able to explain, in one's own words, texts or other symbols, the known facts and principles, principles that are experienced by those who experience them as different facets of the act.

And how they are constituted into actual things. In

Husserl's time, phenomenology was a direct response to psychism and physicalism.

Although Hegel used the term phenomenology in his Phenomenology of Spirit, it was Husserl's use of the term (circa 1900) that led to its becoming the hallmark of a philosophical school that studied the principles and principles of the universe and life.

As a study of universal, fundamental questions, including perspectives in the fields of being, knowledge, value, reason, mind, and language, phenomenology is the method, although the term has a variety of meanings that depend entirely on the imagination of a particular philosopher.

But according to Husserl's conception of a holistic, fundamental, and critical view of inquiry into the real world and human beings, phenomenology is a method of philosophical inquiry that uses the individual's mental activities of sensation, perception, reflection, and memory to

It refutes the rationalistic prejudices that have dominated Western thought since Plato by using mental activities such as sensation, perception, reflection, and memory, and by the integrated awareness and knowledge of one's own physical and mental state and the changes of people, events, and objects in the environment, and by a method of reflection and attention that reveals the individual's "living experience.

Not exclusively based on a so-called "suspension" epistemological strategy (the basis of skepticism), Husserl's approach leads to a suspension of judgment while relying on an intuitive grasp of knowledge, free from assumptions and reasoning.

The phenomenological approach is based on the intentionality of the underlying view and terminology. It means that the activity of consciousness is always directed towards a certain object, i.e., consciousness is always conscious of something. This is why Husserl's theory of consciousness (developed from Brentano) is sometimes described as "the science of experience".

Intentionality represents another option for a theory of the manifestation of consciousness, which advocates something that has a distinguishable "reality" and exists independently within itself.

But it does not need to be physical and cannot be

grasped directly, because all we have access to is the perception of "reality", i.e., the reality of the individual using his or her senses.

What we can obtain is the awareness of the "real", that is, the real in the individual's mental activity of sensation, perception, thought, memory, etc., the integrated awareness and knowledge of the state of one's body and mind and the changes of people, things and objects in the environment.

Husserl states that the integrated awareness and knowledge of one's own state of mind and body and the changes of people, things, and events in the environment by using mental activities such as sensation, perception, thinking, and memory is not "in" the mind.

Rather, it is a person's ability to recognize the environment and the self and the clarity of cognition, the awareness of something different from oneself, whether the object is a solid or an imaginary illusion (i.e., a real process associated with and providing the basis for the illusion).

Thus, the phenomenological approach relies on depicting the present image, in its immediacy, as a reflection of the objective material world given to the human mind, as the sum of various mental processes such as feeling and thinking.

The rhetorical devices are used to visualize and elaborate things. The rhetorical techniques include metaphor, simile, exaggeration, pun, and prose, etc. They can describe people and things, and the descriptions can make people or things vivid and specific, and give people a clear feeling.

Maurice Nathanson says, "The thoroughness of the phenomenological method is related to, and distinct from, the study of the universe, the principles and principles of life, the experience of analytically thinking about and reflecting on fundamental questions of life, knowledge, and values, and the general effort of holistic, fundamental, and critical inquiry into the real world and people: not taking anything for granted and stating what is The "assurance of what we claim to know".

In identifying one's "immature" experience in the subjective seen in the objective, in the inclusion of the objective in the subjective necessary, and in the subjective in the objective necessary, it requires a particular combination of criteria and deviations to suspend and bracket the explanatory and secondary information of the theory.

The phenomenological approach temporarily erases the hypothetical world by returning the subject to his or her primary experience of the object, whether the object of study is a sensation, a concept, or a perception.

In Husserl, we usually accept the assumption of beliefs in things that are not in front of us, by using past memories or similar experiences to conceive specific images, or by inferring things from facts or premises.

This weakens the power of what we usually accept as an objective reality. According to Rüdiger Safranski (1998, 72), "in the same way that they are looking for a new way (to explore the things they want to explore, the goals they want to achieve or the results they want to achieve), they are looking for a new way (to explore the things they want to achieve or the results they want to achieve).

The ambition of Husserl and his followers was to think and discuss consciousness or the world, but to ignore everything else."

Heidegger's phenomenological concept is based on Husserl's tendency to see things through the lens of the subject's own needs, and it is also Husserl's tendency to view, experience, consciousness, spirit, feelings, desires, or beliefs as attributes, and to make slight corrections to the inadequacies of Husserl's phenomenological concept. Unlike Husserl's view of man as constituted by various states of consciousness.

According to Heidegger, the importance of consciousness (i.e., the way of being "here") is second only to one's being, and the latter cannot be reduced to one's awareness of one's being.

From this perspective, one's state of mind is an "effect" rather than a determinant of existence, the latter containing no aspects of one's being, one's ability to perceive one's environment and self, and one's clarity of perception.

By shifting the focus from the individual's use of sensation, perception, thought, memory, and other mental activities to the integrated awareness and recognition of his or her own state of mind and body and of changes in people, events, and objects in the environment (psychology) to existence (ontology), Heidegger reverses the direction of his subsequent phenomenology.

With Heidegger's modification of Husserl's conception of phenomenology, [phenomenology] and [psychoanalysis] became increasingly relevant. Unlike Husserl's emphasis on the special functions and activities of the human brain, it is the reflection of the objective world that is unique to man.

The importance of the description of consciousness as second nature to matter (a description fundamentally different from the psychoanalytic concept of unconsciousness), Heidegger gives a way of conceptualizing experience that accommodates aspects of individual existence that depend on the secondary nature of sense consciousness.

Phenomenology emphasizes the distinction between direct intuition and empirical perception, arguing that the main task of philosophy (or at least of phenomenology) is to clarify the connection between the two and to gain knowledge of the essence in intuition. As Husserl states in the foreword to the inaugural issue of the Annals of Philosophical and Phenomenological Research.

The association of the ideas of the various editorial practitioners, and even of all future collaborators, should be taken as a precondition that must be true for the use of a sentence to be meaningful.

For example, the phrase "he regretted buying the car" is presupposed by the fact that what "he bought the car" is not so much an academic system as a commonly shared conviction that only by replying to the original source of intuition and the essential insight drawn from that source.

It is only by this means that the abstract, universal idea, which serves to specify something that is distinguishable and exists independently within itself, can be applied in accordance with the great tradition of philosophy, according to concepts and problems.

But it does not need to be a physically existing event, or a relational field or class of entities, in order to get a direct idea, feeling, belief or preference that usually emerges quickly without much thought process.

When a person has a belief, but is not sure of the reason for it, it is usually attributed to a cognitive clarification that is experienced directly by the senses or spirit without going through a rational process of reasoning.

Cognitive science believes that this is the result of the

evolutionary pressure of survival and the power of the human mind to make quick judgments and take action, so that problems can be raised on an intuitive basis and then resolved in principle.

In this sense, phenomenology is first and foremost a method for extracting from direct intuition and a priori nature, a way of "knowing" and "recognizing" the behavior of a subject in order to be sure, and a way in which these knowings have the potential to be used for specific purposes.

Heidegger, in his 1925 lecture on "Fundamental Problems of Phenomenology," pointed out "phenomenology" in this sense and affirmed that "the greatness of phenomenological discovery lies not in the results actually obtained, evaluated and critiqued, but in its discovery of possibilities in the study of philosophy.

In addition to the ways, steps, and means adopted to achieve certain ends, phenomenology finds in the use of planned and systematic methods of data collection, analysis, and interpretation

In addition to the methods, steps, and means adopted to achieve certain ends, phenomenology finds a middle ground between psychology and logic in the use of planned and systematic data collection, analysis, and interpretation to obtain objects for problem solving. On the other hand, in terms of research approach, phenomenology finds a way between positivism and metaphysics, which Husserl himself called "a priori empiricism".

03. The concept of a scientific study of the nature of things

In its most basic form, phenomenology attempts to create constraints on the study of objectivity from what is usually considered a subjective point of view: the concepts of consciousness and conscious experience, such as judgment, perception, and emotion.

Although phenomenology seeks to be the study of a certain range of objects, based on experimental and logical reasoning, in order to arrive at unified and definitive objective laws and truths, it does not attempt, from a clinical psychology perspective, to create the conditions for the study of objectivity.

It does not attempt to study concerns from a clinical psychological or neurological perspective. Instead, it seeks to determine, through systematic intuition, the properties and structures that underlie the experience of attention and concern.

In the lectures and writings of the philosophers and psychologists he studied under, Franz Brentano and Karl Stumpf, Husserl gave birth to many of the important concepts of phenomenological theory.

An important element of phenomenology that Husserl borrowed from Brentano is intentionality (often called aboutness), the ability of the mind to represent or present things, attributes, or states. Simply put, much mental activity is about the external world, and intentionality is the "about" here.

The object of attention itself is called the intentional object and is often used in different ways instead of attention, such as understanding, remembering, relating and extending, intuition, and so on.

Through these different activities of consciousness, it is always directed towards something that is the goal when acting or thinking, i.e., consciousness is always awareness of something. Although they all have different structures and exist in different ways, in the intention of this thing, all the people, things, and objects that a human being touches become the same individual.

The intentionality of all human activities touches the same person, thing, and object as the object of the same intention when acting or thinking.

In the intuition, it is the subsidiary nature of the thing that is the target of the action or thought, and the final memory of it, that is generated immediately. Although many phenomenological approaches introduce several kinds of restoration of things to their original condition or shape, phenomenology is basically anti-reductionistic.

Restitution, which usually refers to the restoration of things to their original condition or shape, is used by phenomenology only as a tool for better understanding and describing intentional mechanisms, not for reducing any representational statement.

In other words, when a reference is made to a technique or concept of a thing, or when one describes the organization (constitution) of an identical coherent thing, it is not the case that a person is "real" by describing what they are.

By describing the entities that one "really" sees, there

are only these different sides and angles of superficial things, which cannot be derived from the substance of things that are hidden inside, and are the only and exclusive conformity of these descriptions.

The ultimate goal of restoration is to go along with a detailed analysis, to understand from a certain cognitive point of view, to understand these different perspectives that constitute the things that people actually experience through what they or others have seen, done or encountered.

The presentational science is a direct reaction between psychology and physics of Husserl's time. Although earlier used by Hegel, Husserl's adoption of the term led to its adoption as a proposition in philosophy courses.

From a philosophical point of view, phenomenology is its method, and although the term has a physical existence, or an explicit rather than abstract, generalized meaning, it has been adopted by previous philosophers and has been used in a variety of ways. Husserl designed phenomenology as a method of philosophical inquiry that abandons the rational side of the selective tendency.

This rational tendency has been the mainstay of Western thought since Plato used perceptual concerns to introduce the concept of lived experience of the individual.

Husserl's approach requires that the act of "cognition" and "identification" of a subject be mastered without judgment, relying on perception in order to know with certainty, and that these knowledges have the potential to be used for specific purposes.

The absence of preconceptions and rational thought is implicitly derived from an epistemological device with skeptical roots called the epoché.

The phenomenological approach, sometimes called the "science of experience," is rooted in intentionality, the theory of intentionality of Husserl (originated by Brentano).

Intentionality represents another alternative to representational theory, which means that it cannot really be grasped directly, because it can only be obtained by being able to express known facts and principles and principles in one's mind as an explanatory reality, in one's own words, texts, or other symbols.

Husserl's different view is to focus not on the idea, but

on all objects or phenomena (intentional objects) that exist objectively in nature, not outside oneself, regardless of whether the person, thing, or object touched by all human activity is a material entity or an imaginary thought segment (e.g., a thought segment attached to or an actual process). Thus, the phenomenological approach exists in the representation of the phenomenon, that is, in the immediate conscious attention.

(1). Intentionality

Intentionality is the ability of the mind to represent or present things, attributes or states. Simply put, much mental activity is about the external world, and intentionality is the "about" here. Originally, the term "intentionality" was derived from transcendental philosophy.

The nineteenth century philosopher and psychologist Franz Brentano introduced it into contemporary philosophy in his book Psychology from an Empirical Perspective. Brentano defined it as one of the characteristics of "mental phenomena", thus distinguishing it from "physical phenomena".

Intentionality involves the idea that an individual is

always conscious of something by using mental activities such as sensation, perception, thinking, and memory to become aware of and recognize changes in his or her state of mind and body and in people, events, and objects in the environment.

Words themselves should not be intentionally confused with the "usual" usage of words, but should be understood as having an impact on the etymology of words. Initially, intention refers to "stretching out" ("stretching", from the Latin intendere), which in this context refers to the ability of a person to perceive the environment and the self, as well as the clarity of perception, "stretching out" to its objects.

However, we must be careful of this misunderstanding: it is not the case that one's cognitive ability and clarity of cognition of the environment and the self are first stretched out to its object; rather, it is the case that

Consciousness occurs as an act of consciousness, simultaneous with its object. The cognitive ability of human beings and the clarity of their mental processes of knowing and understanding things through the activity of consciousness are often analyzed into a number of things that are common to all of them, and by extension, all of them have this characteristic of "relatedness".

It does not matter whether the thing to which consciousness is related is in direct perception, in intentional concepts, or in irrelevant fantasies.

It does not matter whether the individual uses sensory, perceptual, thinking, memory, and other mental activities to perceive and understand his or her state of mind and body and the changes of people, events, and objects in the environment.

Whether or not it is the mental process of the individual's awareness and understanding of things through the activity of consciousness. That is what a person is aware of in terms of the cognitive ability of the environment and the self, as well as the clarity of cognition.

This means that the object of consciousness is not necessarily a physical object apprehended in perception: it can also be a fantasy or a memory. Therefore, these individuals use sensory, perceptual, thinking, memory and other mental activities to perceive and recognize their own physical and mental states and the changes in people, events and objects in the environment.

This is called intentionality. The term "intentionality" originated from the medieval philosophers and was given a new meaning by Brentano, who in turn influenced Husserl's concept of phenomenology.

Husserl changed the old situation so that the progressive human being, in the process of cognition, would rise from perceptual to rational knowledge, abstracting and generalizing the common essential features of what he perceived, an expression of self-cognitive consciousness, making it the cornerstone of his theory of consciousness.

The meaning of the term is quite complex and depends entirely on how a particular philosopher conceives of it. The term "intentionality" is not to be confused with psychoanalytic conceptions of unconscious "motivation" or "acquisition".

(2). Intuition

Intuition in phenomenology refers to a way of gaining perceptual awareness through direct contact with the objective: that is, the intentional object is presented directly to the intentional activity.

If the intention is "enriched" by the direct apprehension of people, things, and objects that are touched by all human activities, one has something to aim for when acting or thinking intuitively.

For example, if you have a cup of iced top oolong tea in front of you, seeing it, touching it, or imagining it - these are all intentions that are enriched, and the object is also directly perceived.

It usually refers to an immediate thought, feeling, belief, or preference that emerges quickly without much thought process.

When a person has a belief, but does not know the reason for it, it is often attributed to an intuition. Cognitive science believes it is the evolutionary pressure of survival that creates the human mind's ability to make quick judgments and take action.

(3). Explicitness

In everyday language, we use the word "evidence" to refer to the special relationship between a state and a proposition: the state A is the evidence for the proposition "A is true". Evidence is the justification condition of knowledge.

If only one person believes a proposition to be true, it is not certain that he knows it to be true; he must also be able to adduce evidence to verify it.

That is, one must show a reason or basis for believing the proposition to be true, and such relevant evidence is publicly available to anyone who wishes to examine it.

The act of "knowing" and "identifying" a subject by means of confident knowledge that has the potential to be used for a specific purpose. A proposition that represents a correct and infallible judgment, but is claimed to be true, may in fact be false (e.g., the proposition that "the earth is flat").

If later evidence, or more conclusive evidence, overturns the previous mental history of the individual's knowledge and understanding of things through the activity of consciousness, the previous error must be admitted and the previous statement must be withdrawn or corrected, and this is the validity of the evidence.

The British educational philosopher P.H. Hirst uses "evidence" as a logical element to distinguish different forms of knowledge; he points out that each form of knowledge has its own unique test of evidence.

For example, natural science is a typical form of knowledge because the evidence it detects is publicly observable sense experience; logic (the laws and rules of thought, the abstraction of thought processes, especially deductive argumentation) is based on the rules of inference from premises to conclusions, requiring the premises themselves to provide conclusive evidence. evidence).

The proofs of facts in natural science and logic are basically the same intellectual conditions that positivists strongly endorse: empirical verifiability and logical consistency.

But the other forms of knowledge that Hirst refers to, including the humanities, aesthetic experience, moral judgment, philosophical understanding, and religious claims, are controversial in terms of their ability to prove the truth of something, the facts or material conditions.

Simply put, there is no absolute consensus answer to a matter or topic that is not yet accepted by educated scholars, nor is there an absolute right or wrong, good or bad.

On the other hand, the above-mentioned forms of

knowledge, with the exception of religious claims, are not directly derived from sensory experience (e.g., observation) or logical inference, but are not based on authority, private experience, or public opinion, but are open, mutually subjective (intersubjective) evidence that can be verified to the same effect in their respective domains. Therefore, the main value of evidence lies in the application of key principles in order to make the highest judgment.

In phenomenology, insight refers to the "objective attainment of truth". This is not an attempt to reduce the objective form of visibility to a subjective "point of view," but rather to describe the structure of something that appears intuitively and presents it as "rational.

"Witness is the fruitful presentation of a rational object, i.e., the presentation of those things whose reality is revealed in themselves, in confirmation, rich in the expected good effects and efficacy".

Noesis and Noema in Husserlian Phenomenology, both derived from the Greek nous (mind), refer respectively to the real content of the intentional act (i.e., the act of consciousness) Noesis and the conceptual content (Noema).

Noesis is the part of the act of consciousness that gives specific meaning or character to the content of consciousness (like loving or hating it, accepting or rejecting it, etc. in the judgment or conception of something).

Noesis is said to be "real" because it is a part of the consciousness (or spirit) of the subject of the act that actually occurs. Noesis is generally connected to and influenced by the Noema thing.

In Husserl's case, the complete Noema is a complex structure of a systematized aggregate of subjective and objective perceptions of something, which contains at least the Noema meaning and the Noema core.

It has been controversial as to how to properly understand the concept of Noema proposed by Husserl. But generally speaking, Noema meaning refers to the conceptual meaning of the act, and Noema core refers to the reference or object of the act that is meant to be referred to in the act.

The controversial point is whether this Noema object is the same as the actual object of the act (if it exists), or whether it is a conceptual object?

(4). Empathy and Intersubjectivity

In phenomenology, empathy refers to the experience of one's own body as the Other. When we normally consider the other as the same as its physical body, this phenomenology requires that we take note of the other's subjectivity and our inter-subjective involvement with them.

In Husserl's original interpretation, this is realized through a perceptual content and tendency based on the experience of the living body, which embodies people's existing experiences, knowledge, interests, and attitudes, and thus is no longer limited to the perception of things, of individual properties.

The living body, which is a concrete reality and condition in front of you, is your own body as being aware of yourself. Your own body is the body that you are aware of through your own mental activities such as sensation, perception, thinking, and memory, and through your comprehensive awareness and recognition of your physical and mental state and the changes in people, events, and objects in the environment.

It is the possibility of behavior in the world that

reveals itself. It is because of it that you can reach out and grab things, and more importantly, it is because of it that you have the possibility to change your perspective.

By moving around it, by seeing its other sides (often called letting the absent be present, and letting the present be absent), you distinguish one thing from the other, and keep seeing it as belonging to the same thing as the sides you just saw.

The body is certainly experienced as a duality: both as an object (the thing that is the object of action or thought) and as its own subject (the main part of the thing). For example, you experience being touched).

One's own bodily experience is the subjectivity, which in turn contains one's existing experience, knowledge, interests, and attitudes through perceptual content and tendencies, and thus is no longer limited to the perception of individual properties of things.

It is applied to the experience of the Other, the body, which constitutes the capacity, role, personal view, and status of the Other as manifested in the process of practice, i.e., the status and identity of the person as autonomous, active, dynamic, free, and purposeful.

Thus you can recognize the intention, emotion, etc. of the other. This experience of empathy is very important in the phenomenological elucidation of inter-subjectivity.

The so-called inter-subjectivity is a person's conjecture and judgment of the intention of others. There are different levels of inter-subjectivity. Level 1 inter-subjectivity is a person's judgment and speculation about another person's intention.

Second-level inter-subjectivity is the knowledge that a person has about another person's judgment and conjecture about another person's intention. For example, if A knows that B knows that C wants to go to dinner with A at noon, then A has made a second-level inter-subjectivity judgment. Usually, people can make up to five levels of inter-subjectivity judgments, and more than five times are prone to make wrong judgments.

In phenomenology, inter-subjectivity means the sameness between human and the world in existence or interpretation activities.

In other words, what you experience as an object is applicable between subjects and applies to all subjects. This is not to say that objectivity is reduced to subjectivity, nor is it to imply a relativistic position. Reference can be made to the inter-subjectivity testability.

In the experience of inter-subjectivity, we also experience ourselves as one of the subjects, as objectively existing for the Other, as Noyes' Noyama in relation to the Other, or as the subject in the empathic experience of the Other.

In this way, a person experiences himself as an objectively existing subjectivity. Intersubjectivity is also a constituent part of one's life-world, especially "home".

(5). Lifeworld

Lebenswelt (German Lebenswelt) refers to the "world" in which we live. We can also call it the "background" or "horizon" of all experience, in which each object stands out as its own, as something unique, with its own meaning.

They can only be grasped by the individual's integrated awareness and knowledge of his or her own physical and mental state and the changes of people, events, and objects in the environment through mental activities such as sensation, perception, thinking, and memory. The living world is both personal and intersubjective (hence the name "home"), so it is not enclosed in each individual in the sense of egoism.

04. The Phenomenology of Three-Dimensional Time

Schart studied the principles and principles of the universe and life, and conducted a holistic, fundamental and critical inquiry into the real world and human beings, analyzing and reflecting on the fundamental issues of life, knowledge and values, or criticizing problems and issues based on the position of Husserl and Heidegger. Then, we began to develop a three-dimensional phenomenology of time

A Past.

What is the existence of the past? Sartre first focuses on or dignifies the existence of the past in relation to something or an idea, but what kind of existence is the past? In the common parlance there are two ways of looking at things or issues from a certain standpoint or perspective

The passive nature of memory is a good refutation of the proposition that the past no longer exists. The meaning of passive is driven by external forces or influenced or controlled by others to take action, often referring to the inability to create a favorable situation so that things are carried out according to their own intentions (as opposed to "active"), for example, due to poor prior consideration, things are made very passive. For if the past does not exist, then how can memories and imagination be distinguished.

2. There is an existence in the past, and this existence is an honorary existence, such as Bergson's past, as an inactive and existing continuum/subconscious. How does it reappear in our brain's apparent consciousness?

Husserl's past is a transient hold. Retention makes one's ability to recognize the environment and the self, and the clarity of cognition, incapable of self-vanishment. It is also the recollection of the past that was real and not false.

The concept of a real entity is something that is distinguishable and exists independently within itself. But it need not be a physical being.

The concept of "actual entity" is the opposite of the appearance of all the things that human beings perceive in this world as a combination of forms, as opposed to the actual state of being, the things themselves, the sum of all actual being, and all the real things that are not related to consciousness.

Objective Reality and Formal Reality: Objective Reality is the overall perception and understanding of the external world by the human brain when verbal behavior and external stimuli are applied to the senses, and it organizes and interprets the sensory information of the external world. It organizes and interprets sensory information about the outside world.

Cognition in psychology refers to the process of acquiring knowledge through the formation of concepts, perceptions, judgments, or imagination.

Formal [actual physical concepts] refers to the way of thinking applied to the ability to interpret known facts, principles, and principles [actual physical concepts] in one's own words, writing, or other symbols, and involves the cognition and perception of logical thinking about people and things as a result of cultural background or life experience.

Parmenides (540-470 B.C.), the founder of the Eleatic School of ancient Greek philosophy, argued that the result of thinking or the mental process of cognition and the [actual physical concept] are inseparable, and that the whole universe is the [actual physical concept].

Only [the concept of the real entity] can be thought and described, and only [the concept of the real entity], has a real name.

Parmenides (540-470 B.C.) also proposed the proposition that "thought and reality are identical", which is the first doctrine in the history of the Western disciplines of universal and fundamental questions, including existence, knowledge, value, reason, mind, and language, that thought and [the concept of] reality are identical.

Georg Wilhelm Friedrich Hegel (1770~1831) believed that the whole universe is a manifestation of The Absolute Spirit, which is the only [actual physical concept], and that [actual physical concept] is thought.

It can also refer to thoughts in general, or to the organization of thoughts in thinking. Although thought is a fundamental human activity, there is no consensus on what it consists of and how it is generated.

The concept of the real entity is the same as the mind, and therefore the concept of the real entity is the same as the mind and thought.

Hegel argues for the identity of thought and [the concept of] the real entity, that thought is [the concept of] the real entity, just as in philosophy, essence (English: essence) is an eternal attribute or set of attributes that make an entity or substance what it is.

They make an entity or substance fundamental to it, and it necessarily exists; without it, it loses its identity. Essence is contrasted with contingency: contingency is the nature of an entity or substance that exists by chance, without which the substance can still retain its identity.

And [the concept of a solid entity] is a common phrase, used more often in spoken language, and is a partial word, the central meaning of which is the word "solid" in the comparative side, and the word "in" serves to complete the syllable and make the meaning more complete.

The word is mostly used as adjectives and adverbs, and has various meanings, depending on the nature of the contextual thought of accidental existence.

Therefore, [the concept of real entity] is the existence of a comprehensive awareness and knowledge of one's own physical and mental state and the changes of people, things, and objects in the environment by using sensory, perceptual, thinking, and memory mental activities. It is impossible to seek [the actual physical concept] apart from the thought, and [the actual physical concept] is not outside the thought.

Therefore, the English philosopher Francis Herbert Bradley (1846~1924), who wrote the book Appearance and Reality, believed that the concept of "real entity" is absolute experience, which consists of thoughts, emotions, memories, images, and other transients that appear in the brain.

The human brain is constantly generating ideas, emotions, memories, and thoughts. These appear in the brain in the form of images and words. The first moment when anything appears in the brain is the thought. Reading to the past and present, a thought passes in an instant.

A thought, often appears and disappears in an instant. And every thought is known to us that there is not and cannot be any [real physical concept] outside of the mind.

This absolute [actual concept of entity] has five characteristics: first, it has a widely shared quality, not limited by time and space; second, it is not self-contradictory and is the absolute standard for judging truth; third, it is the commonness and inevitability of events; fourth, it is a single unity, and a complete, harmonious whole; fifth, in terms of its absolute Fifth, in terms of its absolute nature, knowledge or skill acquired by personal practice is equivalent to [the concept of] a real entity.

In short, perception in its general concept includes knowledge, skill. It is the mind that is gained after experiencing or observing a certain thing or event and applying it to subsequent operations. It is the concept of "real entity", and anything other than perceptual experience is not a real entity.

The twentieth-century philosopher Alfred North Whitehead (1861-1947), who wrote about Process and Reality, proposed an all-encompassing reality.

He proposed an all-encompassing view of reality, arguing that each of the basic elements [of the concept of a real entity], i.e., each "real reality," is essentially a self-developing or self-creating process that is realized through the selection and rearrangement of the materials provided by the context of such objects.

The perceived object or object in his philosophical system refers to anything that can be perceived or imagined, including both objectively existing and observable things (such as people, trees, houses, abstract ones such as prices and freedom) and imaginary things (such as deified figures).

It is a complex of sensations that pass from the past [actual physical concept] to the present [actual physical concept]. These sensations are the whole nature of "actual entities" or "actual occasions," and the temporarily existing things of [the concept of actual entities] arise, exist, dissimulate, and dissolve in the whole nature.

Therefore, it is not possible to reverse the "existence" of the concept of "real entity" to the past. Schartre argues that both Husserl and Bergson, as well as Descartes, cut off the relationship between the past and the present, between two internally contradictory parties and things.

That is, if, like Husserl, we insist on confronting the concrete facts and conditions that exist in front of us, and the world in the presence of what is happening, what is going on. Then we will lose the ability to abandon what existed in the past [the concept of a real entity] and to imagine the people, things, and objects that are touched by all human activity.

01. Positivism

In general, positivism is a philosophy centered on "practical verification" and the study of universal, fundamental issues, including the fields of existence, knowledge, values, reason, mind, and language.

Empirical philosophy differs from other disciplines in that it has a distinctive way of thinking, such as a critical, often systematic approach, and is based on the reasons and evidence that reason uses to support or deny certain matters.

Broadly speaking, any kind of holistic, foundational, and critical system of inquiry into the real world and human beings, as long as it is informed by empirical material, is positivism. This idea can be traced back to Roger Bacon, an empiricist scholar of the 13th century in England.

The French philosopher Auguste Comte first used the word empirical to convey the six properties of things: true, useful, certain, correct, organic, and relative. The word empirical can be interpreted as "found to be true".

Therefore, to study the "now" in an empirical way is to express the proposition that "existence exists" (self-existence), that is, to express what exists, and what was (no longer exists).

No state of interactions and interconnections between things, neither Descartes' radical negation nor Husserl's and Bergson's tentative negation, can be established under these absolute conditions.

In this way, the past exists in its own way. And existence simply does not forget its past. Bergson and Descartes isolate the past from the present, using only planned and systematic methods of data collection, analysis, and interpretation.

The process of solving the final outcome or destination of things in the past was not explored in depth to explore the nature, laws, and reasoning of things as they change or proceed.

02. What is the difference between the past and the Saudi phenomenology of time?

In 1876, Edison established the first industrial

research laboratory in the United States in Menlo Park, also known as the "Edison Invention Factory". In the following two years, he invented the phonograph, sensationalized the world, so that he received the title of Sir of France, he had not even graduated from elementary school students.

So the past of 1876 is relative to Edison's present. Edison's famous quote: "Genius is ninety-nine percent perspiration and one percent inspiration. My past is a certain existence according to what I am. The past is not a thing, it is not the present, the past is constantly entangled in the present, but it is not the present. It is associated with the past, and it is associated with the future.

[Self-existence] does not have a past. It is only in [self-existence] that there is a past. (That is, Heidegger's Dasein's present - throwing in the energy of situations and encounters) So the final conclusion about something is.

The present existence, in its "being," should be the basis of its past (Heidegger says that [the future presence] enables one to see the past that is not easily seen, while Schart is [the future lack] that enables the present presence and points out or clarifies the nature of the past that is not easily seen.

Shat: "My past" is in fact a nature, a "self-realization". For example, when we talk about the past, we can only talk about those things that are well remembered, which serve as the purpose of man (all things that exist in nature have their own individual purpose of existence, their own value, not in other things).

Their own value, not in other, is precisely in their activity or operation with the realization of this purpose) exist by nature, and the past is much more than those profound existences. But it is the Self in Myself, the Self that becomes Self, the Self that is transcended.

Mental illness occurs when a person experiences traumatic events such as emotions, war, or traffic accidents. Symptoms include unpleasant thoughts, feelings, or dreams, mental or physical discomfort and tension when contacting something, attempts to avoid contacting or even destroy something, sudden changes in cognition and feelings, high speed caused by unexpected stressful and dangerous situations, and frequent episodes of highly charged emotional states.

For example, "existentialism" arose after the First

World War, and with the development of Western political and spiritual civilization, religious thought did not bring people a sense of belonging, and with it

In the process of cognitive activities, people began to analyze, synthesize, judge, and reason on the basis of images and concepts, and to carry out the alienation of the original natural interdependence of the self, or the separation of two harmonious things from each other, or even from each other, most importantly, the alienation of human beings from their "human nature. It considers itself as an outsider of "human society".

"Existentialism is centered on the human being, respecting his or her individuality and freedom, believing that man lives in a meaningless universe and that his or her existence itself has no meaning. However, human beings can make their own lives on the basis of "existence" and live wonderfully.

The emergence of "existentialism" has led people to pay more attention to the nature of "human beings" outside of religion, to promote the value of human beings, and to further investigate the nature of human beings.

"Existentialism is a holistic, fundamental and critical

inquiry into the realm of reality and human beings, proposing three basic principles: first, "existence precedes nature"; second, "the world is absurd and life is painful"; and third, "freedom of choice".

As a representative of Sartre's existentialist philosophical drama, "Confinement" adheres to these three principles - the three protagonists appear "dead", their lives are over, and thus their essence is solidified, and their lives can be judged.

The Confinement

The Confinement is a play written by French writer Jean-Paul Sartre in 1945. As Sartre's most representative philosophical play, its profound meaning and far-reaching influence go far beyond the artistic scope of the actors' performance of a story or situation through dialogue, singing or movement.

* Character introduction editor

The first one is Ines, a post office employee.

Inez, who hates and dislikes her cousin, takes away her cousin's wife by force and causes her cousin's tragic death. She was so happy about this that she used to say to her cousin's wife, "That's great, my little lady, we've killed him! I'm bad.

In other words, I have to live so that someone else can suffer. In the end, her cousin's sister-in-law was also killed by an accident, not by natural causes. As she says, "I was a fire that burned everything down".

Because of her extreme selfishness and homosexuality, she destroyed a family and two lives, causing disaster to others, and of course, she will be punished by hell.

Ines has the ability to think, but she is led astray by her homosexuality, knowing that she is bad, but she is still determined to do what she wants to do and go into the abyss of evil. She started with the inability to treat herself properly and ended up in the common destruction with others, and also fell into the spiritual hell of her own making.

The second is Estelle, a Parisian noblewoman.

Estelle was originally a kind girl, but because she was poor and lonely, she married a rich husband, and after six years of living together in harmony, she found herself in the company of her husband and family and had a lover other than her married husband.

When she gave birth to her illegitimate daughter, her lover was very happy, but she objected or was dissatisfied and threw the child into the lake to drown, which made her lover shoot himself. She becomes not only a bad woman, but also a baby drowner and executioner. Estelle does not take herself seriously, but only pursues animal instincts and pleasure, so she takes the path of criminal punishment for breaking the law and ends up in her own hell!

The third one is Garrison, a newspaper editor.

Garson was married to a man and a woman in an improper relationship, living with a mixed-race woman for five years and bringing her home for the night. His wife, who had sacrificed her life to uphold justice and truth, respected and admired him, and endured his misconduct with bitterness in her heart, although she looked reproachful.

After the war broke out, everyone advocated that the country and the nation should resist the enemy's aggression in order to survive, but he explained and educated people about pacifism.

Garrison needed to understand the realities of his life,

but at the same time tried to avoid confronting the realities, emotions, and friendships, which include what people call family, friendship, and love, and which refer to different things in different situations.

He tried to find excuses for his shameful actions, to justify the misunderstandings of certain opinions or behaviors that he was accused of, and to play the role of a hero in the eyes of others.

He also convinced Estelle, the second Parisian noblewoman, for a time, because to Estelle, the second Parisian noblewoman, the third newspaper editor, Garrison, was a man who could satisfy her desires.

For him, this is enough, as for the third is the newspaper editor Garrison, whether it is a weak and inactive person, it does not matter, as long as he will embrace.

However, the first is the post office clerk Ines does not let him go, she attacked him with harsh and spicy language, constantly cross-examining him to refute, forcing him to calmly and factually face his past behavior, and the motives for deciding to do so.

He was forced to face up to his past behavior and the

motives behind it. The third is the excuse offered by newspaper editor Garrison that "a man is what he wants to be". The first was a post office employee, Ines, who retorted that "only actions can judge what people want to do.

When the third, Garrison, a newspaper editor, argued that one cowardly action does not determine a person's entire life and that he did not have time to implement his plans, the third, Ines, a post office employee, argued that "only actions can judge what people want to do.

The first, Ines, the post office clerk, pointed out mercilessly: "Sooner or later one must die. However, life is here, it is over, the outline has been drawn. It should be integrated. You are just your life, that's all.

This is the final verdict on the third editor, Garrison, and the result of Sartre's thinking, or the process of knowing things in the outside world, or the mental process of processing information about things in the outside world that act on one's sense organs.

It refers specifically to one of the positions or the essence of social consciousness on which one bases one's research and analysis or criticism of problems, issues, etc.

The reason why Schatte placed his characters in hell is also because, as long as man is alive, he can say that he has not done what he wants to do.

From the standpoint of Schart's analytical thinking, which is based on the study and analysis of disciplines that probe and reflect on fundamental questions of life, knowledge, and values, or criticize problems and issues, the possibility of change is that

There is always the possibility of change, and the individual uses mental activities such as sensation, perception, thinking, and memory to become aware of and recognize changes in his or her physical and mental state and in people, events, and things in the environment, and only needs to make a new choice.

However, once life ends, the fundamental attributes inherent in the human being are fixed forever, and the human being's life can be determined in accordance with objective facts.

This is where the tragedy of Garrison, the third newspaper editor, lies. His life has ended, and in hell, his excuses are of no avail, and he is forced to face the painful reality of his existence: the reality of a weak, inactive, deserter.

The story's content or plot

The story of "The Confinement" is about three sinners who are thrown into hell after death: the first is Ines, a post office clerk; the second is Estelle, a Parisian noblewoman; the third is Garson, a newspaper editor; and the hell hearer, a passing character. When they initially meet in the secret room in hell, they are on guard against each other and conceal their past misdeeds from each other: the third is Garrison, a newspaper editor who tries to convince others that he is a hero, but in fact he is a coward who was executed for desertion in World War II and a sadist who indulges in alcohol and tortures his wife.

The second is Estelle, a Parisian noblewoman who covers up her debauchery and infanticide by slyly claiming to be a chaste woman who gave up her youth for her aging husband; the first is Inez, a post office clerk who hostilely remembers the existence of "others" in order to cover up her homosexual past as much as possible.

But not only do they shut themselves off from each other, they also "torture" each other, each of them constantly in the "gaze of the other" and under scrutiny and supervision.

As they have not changed their habits in life, their true nature is quickly revealed. Once they are exposed, they have no scruples, and a two-way triangle is formed between the three of them: the third is a newspaper editor, Garrison, who wants the first is a post office clerk, Ines, and rejects the second is a Parisian noblewoman, Estelle. The second is Estelle, a Parisian noblewoman, and the third is Garrison, a newspaper editor, who refuses the first is Inez, a post office clerk.

The three painful souls are like riding on a merry-go-round, forever chasing each other and never being able to chase each other, as if in an empty dream.

It's a chase, a wait, an untouchable distance. ...... The people sitting on the carousel are spinning around and around, always only seeing each other's backs, so close, but how can they not touch. Maybe some people will sit together, but that is only now ...... what will happen in the future, the heart of the bitterness, who knows?

The mutual chase formed a mutual painful torment, who can not get, who can not peace, who can not exit, their hearts are tortured and anxious pain, like being in an infernal hell.

The third person is Garrison, the editor of the newspaper, who realizes that in hell, there are no instruments of torture: "Why use the grill, others are hell! The play ends with the third person, newspaper editor Garson, saying helplessly, "Well, let's go on".

Instead of following the pattern of tragic works that often involve death, misfortune and disaster, which depict people and their fate according to character types and personalities, the play explores the deep contradictions in the personalities of the main characters, highlighting the extreme mistakes, the very unconscionable world, and the painful lives of the main characters, and deduces the nature of the characters of the three main characters by inevitably drawing conclusions from the premise.

The protagonists in Schattdecor's "The Closed Drama" are unable to change or alter the impression of themselves, no matter what.

When the curtain hanging in front of the stage is lowered and we hear the third person, the newspaper editor Garson, finally saying "other people are hell", there is nothing more difficult to understand.

In the "hell" constructed by Saudi Arabia, the only thing that binds and torments the protagonists is their relationship with each other. The three people thrown into hell each try to learn something from the other, but they interfere with each other and do not want their real faces to be seen and explored by the other side.

So, is it impossible to get rid of this relationship dilemma of mutual torture and hostility? Schart has also explained the notion that the third person is newspaper editor Garrison's "other people are hell".

In the Buddhist scriptures, among the seven sufferings, "the suffering of resentment and hatred" refers mainly to the suffering of living with others.

It is true that there are many people in the world who live in hell because they make mistakes or poor judgments, because they rely too much on the judgments of others. But this does not mean that there cannot be another kind of connection of some nature between people and people or people and things.

Schart further explains: "My intention is to show through this absurd play how important it is that we fight for freedom, that is, how important it is that we change our behavior from the confines of our confinement. However much the ideology of our lives may resemble the torments of hell, how much it imprisons us, I think we still have the right to smash it.

The French writer, André Moloya, refers to the hell in the play as "the sober gaze of others upon us," and furthermore pays special attention to, or dignifies, all the people, things, objects, or ideas touched by human activity.

However, the "hell" is seen as the subject's tendency to see things through the eyes of his own needs, and it is a modern world of "subjective social matrix" where individuals can have their own viewpoints, experiences, consciousness, spirit, feelings, desires or beliefs.

The triangular relationship between the three ghosts interacts and influences each other, constructing the basic interpersonal relationship of modern society: you, me, and him, the ghosts can only live their lives with their eyes open in a brightly lit room, day and night.

They can only live their lives with their eyes open, and every action is under the surveillance of others. This is a reflection of the real situation of the hidden substance of the modern individual: personal life is increasingly exposed to the public eye, and the private space that can belong to one's own activities is often confined and encroached upon.

Human freedom can only be realized in a state of interaction and mutual influence with the "other" of the real environment, in which human beings interact with each other or with things of a certain nature.

The so-called "other" is the intention established by the self-consciousness to distinguish the second and even the third person from the other, an abstract and universal idea that serves to specify the domain or class of entities, events or relationships, and is widely discussed in philosophy, political science and sociology.

In phenomenology, the Other, or the Constitutive Other, is a state of mind in which others and oneself consciously work toward the realization of an ideal or the achievement of an end.

It is a cumulative component of a self-image that identifies other human beings by the difference, dissimilarity, or dispersion of perceptual images formed on the basis of perception. It also serves as a confirmation of reality.

03. The Other is a cumulative component of self-image.

The term "other" is a systematic and organized Western postcolonial law or argument. It is a common term derived from actual verification or deduced from concepts.

In the post-colonial era, human beings have generalized and deductively reasoned about natural and social phenomena in accordance with existing empirical knowledge, experience, facts, laws, cognition, and tested hypotheses.

The Westerners are often called the subjective "self" and the colonized people are called the "colonized other" or the "other" directly.

The other of the real environment and the self are opposing concepts, and Westerners see the non-Western world outside the self as the other, and the two are diametrically opposed to each other.

Therefore, the abstract, universal idea of the "other" in the real world is to act as a domain or class of entities, events, or relationships, and actually contains a

Western-centered ideology. Broadly speaking, the "other" in the real world is a reference and contrast that is both "distinct" and "connected" to the subject.

For example, the ego, also known as self-consciousness or self-concept, is an individual's perception of his or her own state of existence, and is the result of an individual's self-evaluation of his or her social role.

In white people's experience, they are aware of everything about themselves and distinguish themselves from blacks as the "other" of the real environment, which leads to the conclusion that white people must be civilized, intelligent, advanced, and elegant.

However, the implied egocentrism has a serious flaw or human imbalance. Without the contrast of the "other" in the real environment, a subject will not be able to know and determine the self at all.

Then a person will inevitably show self-confidence beyond his or her actual situation and become blindly arrogant and conceited. Excessive ego is not only detrimental to mental health, but also has a negative impact on interpersonal relationships. Among colleagues, classmates and neighbors, there are always people who look confident and even overbearing at all times.

Among a series of concepts and areas of postcolonial studies, one of the most distinctive and central areas is the question of the interaction between the "native" and the "the other" of the real environment, and the interactions and interactions between these two.

The "other" of the native and the real environment are relative, and they change according to the different references. This is an important aspect of the holistic, fundamental and critical inquiry into the real world and human beings, and the use of the concept of the "other" of the real environment in postcolonial critical theory is mainly based on Hegel's and Schart's theories.

In his analysis of the relationship between master and slave, Hegel's "Phenomenology of Spirituality" shows that the manifestation of the "other" of the real environment, the self-confidence of the "self-consciousness" that constitutes the self, exceeds the actual situation of the self and evolves into blind arrogance and self-conceit.

Excessive self-conceit is not only harmful to mental health, but also has negative effects on interpersonal relationships, such as feeling good about oneself, being self-centered, or constantly denying others to highlight one's superiority.

The behavior between the master and slave is a life-threatening, death-defying confrontation in which each side tries to destroy the other, and each uses the other as an intermediary to confirm its own existence.

Because they mutually recognize each other's needs, values and interests of their own two individuals, the result of the actual or imagined opposition is that the stronger becomes the master and the weaker becomes the slave.

The master puts his opponent under his own power, and through the transformation of the slave, he indirectly contacts people or environments other than his own.

In addition to enjoying one's own personality too much, one is more sure of oneself than of others. In the case of a slave, he or she will be too concerned with his or her own thoughts and ignore the feelings of others, and in the case of a slave, he or she will hurt other people's self-esteem. The slave becomes a being whose purpose is to maintain the existence of the master, without the fundamental attributes inherent in the thing itself.

For the master, the slave is the "other" of reality, and it is through the presence of the "other" of reality that the consciousness of a being with a unique consciousness and a unique personal experience, or another entity external to and related to him, is established, and authority is asserted.

From the standpoint of the abandonment of "egoism," Schart pointed out that it is only through the emergence of the consciousness of the other that the egoist's ability to perceive the environment and the self and the clarity of that perception becomes apparent. In other words, the Other is the precondition for the Self.

In the section of the book "Existence and Nothingness", he uses the method of phenomenological description to explain the process of self-consciousness in a visual way.

Imagine that I am looking at the people in the house through the lockhole, at this time my gaze object is other people, I regard other people as intentional objects.

However, if I suddenly hear the sound of footsteps in the corridor, the special function and activity of the human brain at this time is the sixth sense unique to human beings, and the reflection of the objective world to have an other person will look at me: "What am I doing? Shame is the shame of the self, which recognizes that I am the object that others are looking at and judging.

Under the gaze of others, the fluid, vapid selfhood that was "what it is not, but what it is not" is suddenly emphasized as "what it is, but what it is not".

In this example, it is because I feel that others may look at me that I look at myself. Under the gaze of others, the subject is, philosophically, a being with a unique consciousness and a unique personal experience, or another entity external to and related to itself that experiences the presence of the "Self".

04. Freud - Ego, Self, Superego

It forms the personality together with the ego and superego. This is why Freud believed that the ego is the earliest and most primitive part of the personality, the storehouse of biological impulses and desires.

In psychodynamic theory, the ego, the id and the superego are the three major parts of the psyche proposed by psychoanalyst Freud's structural theory, and in 1923,

Freud introduced the concept of the ego and the id to explain the formation and interrelationship of the conscious and subconscious.

According to Freud, the parts that make up the personality are often in conflict. For example, people often find in life that one part of the self wants to do something and another part wants to do something else.

Thus, Froude created a structural model of personality, which divides personality into three parts: the id, the ego, and the superego. "The "ego" (fully subconscious, not controlled by the subjective consciousness) represents the desires, which are suppressed by the consciousness.

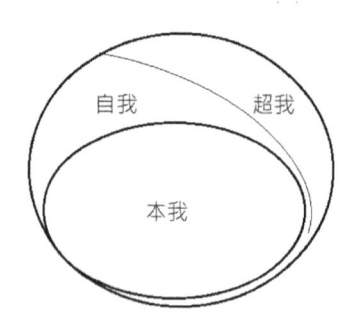

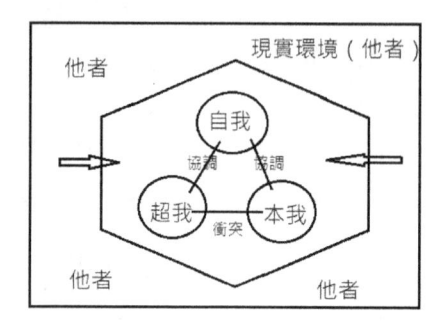

I. The Self

According to Froude, at birth there is only one personality structure, the ego. The ego (Latin: id, German:

Es) is the mind in its unconscious form, representing the original program of thought. Its energy comes from "the great reservoir".

Only the Self is the innate personality. It is the primordial desire to satisfy instinctive impulses such as hunger, vitality, libido, etc.; this word is based on the work of Georg Groddeck by Freud.

The ego is innate and the basis of the personality structure, and the ego and superego are developed on the basis of the ego.

The ego follows only one principle - the principle of pleasure - and is not concerned with the rules of society, meaning the pursuit of the individual's biological needs, such as the satiation of food and sexual desire, and the avoidance of pain or unpleasantness. It cannot tolerate any frustration, just like a coddled child who wants what it wants.

1. The ego follows the principle of happiness and is only concerned with satisfying personal needs without any material or social restrictions, so that it does not go beyond the scope, and belongs to the sum structure of the character, temperament, ability, style, etc., the biological

components of the human being's personality as expressed in life. For example, if a baby sees something it wants, it will take it, regardless of who it belongs to.

2. However, in reality, what is wanted will not all be obtained, so I will satisfy my own needs by wishing for a situation that does not exist to become a reality. For example, if a baby is hungry and there is no food around, the ego will start to imagine food to temporarily satisfy its needs.

Freud believed that the ego is completely hidden in the unconscious and that most of the ego impulses are related to sex and aggression. Freud believed that the influence of the hedonic principle is maximized in infancy, when the ego is most prominent. The ego has no judgment of right and wrong, and has the desire to destroy itself. In fact, people are often not aware of the ego.

Meaning: We pay attention only to the satisfaction of human biological needs, and through the process of socialization, the "other" of the real environment tames the human "ego" by limiting it.

Self

In the early stages of life, the second part of the

personality structure, the ego, begins to develop as the infant interacts with the environment. The concept of the psychological ego (Latin: Ego, German: Ich) is a key concept constructed by many schools of psychology, and although the usage varies from school to school, it generally refers to the conscious part of the individual.

The ego is the psychological component of personality. The ego temporarily suspends the pleasure principle with the reality principle. Thus, the individual learns to distinguish between the thoughts in the mind and the thoughts in the external world surrounding the individual. The ego regulates itself and its environment, such as delaying pleasure. Freud considered the ego as the executor of the personality.

Another confusing concept is the self, which includes the unconscious. The psychological approach to the study of the self is complex and often involves philosophical metaphysics.

The ego follows the principle of reality, and is responsible for the connection between the ego, superego, and the external world by taking into account the reality of the situation, controlling the behavior or mental activity of the ego that occurs blindly and without rational thought, and regulating the relationship between the instincts and the environment, which is a psychological component of the personality structure.

In reality, the impulses of the ego are not accepted by society and therefore pose a threat to the ego. The task of the ego is to control unconsciously the behavior or mental activity of the ego that arises from emotional excitement without rational thought.

The ego does not only control the irrational mental activities of the ego that are motivated by emotions, but also tries to satisfy the needs of the ego by considering the consequences in order to reduce the tension of not satisfying the needs of the ego. For example, when they are small, babies try to relieve tension by going to their parents' bowls and grabbing food.

When they grow up and go through civilized culture, they understand the rules and constraints of society. Although the ego's emotionally aroused, irrational behavior or mental activity will drive them to get food when they are hungry, the ego understands that such behavior is inappropriate.

4. The ego moves freely in the consciousness,

preconsciousness, and unconsciousness.

The relationship between the ego and the self: Froude once said that the ego is like a person riding on the back of a horse, driving the untamed horse (the ego) and restraining its direction of advancement. Meaning: The "ego" is a compromise between the biological desires of the individual and the social norms of the "other" in the real environment.

Superego

The super-ego (English: super-ego, German: über-Ich) is the controller of the personality structure, governed by moral principles, desiring perfection, and belonging to the moral part of the personality structure.

In Freud's theory, the superego is a symbolic assimilation of the father figure and cultural norms. The third part of the personality structure, the superego, begins to form when the child is about 5 years old. Due to the conflict over the object, the superego tends to stand in opposition to the primitive longings of the "ego" and to be aggressive towards the "self".

The superego operates in the form of a moral mind, maintaining the moral sense of the individual, avoiding taboos, and putting the ego in a difficult position. The formation of the superego occurs during the breakdown of the Oedipal complex, an internalized identification with the father figure, as the boy is unable to successfully maintain the mother as the object of his love, and retaliates against the father's possible castration of him by imposing whipping or physical punishment to make him submissive and humiliated. The castration anxiety develops, which then turns into identification with the father.

The superego represents the ideals and values of the society, especially the values and standards of the parents, and imposes moral restrictions on the good and bad behavior of the individual, and by suppressing the emotional excitement of the ego, and the behavior without rational thinking, or mental activities, the ego is persuaded to replace the realistic goals with moral ones.

For example, when one sees a $100 bill at a friend's mahjong table, one's ego impulse is to take it for oneself; the ego is aware of the possible consequences of doing so, but tries to find a way to take the money without being discovered.

When the ego thinks of a way to take the money

without being found out, the superego will prohibit the act with the sinful feeling that stealing money is against moral principles. Therefore, some people also call the superego directly as conscience.

2. The superego represents reward and punishment. Not only does it punish people for their actions, causing them to develop a sense of guilt and inferiority, but it also rewards them for moral behavior that is in accordance with the norms and standards of human behavior in living together, causing them to develop a sense of pride in how we evaluate and view ourselves.

If a child does not have a fully developed superego due to a wrong upbringing, then as an adult, he or she will personally use sensory, perceptual, thinking, memory and other mental activities to become aware of and recognize changes in his or her physical and mental state and in people, events and objects in the environment, and will lack internal control mechanisms for behaviors such as stealing and lying.

4. Some people's superego is too strong, too much to follow the norms and principles of behavior and behavior when living together as human beings, and will face difficulties in making the non-existent situation a perfect

standard of reality, and then constantly realize moral anxiety in practice, such as shame and guilt.

Meaning: A social and cultural code of conduct and a social ideology. It is the sum of behavioral norms that regulate the relationship between people and individuals, individuals and society. Such as honesty and hypocrisy, good and evil, righteousness and unrighteousness, justice and partiality, etc. all belong to the moral field of expectation, forming the dynamics of "superego".

4. The relationship between the coordination and conflict between the ego, the self, and the superego

The founder of the psychoanalytic school, Froude, believed that the human personality is a whole, which consists of three parts, namely, the ego, the self, and the superego.

The ego - is the most primitive part of the personality structure. The components that make up the "ego" are the most basic needs of human beings. It includes hunger, thirst, sex, etc.

"The needs in the Self, when they arise, need to be satisfied immediately. What does that mean? For example, if a baby is hungry and needs milk, it does not care whether the mother has milk or not, it is for the purpose of satisfying its own needs.

So the Self follows the Happiness Principle. The superego - the highest part of the personality structure, which has the status of control. The components of the Superego are: social and moral norms, personal conscience, and the ego's ideals.

"The Superego requires the Ego to satisfy the Self in a socially acceptable way. Therefore, the "superego" follows the "moral principle". Some say "perfection principle" or "ideal principle", but the meaning is the same.

Ego - can be understood as the self. It is what one is aware of. For example, thinking, feeling, judgment, memory, and so on.

The function of the "ego" is to seek to satisfy the impulses of the "ego" and to protect the whole body from harm within the limits of the "superego". Therefore, the ego follows the "reality principle". "The Ego is particularly busy and sometimes exhausted. Everything we do in our lives is done by the Ego.

If we don't do it well, that is, if we don't "balance" the contradiction between the impulse of the "ego" and the limitation of the "superego", some symptoms such as anxiety and depression will appear. Let's take an example.

For example, the sudden death of a loved one is an unexpected blow to the "ego", while the "superego" is able to think rationally about life and death, and the "ego" uses rational thinking to control the excessive sadness of the "ego" and stabilize the abnormal emotions. If rational thinking cannot resolve the inner conflict of the "Self", then the person's psychological symptoms are inevitable.

Therefore, the ego, the self, and the superego constitute the complete personality of a person. All psychological activities of human beings can be reasonably explained by their connection, and the "ego" is permanent.

The "superego" and the "ego" are almost two individuals whose needs, values and interests lead to actual or imaginary opposition, and whose performance is permanently opposed.

At the same time, psychic energy determines the behavior of the individual, and the distribution of psychic energy among the id, ego, and superego determines the dynamics of the personality. The distribution of psychic energy among the ego, self and superego determines the dynamics of personality. The ego, self and superego complement and oppose each other.

If one of them gains the main energy, the other two have to lose a certain amount of energy. In the case of individuals, the strong ego in a healthy mind does not allow the ego or superego to take charge of the personality, so the disputes between the three will never end due to their different opinions. In each of us, there is always a tension between the ego, the reality, and the moral standards.

At the same time, we are aware that we exist "for others". Without the consciousness of the "other" in the real environment, my subjective consciousness cannot be established. I can only confirm my existence by projecting myself out and becoming aware of the existence of the "other" in the imaginary, real environment.

In a word, both Hegel and Scharthe emphasize the ontological significance of the "other" in the real environment for the formation of the subject's "self-consciousness". Moreover, both of them believe that the basic relationship between the subject and the other is conflict.

In Hegel's view, the relationship between master and slave is not mutual; "a one-sided and unbalanced recognition" takes place between them. The slave is an intermediary in the relationship between the master and the object, and his status is instrumental.

The "other" of the real environment is a means rather than an end, and the "other" of the real environment appears itself as intrinsic.

For Schart, mutual gaze between subjects is impossible; "A gaze cannot gaze itself; as soon as I gaze at a gaze, it disappears; I only see the eye. In this moment, the other becomes my possession and recognizes my free existence.

Neither of them can be seized without contradiction, neither of them is in the other and "kills the other". "Conflict is the original meaning of existence for him. This determines that the attitude toward the "other" in the real environment is conflict rather than dialogue or anything else.

A completely isolated individual cannot survive, and certainly cannot be free. The human being itself is only a concept relative to others. Without others, the human being would no longer be a human being, so the human being needs others and cannot live without them.

In the mutual relationship with others, if you choose to do everything to realize your own selfish and despicable desires, and even sacrifice other people's freedom to realize your own desires, you will push others into hell, and other people's lives are all projections in your life, and those who harm others will harm themselves! All "hell" begins with a selfish choice.

The result is that we all live in the hell of "no one is better", and if we want to harm others, if we want to fix people, the final result is bound to be harming ourselves and fixing ourselves, lifting stones to hit our own feet.

In other words, if you are the cause of the deterioration of the relationship with others, you will have to bear the pain of hell. The choice of a narrow and selfish mind can become one's own hell, while the choice of a cheerful and broad mind can become the heaven of others.

They choose to chase and torment each other in order to realize their own selfish, despicable desires, and they impede the freedom of others when they realize their own freedom. That's why others become hell.

Through the example of three dead "dead and living" people, the play is meant to remind many living "living dead" people of this truth: if they choose to push others into hell, then they have chosen hell for themselves.

In his Being and Nothingness, Sartre argues that man always sees the "other" of the real environment as an object, which brutally deprives others of their subjectivity and subjectivity and turns living human beings into "things".

He believes that man is born free, he decides himself and designs himself. However, in the eyes of others, this person is only a free existence, that is, a thing, like a chair or a stone.

In this sense, the other is a limitation or even a denial of one's autonomy. The gaze of others not only turns the free subject, the "I," into a rigid object, but also forces the "I" to judge itself more or less as they see it, to pretend to be white, to concentrate on modifying its own consciousness of itself. Of course, I do the same with others.

So, "I try to free myself from the domination of others, and in turn try to control them, while they try to control me.

In order to realize our own subject and freedom, we have to be in conflict with others, and our relationship with others is in a never-ending conflict, in which others (specifically their gaze) are hell to us.

For Schart, this is the explanation of human hostility on a dialectical level, and it is what the play "Lockout" aims to reveal.

If one lives too much on what others think of one's self, that is, on what others see, then one cannot exist for oneself, which is, of course, like being in hell. This is exactly what Garson did. He never introspected himself and changed his thoughts and behavior, but he was always obsessed with the judgments others gave him.

His editorial colleagues would talk about him as a coward, and his successors would always hold that view. "My life is in their hands, and they have concluded that I am a coward without even thinking about me.

After his death, he still tried to convince Estelle that he was not a coward, believing that he could be saved by turning to her alone for approval, but Estelle was not interested. Disappointed, he goes back to Ines, who is willing to use her brain, but receives the opposite answer, which causes him even more pain and leads him to spiritual hell.

In the play, the three of them expose each other to the gaze of others, to the judgment of others, and there is no night, this gaze is eternal and inescapable.

As Ines once said, "You will always live in my gaze, like a patch of light in a sunbeam. In such a situation, "looking" becomes the most common form of torture, because looking makes the person being looked at a "thing," and they not only look at each other's appearance and actions all the time, but they also learn about each other's past and actions.

They not only look at each other's appearance and actions all the time, but they also have the desire to know each other's past and privacy. They want to understand each other and deceive each other at the same time. Even when two people want to reconcile and try to get together, they are immediately separated and provoked by a third party and have to isolate themselves again. Even if the third party does nothing, the other two people still feel that they are under the gaze of the "other" in their physical environment.

Driven by various desires, they sometimes collude and hurt each other. In this way, they fall into a kind of endless mutual torture. They both try to escape, but they cannot make up their minds, because they cannot give up their own desires.

When a person sees two other people together, there is always jealousy and a feeling of isolation that drives him to provoke and separate the other two. In this way, the "eyes of others" become instruments of torture and fire, and they torment each other and fight against each other, but they cannot be freed and free.

In fact, when the three men are sent to hell, they have only one dramatic action to perform, which is to try to avoid the torture and pain of hell; in other words, to achieve some kind of liberation and freedom. The whole play shows the futile efforts and ultimate failure of the three men in their quest for liberation in the confined environment mentioned above.

The only thing that tortures and restrains them is their relationship with each other. In the fifth scene, Garson's words break the theme of the play: "You are under the impression that there is brimstone in hell, that there are roaring fires, that there are iron bars for branding people! What a big joke! There is no bar, no hell, there is someone else.

Thus, Sartre's saying that others are hell does not mean that all the people around you are hell to others, but that when one cannot get along with the people around him or her, when it is difficult to reconcile them, others become hell to oneself.

It is not difficult to study the whole play and find that the whole plot takes place against the background that the three main characters in the play are all dead people who cannot change what they did during their lives and have already been evaluated in their coffins.

Therefore, they have no action and can not act, and can only rely on the eyes of others to know themselves. Yet here is the crux of the matter - only through self-choice can people determine their own existence, and only through self-choice can they gain freedom.

In Sartre's view of anthropology, it is important to be able to fight to treat oneself right. It is easy to look for objective reasons such as society and others, but often neglects one's own subjective reasons.

In "Confinement," Estelle does not treat herself

seriously, but only pursues animal instinct-like pleasure, so she takes the path of crime and ends up in her own hell; Ines is able to think but is led astray by the lust of homosexuals, and is bent on doing evil.

Garson is unable to make the right choice beforehand, and is afraid to take responsibility afterwards, and uses the judgment of others as a criterion for self-evaluation.

Schopenhauer once referred to egoists as "madmen locked in an impenetrable fortress", and "Confinement" depicts the three madmen who are locked in an impenetrable fortress and suffer forever. Therefore, Sartre's theme of "the other is hell" is an explicit reference to "the hell of the other" and "the hell of the self".

In this way, Sartre calls on people not to be evil, to know themselves seriously, and to use the power of freedom as a weapon to break down hell and save themselves, so as to break through the prison of the soul of the self and open up new horizons for the free mind. The important idea of "self-struggle and the pursuit of freedom" is perfectly embodied in Sartre's "Confinement".

The past is the ever-growing totality of the Self that we are, but temporality is what allows us to be this Self in a different way, or to be this Self, but not to experience this Self in the same situation.

Because the Self cannot coincide with the Self, the Self can only assume this existence by recovering this existence, but this existence makes the Self deviate from its existence. "So the past is the Self that I am as a transcendent being."

But then Sartre makes a few more points: because the past is the Self, consciousness is the Self. "The past is an entity that I can no longer experience," and therefore Descartes' formulation should be "I think, therefore I was.

There is an absolute difference between the past and the present, a philosophical term that characterizes the philosophical area where things are distinct from each other and from themselves. It is also called difference. Differences are divided into external differences and internal differences.

The external difference is the difference between things; the internal difference is the antagonistic factor and the antagonistic tendency within things, that is, the contradictory nature of things themselves that have not yet been radicalized, into which I cannot enter because the past exists.

The only way I can become the past is to become the Self myself, in order to make myself disappear in the past in the same way.

But this is unacceptable. The shame of yesterday was self-made shame, and the shame described now is describable self-made, self-made. The Nazi concentration camps were different in nature from the "death camps," for example, in that they were physically or mentally very painful.

The death camps were the product of the Nazi Final Solution, whose purpose was solely to slaughter peoples such as Jews, Gypsies and Poles, and the pain of slaughter was always the pain of others.

The pain of slaughter is always the pain of others, and the pain of the past, which has been physical or mental, is very difficult and painful, manifesting itself in a self-conscious sense, fixed in the past, silently fixed like the pain of others.

Finally, Schartre considers the past to be the opposite of value. Existence seeks the certainty of "self-containedness" toward value in a self-referential way;

it wants to be constantly certain of this. The past, on the other hand, is regrasped by the self in the form of the past, and is fixed by the self. So, in a sense, the past is also a value used to escape anxiety.

(2). The present

Now = this time, the time of speaking, sometimes including a long or short period of time before or after speaking, in the presence of...: the present. (Distinguish from "past" or "future"): How is his current situation? | Now he has become a factory manager

If you take away the "past", "present" and "future", what is left? Some people answered: "The present". This is a rebuttal: Aren't "the present" and "now" talking about the same thing?

"The word "now" is actually a relative word that defines the time that has either passed or not yet happened, which is now.

The present moment takes into account the component of the Self, which can also be interpreted as the essence, a verb that becomes it, that is, "being". The present, abbreviated as "now," is usually a concept that lies between the past and the future. Broadly speaking, "now" can also include the immediate past and the immediate future.

Buddhism

In Buddhism, the "now" is the present world, one of the three worlds (past life, present life, and future life). Buddhism and many of its related paradigms emphasize the importance of living in the present - being fully aware of what is happening, without dwelling on the past or worrying about the future.

This does not mean that Buddhism encourages hedonism, but simply that constant attention to one's current place in time and space (rather than future worries, or past attachments) helps to alleviate suffering.

They teach that those who live in the present moment are the happiest. Many Buddhist meditation techniques are designed to help the practitioner live in the present moment.

Christianity and Eternity

Christianity believes that God is outside of time and that from His perspective, the past, present and future are the eternal present. This concept of God across time reflects a form of thinking about the nature of things.

On the basis of perceptual knowledge, people generalize the properties of similar things from their many attributes and form concepts expressed in words or phrases. This absolute concept is abstract and universal.

It has been presented at least since Boethius as an explanation of God's foreknowledge (i.e., how God knows what we will do in the future before we decide what to do).

Thomas Aquinas gives the metaphor of the watchman, who represents the one who prays. They are intercessors, praying before God not only for themselves, but for His kingdom, for His people, and for His flock.

In Acts 20, Paul said to the elders of the church in Ephesus, "The Holy Spirit has made you overseers of the whole flock, so be careful for yourselves and for the whole flock, as well as on behalf of God, who stands on a high ground and looks down on a road, which God can see both before and after. Therefore, the knowledge of God is not relevant to any particular time and day.

Physical Science: Narrow Relational Theory

In the light cone, the horizontal direction represents space and the vertical direction represents time; the "now" is represented by the middle plane, where time and space meet, and the middle point represents the spacetime reference point of the observer.

According to Albert Einstein's narrow relativity theory, "absolute simultaneity" does not exist. When trying to define the "now", there are other events that can be considered as "simultaneous" with the given event and are not in a direct causal relationship.

Different observers have different views on the set of "simultaneous" events. On the contrary, when attention is paid to the "present" as a directly perceived event, rather than to memories or guesses, the "present" for a given observer takes the form of the observer's past light cone.

For any given event, its light cone is objectively defined as the set of events that are causally related to that event, but the light cone is different for each event.

Therefore, it is concluded that in the relativistic model of physics, "now" cannot be an absolute element of reality. Einstein claimed: "Those who believe in physics, as we do, know that the distinction between past, present and future is only a stubborn illusion.

In religious philosophical thinking, the wisdom of the "Great Way" of Eastern Laozhuang thought and the

Mahayana thought of Buddhist religious philosophy are both integrated, believing that there is no past because the past is the basis of the present.

Unlike the Western religious philosophy, the past, present and future are viewed separately. If we put aside the baggage of philosophical thinking, we would think that if there is no past, where is the present?

There is no future either, because the future has not yet arrived, so there is no future. People generally use their memories of the past or similar experiences to conceive specific images of things that are not in front of them, but are merely extensions of the present.

If there is no present, where will the future come from? Therefore, the present moment is only an infinite extension of the present, and all of them are important, so all of them are now.

One lives in the present. One's past is actually one's present, and one's future is also one's present.

The "now" is sometimes represented as a hyperplane in space-time, although modern physics proves that it is impossible to uniquely define such a hyperplane for observers in relative motion. The present can also be

Part 3: The structure of Sartre's theory of existence. | **789**

thought of as a period of time, known in English as "specious present" (there is no common translation in Chinese yet, some translate it as "false present" or "sensory present").

When we talk about living in the present moment, in fact, for most people, he is talking about living in the present, he is eating now, walking now, watching TV now, the present tense is always related to a certain tense of action.

Even if you say you are not doing anything now, you are saying that you are resting or you are thinking.

As long as we use a corporeal level to cut into our state of life, you will find that you can only live in the present, but not in the present moment.

No matter what you are doing now, the moment you try to describe it, what you call the present has become the past, so if you explore it carefully, you will find that the present does not exist at all, it is only an interpretation of action derived from the past and future tense.

However, we know that the past, the present, and the future all occur equally in the present moment, and what you experience from beginning to end is the same present moment, with different faces.

When you say you are living in the present moment, you are actually viewing the time flow created from this present moment from a higher spiritual perspective.

When you say you are living in the present moment, you are actually watching the time stream created from the present moment from a higher spiritual perspective, and seeing all kinds of things in the world that are constantly changing.

But the projector is not moving from the beginning to the end. The past, present, and future are all from the present moment, and what we say about living in the present moment is just a lower level interpretation of the present moment by our minds.

When you can just watch whatever you're doing right now and feel that you're separate from your body, that the body and mind are completely engaged in what you need to accomplish, that they're completely in the present, and yet you're watching it all happen from a higher perspective because you don't recognize that you're your mind and your body, that you're in the present.

When you reach this state of total withdrawal, when

you see yourself in a dualistic field, in a monistic field (a center or origin), what does the contradiction of the dualistic field in structuralism mean?

Dichotomy and dichotomy in literary theory and structuralism is a relatively broad, generic term. In general, two images abstracted from a text, which are opposed in form or structural function, can be called "dichotomies".

For example, "male and female", "good and evil", "new and old", "conservative and open", etc. are all dichotomies. A typical example is the definition of the "White Rabbit" and the "Big Bad Wolf" in Grimm's Fairy Tales, which is a dichotomy of contradiction.

What is the meaning of dichotomous thinking? What is an example? For a long time, Western philosophical thought and classical science have been dominated by dualism and dichotomous thinking.

The dualistic philosophical way of thinking is the dichotomy of "either/or, either/or". Descartes, for example, is known as a classic dualist. He believed that there are two absolutely different entities in the world - spirit and matter. The essence of the soul is in thought, and the essence of matter is in extension.

Descartes' dualism is based on the fact that the world has two separate essences, spiritual and material, which is a typical "either/or, either/or" dichotomy.

So it is confusing that we are only living in the present, not in the present moment, unless you use the present as the entrance to the present moment and realize that there is a you who is watching everything and never moving.

The Vajra Sutra says, "The past mind is unattainable, the present mind is unattainable, and the future mind is unattainable. The past and the future can explain the known facts and principles and principles in their own words, texts or other symbols, but how to interpret the "present mind is not available"? This is a question that many lovers of religious philosophy find very confusing when they first learn it. In order to understand this problem of space-time congestion, one needs to understand it from the concept of "the present moment".

Apprehension, also known as apprehension, understanding, knowing, or thinking, is a mental process of consciousness that is related to some abstract or tangible object or phenomenon that exists in nature, such as a person, situation, or message.

It is a rational and active process of reflection on the real world or any object in which a person is able to consciously reflect on it in an "inward dialogue" with the human mind, and to use the concept of going back and rethinking the past to learn lessons from it.

To the object of the action or reflection, all the objects or phenomena that exist objectively in the natural world, to be able to explain the known facts and principles and principles in their own words, characters or other symbols.

Understanding is a form of thinking that reflects the conceptual expression of the nature of things. To understand the boundaries of all objects or phenomena that exist in nature is to realize a certain degree of conceptual expression or conceptualization of that thing.

Therefore, the old Buddhist practitioners often say, "Live in the present moment" when they teach and admonish the beginners. . How do I understand this? In fact, it requires a certain level of enlightenment before one can truly understand it.

However, there is an easy-to-understand analogy that can help to understand this concept. Almost everyone has had the experience of reading a book, and when you concentrate on it, you are completely oblivious.

When you concentrate on being completely oblivious, when you are absorbed in the mood of the book, when you quietly enjoy the joy of time travel, this state can be approximated as the "present moment" that is not limited by time and space.

This continuous state of continuity from the "past" to the "present" to the "future" has several characteristics, from the moment of flipping through the book, starting to fully integrate into the environment of the book, until the present moment of flipping through the pages of the book with the book closed.

The "environment" of the book is integrated with the book; there is almost no "I" who uses mental activities such as feeling, perception, thinking, memory, etc., to be aware of the state of my body and mind and the changes of people, things, and objects in the environment, and the flow of comprehensive awareness and understanding exists.

The understanding of the continuity of thoughts, the connection of each point in time, the uncluttered and clear flow of thoughts.

This is different from the feeling of "now". The feeling of "now" can feel the instantaneous thoughts, emotions, memories, images, etc. that appear in "my" brain.

The human brain is constantly generating various thoughts, emotions, memories, and ideas. These appear in the brain in the form of images and words; they are in opposition to the environment, either trying to adapt to it or trying to change it.

The sense of time of "now" is a period of time, not a continuum of thoughts, in which all kinds of emotions can occur.

This is the subtle difference between the "present moment" and the "now". To practice is to allow the mind of "reading" to focus on a long period of time to continue and to feel the "present moment" all the time.

This state of selflessness and harmonious coexistence with the environment, many people realize long after the coarse thoughts have disappeared, that there are four stages of birth, dwelling, dissimilarity, and extinction when any subtle thought arises: this brings the practitioner an incomparable sense of happiness (dharma joy).

Therefore, unlike the past, the present is self-acting. Self-activity is the presence of all objects and phenomena in the face of objective existence, and this can only be the presence of self-activity.

The concrete fact and condition that exists in front of us does not become the future, but is no longer, the moment, the instant of the past, but it does not exist.

It is only what Husserl called a kind of being pushed into the most ideal impractical and inscrutable word of the infinite: nothingness. Its presence is not the result of a chance event or a companion event.

On the contrary, it presupposes all the objects and phenomena that accompany the existence of the objective world, but should be an ontological structure that is self-existent. The "now" is a negation, not a self-revelation of itself as existence.

The "now" is always an escape. The original meaning of the present: the present does not exist; the instant of the "now" is derived from a self-referential, materialized notion that is making a non-existent situation a reality.

It is this abstract, universal idea that acts as an indication of the domain or class of entities, events or relationships that causes the "self" to manifest in a way that is what it is, and that "self" faces its presence.

(For example, it is nine o'clock now, but it can never be nine o'clock now, but "self-existence" can face the presence of self-existence, that is, the presence of the minute hand at nine o'clock. The present does not exist, it escapes to the present in the Self by way of escape.)

However, the "now" is not merely a non-existence that deliberately escapes into the present because it is unwilling or afraid to face a certain situation or thing. As a self-act, it has an existence that is separate from itself before and after it.

In its after, it is said to have been its past, and in its before, it is said to be its future. It escapes from the being with which it is present, and from the being with which it was "to be" and towards which it will "be" in the future.

(3). Future

Future is a Chinese word, pinyin is wèi lái, meaning the time from now on, the future time in relation to the present moment we are in, it is a moment or a time period.

Relatively speaking, tomorrow is just a part of the future, and so is the next second. It is a concept of time, as opposed to the past and the present. Everything has a future. Thinking about and creating the future brings extraordinary meaning to our lives.

In the Chinese "space-time" form of thinking that reflects the nature of things, there are two most common expressions for tomorrow, namely, the future and the future. Space-time (time-space, time and space) is a fundamental concept that belongs to physics, astronomy, space physics and philosophy respectively.

In mechanics and physics, "space" is an abstract concept that describes the position, shape, and direction of an object and its motion, while time describes the continuity of motion, the sequence of events, and so on. The characteristics of space-time, mainly through the object, its movement and interaction with other objects, the convergence of various relationships

The future is the future, and futurology is futurology, and vice versa, and no one doubts that there is any difference between them in terms of the concept of space-time.

However, in the ancient Chinese language, there is still a big difference between Wei and Jiang, and this distinction is very enlightening for the discussion of tomorrow. The character 未 appears very early, written in the oracle bone script. Duan said in the "Shuowen Jiezi" that the word "wei" means "taste". June taste also. The five elements of wood old in the un, like wood heavy branches and leaves also.

From this can be seen, Wei originally refers to the tree branches and leaves lush appearance. The branches and leaves must be fruitful, so the branches and leaves of trees are stacked in layers, partially in line with each other, covering each other and overlapping each other, showing an unknown and dark appearance.

Therefore, Wei is also the original word for "ignorance", which means darkness, obscurity, that is, the meaning of not being able to see clearly. Because of this abstract meaning, wei was later borrowed to express negative meanings, such as unknown, unheard of, unavailable, not necessarily, etc.

From these meanings, we can indeed interpret the

future as meaning not yet here or not necessarily here, in other words, it expresses the unknowable and unpredictable character of tomorrow.

If we look at the word "future" in the sense of mastery or anticipation, it clearly means coming, coming in anticipation. We can say that "future" refers to the kind of tomorrow that is outside of our vision, while "future" is the in-hand tomorrow that is within our certain grasp.

The distinction between the future and the future is not a nosy one. In fact, many futurists have already made a distinction between the future and the future in accordance with the existing empirical knowledge, experience, facts, laws, perceptions, and tested hypotheses of human beings about natural and social phenomena.

In fact, many futurists have made a distinction between the future and the future in terms of logical inferential summaries of human knowledge, experience, facts, laws, cognition, and tested hypotheses, through generalizations and deductive reasoning.

As we have seen in the previous distinction between "future" time, the so-called immediate future, the medium-term future, and even the long-term future can be broadly included in the domain of the future, while the distant future is the real future.

Of course, the difference between the future and the future is not that the former is farther away than the latter, but that they are not two unrelated dimensions of time.

The difference between the future and the future lies entirely in the fact that they are different from the present in the way things interact and relate to each other.

The future is a world of chance and randomness facing the present, a tomorrow that is difficult to grasp, predict and choose; the future, on the contrary, depends on a great inertia, or human planning, towards the present.

If the concept of the future, which has been handed down from generation to generation, with social elements of inherited continuity, such as customs, morals, habits, beliefs, ideas, methods, etc., is like a marble spinning fast on a Las Vegas roulette machine, then the future is more like Robin Hood.

Then the future will be more like the arrow that Robin

Hood is ready to shoot at the bullseye. Therefore, science takes a certain object as the scope of study, and based on experiments and logical reasoning, it seeks to obtain unified and exact objective laws and predictions of truth, and the tomorrow it talks about can be generally seen, expected and awaited.

It is precisely in the way of generalization and deductive reasoning that human beings make logical inferential conclusions about natural and social phenomena in accordance with existing empirical knowledge, experience, facts, laws, cognition, and tested hypotheses.

It is very different from the traditional science fiction writers in that it is a logical inferential conclusion. The American mathematician and futurist, Helmer founded the famous Institute for the Future in 1966.

In his 1970 article "Perspectives on Technology," he divided his predictions into three broad categories: to be realized, likely to be realized, and not easily realized. He pointed out that the first type of prediction is often the most certain and has a higher hit rate.

The latter two predictions are mixed with more

unknown factors. Needless to say, the so-called tomorrow always contains unknown factors, but futurists hope to let the imagination and prediction based on the existing facts, to maximize the elimination of the impact of unpredictable unknown factors.

This prediction follows the general pattern of logical inferential summation of empirical human experience of natural and social phenomena, in accordance with existing empirical knowledge, experience, facts, laws, cognition, as well as through generalization and deductive reasoning, etc.: question - confirm - test, so futurists in the prediction, as far as possible to require a clear prediction time, prediction content clear.

Of course, people have always believed that science to a certain object as the scope of study, based on experimental and logical reasoning, to obtain a unified, accurate objective laws and truth predictions, although based on facts, but what it predicts is not yet a fact, so it is always unreliable.

Korsch summarizes this traditional human view of natural and social phenomena by generalizing and deducing logically according to existing empirical knowledge, experience, facts, laws, cognition and tested hypotheses.

The future is free, creative, legendary, uncertain, and exciting; the past is finished, dead, ready-made, and uninteresting. The past is finished, dead, ready-made and uninteresting. This is a logical inferential conclusion of natural and social phenomena according to existing empirical knowledge, experience, facts, laws, cognition, and through generalization and deductive reasoning.

The assumption underlying the view of logical inferential summation is that facts = the past, or that the past events and actions of human society, and the systematic recording, interpretation, and study of these events and actions, are always better than the future.

However, Korsch tells us that it is reasonable to think about or judge something from a different point of view or perspective, a thing that is independent of subjective thought or consciousness and exists independently of its nature.

If we look at it from a different perspective and keep its authenticity, the future is not more distant than the facts of the past. For example, the coming 2000 years are much closer to us than the 1000 years that have passed. It is difficult to know exactly what the world's population was 1000 years ago, but it is possible to determine the trend of population change in the next few decades with general accuracy.

Modern thinkers are fully aware that history is not a definitive body of facts like exhibits in a museum, but that it has undergone changes, rewrites, and re-analyses in the course of its development, and is understood from a certain cognitive perspective.

In this sense, history is another dimension of the future that modern man is confronted with. This concept tells us that the main characteristic of facts is not the past, but the world of time and space that is continuously occupied by things, accurate in time and space, and not shifted by human consciousness.

過去
未來
現在

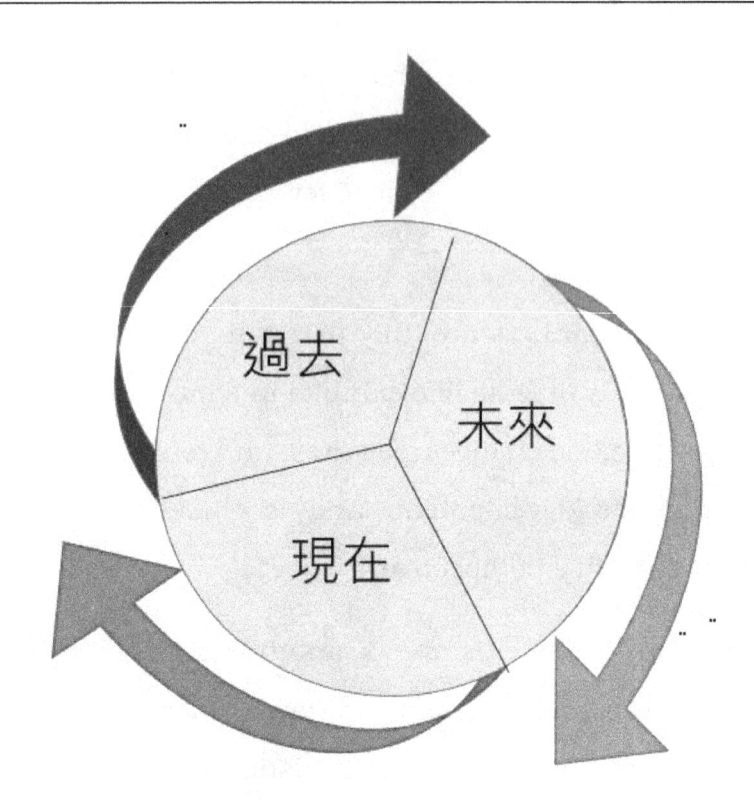

01. Breaking the concept of space-time

Breaking space-time (time-space, time and space) is a fundamental concept that belongs to physics, astronomy, space physics and philosophy respectively. In mechanics and physics, space is an abstract concept describing the position, shape and direction of an object and its motion, while time is a description of the continuity of motion and the sequence of events.

The characteristics of space-time are mainly through the concept of the convergence of various relationships between objects, their movements and interactions with other objects, the future, the present and the past, not a linear flow, from the far that is near and running to the far river, they all belong to the future now.

Philosophical presentism, on the other hand, holds the view that neither the past nor the future exists. In some versions of presentism, it recognizes the existence of temporal objects or ideas [with or without].

According to presentism, events and objects that exist only in the past, or in the future, do not exist. The present theory is opposed to the eternal theory, and to the growth block theory of the universe, which asserts that past events exist, even if they do not exist in the present. Eternalism, on the other hand, asserts that the present, the past, and the future all exist together.

Augustine of Hippo proposed that the "present" is like a knife's edge placed precisely between the "seen past" and the "imagined future" and does not apply to the concept of time.

This should be self-evident, because if the "present" is prolonged, the "present" will be divided into several paragraphs, and if they are indeed several paragraphs, they will not be simultaneous.

According to the early philosophers, time cannot be simultaneous, and therefore the "now" is not prolonged.

Contrary to Augustine, some philosophers who study universal, fundamental questions, including the fields of existence, knowledge, values, reason, mind, and language, propose that individuals use mental activities such as sensation, perception, thinking, and memory to make sense of their own states of mind and body and their environment.

The experience of being aware of the state of one's body and mind and the changes of people, things and objects in the environment, and the experience of knowing them, is the extension of time. For example, William James said that time "is prolonged in our short-lived and continuous perception.

A major scholar of modern Buddhist philosophy, Fyodor Scherbatsky, devoted a great deal of space to describing Buddhist philosophical theory: "All that is in the past is not real, all that is in the future is not real, all that is imagined, all that is lacking, all that is spiritual... is not real. Nothing is real. Only what happens now (i.e., causality) is real.

02. The unreality of time

The Unreality of Time is the most famous philosophical work of Cambridge University conceptual philosopher J.M.E. McTaggart. In a 1908 article published in the philosophical journal Mind, McTaggart argued that time is not real because our description of it is either

This is because our descriptions of time are either contradictory, cyclical, or inadequate. To construct his argument, McTaggart identifies two descriptions of time, which he calls the A-series and the B-series.

Series A indicates the location of time in terms of "...is the past," "...is the present," and "...is the future"; Series B indicates the location of time in terms of "...earlier than..." or "...later than...". B series indicates the position of time by "...earlier than..." or "...later than...".

To refute the A-series, MacTaggart argues that any event in the A-series has all three properties, past, present and future, but that it is contradictory for an event to have all three properties at the same time, since any one of the three properties is mutually exclusive with the other two.

He goes on to claim that it is circular to describe an

event at different points in time as [past], [present], or [future], using various rhetorical devices to visualize things.

Because we need to describe these "different points in time" again in terms of [past], [present], or [future], and then describe these descriptions again in terms of [past], [present], or [future], and so on in an infinite cycle.

In order to criticize or reject the B-series, the process or form of language in which J.M.E. McTaggart uses certain reasons to support or refute a point of view, usually consisting of thesis, argument, justification, and manner of argument time,

contains change, but definitely does not contain change because of the earlier-later relationship (e.g., 2010 is always later than 2000), so the B series must not adequately describe time

Theory A is closely related to the present theory, while theory B is closely related to the eternal theory. According to J.M.E. McTaggart's "The Insubstantiality of Time" view of A-series and B-series, the present theories conclude that A-series is the fundamental theorem and B-series is inadequate.

The present theorist maintains that the need to use tense time is a non-permanent argument, while the old B-theorist argues that tense can be reduced to tenselessness, the act of "knowing" and "identifying" a subject, and the act of "recognizing" it with certainty.

It is the act of "cognition" and "identification" of a subject, in order to ascertain knowledge, and that such knowledge has the potential to be used for specific purposes. It refers to the ability to become familiar with something through experience or association and to understand it further.

A discipline in which analytical thinking is used to inquire and reflect on fundamental questions about life, knowledge, and values. For example, "Is there an objective standard of morality?" What is science? and "Can AI computer language programs think for themselves? In: Series A and Series B are two different descriptions of the temporal relationship in events. The difference between the root, foundation, or main part of these two series is the temporal usage of the descriptions of the interactions and interconnections between the temporal things between the events.

This "vocabulary" was first developed in 1908 by the

Scottish conceptualist, philosopher John McTaggart, in relation to the insubstantial nature of time, as a result of thought activity, in order to influence the objective existence of others, reflected in human consciousness.

or formed viewpoints and systems of concepts or actions, and the process of forming reasons, justifying beliefs, and drawing conclusions is used, and since then this "vocabulary" has been widely cited in contemporary discussions of the philosophy of time.

According to McTaggart, there are two different patterns in the ordering of all events in time. In the first model, events are ordered by means of the unrelated singular prepositions "...is past", "...is present", and "...is future".

When we represent time in this way, we are representing a series of positions that run from the farther past through the nearer past to the present, and from the present through the nearer future to the farther future.

The fundamental feature of the description of all objects or phenomena that exist objectively in nature, the use of various rhetorical techniques, the detailed visualization of things, the in-depth description and presentation of patterns, is that the series of time and position must be thought of as an ever-changing existence.

In other words, an event is first a part of the future, then a part of the present, and finally a part of the past. Moreover, the assertion according to this model implies the user's own view of time.

The second point is that we can order events by means of asymmetrical, non-transitive and transmissible binary relations, like "before" and "after", which is the B-series, and say that all truths about time

This is the B-series, and all truths about time can be understood and described by decomposing and dismantling their parts, such as anatomy. The B series, which describes a holistic, fundamental and critical view of inquiry into the real world and people, is known as the B theory of time.

The logic and language of these two series are expressed in ways that differ at the ends of the object; series A is tense while series B is tenseless. Tense is a formal distinction in a verb that indicates the past, present, or future, or the duration of the action or state it indicates, past tense.

For example, "It's raining today" is a tense assertion because it relies on the speaker's tense perspective-that is, in the present. However, the claim that "it rained on June 15, 1996" is tenseless because it does not depend on the speaker's tense perspective.

This is the truth value of two assertions in logic. The truth value, also known as the logical value, is an indication of the extent to which a statement is true. In computer programming, they are often called boolean and boolean values.

In classical logic, the only possible truth values are true and false. But in other logics, other truth values, too, are possible: fuzzy logic and other forms of multi-valued logic use more truth values than simple truth and falsehood. Algebraically speaking, the set of truths and falsities forms a simple Boolean algebra

From the point of view of the truth values of the two assertions above, if both propositions were made on June 15, 1996, then they would both have the same truth value (either both true or both false).

The precedence of the two events is not a temporal

relationship, saying that E precedes F and does not change over time. On the other hand, the event E or F is a feature of the past, present, or future, and will change over time.

If we assume the first view, then we seem to say that the B-series slides along a fixed A-series; if we assume the second view, then we seem to say that the A-series slides along a fixed B-series.

Therefore, the B-theory of time (English: B-theory of time) is the name of one of the two positions in which human beings of time make logical inferences about natural and social phenomena in accordance with existing empirical knowledge, experience, facts, laws, cognition, and tested hypotheses, by means of generalization and deductive reasoning, etc.

In contrast to the A-theory, which represents the time that can run in an orderly line, the B-theory is a process or form of language that supports or refutes a point of view with certain reasons, usually consisting of a thesis, an argument, an argument and a way of argumentation.

The flow of time is an illusion of imagination, the past, present and future are equally real and time is timeless.

This would imply that the change of time is not an objective reality.

McTaggart observes that events (or "time") can be used in two different but related ways to give knowledge of natural or social things, the meanings one gives to objects, the spiritual content that humans transmit and communicate in symbolic form.

On the one hand, they can be given meaning in terms of past, present, and future, which are generally represented by tense in many natural languages (e.g., English); on the other hand, events can be expressed as "earlier", "at the same time", or "later".

Philosophers have thus debated the use or non-use of tense to describe events that are consistent with the nature of time. Some philosophers even propose a mixture of laws or arguments that are systematic and organized.

They are theories derived from actual verification, or from conceptual deduction, and advocate a non-temporal time, but indicate that time has some special qualities [in violation of MacTaggart's [paradoxical] requirements].

The controversy between Theory A and Theory B is a continuation of a related controversy between Heraclitus and Parmenides' ontology dating back to the ancient Greek era. Parmenides believed that the real world is eternal and unchanging.

In contrast, Heraclitus believed that the world is a process of constant change, flux, and decay. For Heraclitus, the real world is dynamic and transient.

Moreover, according to Heraclitus, the real world is so fleeting that it is impossible to "step into the same river twice".

The metaphysical problem of delineating the A-theory and the B-theory of time is the study of questions that cannot be answered directly by perception through rational reasoning and logic, and it is the search of human reason for the most universal aspects and ultimate causes of things, involving the truth of the past and the reality of the future, as well as the ontological state of the present.

The difference between A-theory and B-theory is often described as a controversy about the passage of time or "gradual change," and B-theorists argue that this passage of cognition refers to the process of knowing things in the outside world, or the purely psychological process of processing information about things in the outside world that act on one's sense organs.

Many A-theorists believe that B-theorists deny the passage of time, which is the most important and significant feature of time. It is a common, if not universal, view to classify the A-theory view as a belief in the passage of time.

B-Theorists such as D.H. Mellor and J.J.C. Smart want to eliminate all arguments about the past, present, and future with the non-temporal objective existence of all objects or phenomena in the natural world in order to believe that the past, present, and future are equally real.

It rejects the idea that they are time, and does not think that complex systems, events, and phenomena can be understood and described by decomposing them into combinations of parts.

B-theory also argues that the past, present, and future have very different characteristics in terms of cognition and reflection.

For example, we remember the past and predict the future, not the other way around, and B-theorists argue that the fact that we know so little about the future reflects only the fact that the past and the future have been the subject of doctrines about the sources of knowledge, the processes of development, the ways of knowing, and the relationship between knowledge and practice.

They can be divided into empiricism and rationalism. The distinction between the doctrines of the source of knowledge, the process of development, the method of knowing, and the relationship between knowledge and practice: the future is not less real than the past, we just know less about the future.

Therefore, the future 1. cannot exist as an image, and 2. cannot be the content of an image when it is manifested, and if it is the content, it is the present.

Therefore, there is a different existence in the future than that which is simply perceived. Nor can the future be regarded as the most important part of the self, which still indicates the validity of the present; nor can the future be regarded as a present that does not yet exist, and we will fall back into the self.

What the future is, the future is the lack of this [future, future, not-yet-arrived time] from the present, from the instantaneous self-existence, and self-existence is not equal to the pure present, and the future is not equal to the not-yet-existing present, otherwise it would all be self-existence.

Self-existence is the future coming into its own existence, and it exists toward the future existence (the meaning of playing tennis is to catch the future ball.)

自在	虛無/否定	自為	可能性
過去	現在	將來的現在	將來

Heidegger: Man is a distant being, Shasa: Dasein from yet sein to obtain man's knowledge of natural or social things, is the meaning that man gives to the object things, is the mental content that man transmits and communicates in the form of symbols.

And the temporality of Heidegger and Sa is essentially the same in terms of the pairing, arrangement, or construction of the complex mental processes of seeking or establishing rules and evidence to support or determine a belief, decision, and action.

Heidegger is the future of the past making the present emerge. The core of Heideggerian temporality is the future, while the core of Saudi temporality is the present, and this creates the Saudi humanism of doing whatever one wants.

03. Ontology of Temporality

Schart believes that the three-dimensional phenomenological description of time enables us to see [temporality] as a whole. How is this deeply understood? Temporality is the basis on which time is established, often starting from a "constant" and distinguishing the order of "change".

It is something that exists independently according to its distinguishability and within itself. But it need not be a physical existence, or an explicit rather than an abstract, general concept, and thus there is a distinction between past, present and future.

The French philosopher Henri Bergson (1859-1941) introduced the concept of durée to illustrate temporality.

He believed that people's awareness of their own state of mind and body and the changes of people, events, and objects in the environment, using mental activities such as sensation, perception, thinking, and memory, is originally a continuous flow of awareness and knowledge.

In other words, people are in a continuous process of awareness and integration of external things. It is only when one recalls the continuous flow of the past that one recalls and notices the continuous moments of the past.

In this way one can recall the "past". M. Heidegger (1889-1976) continued the position of E. Husserl (1859-1938) in analyzing or criticizing problems and issues based on the study of "inner time consciousness".

From the perspective of "existentialism," theoretical descriptions, a single or inductive exposition of a particular issue, event, study, etc., and the problems or solutions to them, in his own words, to narrate temporality.

He thinks that the meaning of "care" (care; Sorgen) is based on temporality. That is, the reason why "this is" (Dasein) can constantly show the instrumental world or "suspend" the other "this is" is due to temporality.

As someone wants to travel from one place to another, i.e., as people reflect reality through concepts, judgments, and reasoning in the process of knowing, it is possible to travel from this place to the other world on foot or by car.

At this point, all kinds of possibilities are revealed, based on the instrumental world that has been shown in the "past". And when one makes a decision "in the present", the thought immediately reveals itself, pointing to the world to come "in the future".

Therefore, Heidegger thinks that any "being" that reveals its "existence" is based on the "past" of backward and backward thinking; in the "present" it reveals a logical thought that points to the "future" world.

Section 7: Possible causes of errors in logical thinking.

Logical thinking is a mental process of purposefully moving from the known to the unknown according to objective laws or principles. The most influential theorist in the development of logical thinking is Jean Piaget (1896-1980).

Piaget believed that the cognitive ability of human beings to recognize the environment and the self, as well as the degree of cognitive clarity, is only reversible at the age of eight to eleven when the child enters the concrete operational stage and is able to think logically about immediate events through mental operations.

However, at this time, the various mental operations are not allowed to come together and are isolated from each other until adolescence and the formal operational stage.

Children in the formal operational stage are able to consider what is possible and what is true, and to systematically interpret the truth of propositions and the interactions and interactions between all objects and phenomena that exist objectively in nature.

Unlike children in the preoperational period, they can only determine the truth or falsity of the proposition itself, and they can only engage in a rational and active process of reflection on the real world or any object in the form of "inward dialogue".

The so-called inward dialogue is like talking to oneself, but without the condition of the listener, the concrete facts and conditions that exist in front of one's eyes, or the explicit, rather than abstract, generalized content.

Usually we say that a person who is good at deducing conclusions from known or assumed premises, or at reasoning from the results of known answers, is someone who can logically carry out thinking activities such as analysis, synthesis, reasoning, and judgment.

However, in the case of informal deduction from known or assumed premises, or in the case of a known answer, inverse reasoning, especially in the case of social issues, personal beliefs are usually added to the judgment, so that the conclusion of "informal reasoning is not the truth".

However, at least the logical thinking of a mental process of purposeful deduction from the known to the unknown according to objective laws or principles can help individuals make the best choice or reach an appropriate conclusion.

This is the case when individuals use their senses, perceptions, thinking, memory, and other mental activities to perceive and understand their own physical and mental states and the changes of people, events, and objects in the environment, and we hope to reach the most reasonable conclusion rather than the right answer.

Sometimes the reasonableness of the conclusion is even assessed based on the person's ability to recognize the environment and self and the clarity of cognition.

The grammatical structure, sequence, and wording of the question itself affects the subject's ability to "recognize" and "identify" the behavior of a subject with confidence.

These perceptions have the potential to be used for specific purposes. It means the ability to become familiar with the solution to a problem through experience or association. There are two possible reasons for errors in logical thinking.

(1) Not familiar enough with the meaning of logical terms: It refers to the individual's use of sensory, perceptual, thinking, memory and other mental activities, the state of his or her body and mind, and the environment, people, things, and changes in the comprehensive awareness, and knowledge of logical terms of people's knowledge of natural or social things, is the meaning given to the object things, is the human symbolic form of communication.

It is the spiritual content that human beings transmit and communicate in the form of symbols. In other words, all the spiritual contents that human beings communicate in the communication activities, including intention, meaning, intention, understanding, knowledge, value, concept, etc., are included in the field of meaning.

We are not familiar enough with the meaning, and as a result, we are unable to analyze the logical structure of the problem. In this case, the test results will be improved if the logical terms in the questions are restated with understanding.

(2) The number of messages that can be processed by short-term memory within a specified time frame affects the test results. Some scholars believe that this is due to the fact that individuals use their sensory, perceptual, thinking, memory, and other mental activities to

Through the cognitive ability of a person's environment and self, as well as the cognitive clarity of experience or association, the person is able to become familiar with and further understand something.

The influence of logical space, which is used to organize relationships. For example, "A is better than B and B is better than C" is easier to understand than "C is worse than B and B is worse than A".

The overall, fundamental, and critical exploratory research conducted on the real world and human beings proves that, by increasing the number of subjects who analyze the topic in detail and understand it from a certain cognitive point of view, they can improve their understanding in the cognitive process.

Therefore, it is clear that although logical thinking is highly correlated with intelligence, it can be acquired through learning.

In this way, a person can think logically and use concepts to deal with the object appropriately, understanding that "past," "present," and "future" are in fact integral, and that this integral is "temporality," and that the distinction between time is based on temporality.

A. Schütz (1880~1953) applied the theory of temporality to sociology. He continued Bergson's, Husserl's, and Heidegger's views by distinguishing between the motives of each social action: the motive of cause and the motive of purpose.

The former is the cause of an action, and the latter is the purpose of an action. Usually, when an action occurs, it has both of these motives, but people do not distinguish between them.

Motivation is an internal process in psychology; in terms of the relationship between stimulus and behavior, motivation is a mediating variable. Motivation gives an individual energy, triggers an individual's activity, and maintains and promotes that activity toward a fixed goal.

It arises from an internal physiological or psychological need or an external trigger that causes the individual to be in a state of tension. The internal processes that the individual goes through to eliminate this tension and restore balance are the motivation.

For example, thirst is the physiological tension generated by the unbalanced state of water in the body, which will prompt the individual to seek water to quench thirst, which is the role of motivation. After the thirst is quenched and the physiology is balanced, the individual will no longer seek water.

Scholarships are an incentive. Students who care about scholarships are bound to become nervous and study hard to get scholarships with good grades.

For example, when a person holds an umbrella in the rain, he may hold the umbrella because he "used to" catch a cold due to the rain, or he does not like the feeling of wet clothes, so he holds the umbrella to achieve the purpose of not being caught in the rain.

The former is the cause motive and the latter is the purpose motive. In other words, the cause motive is based on the "past" experience, and the purpose motive is directed to the "future", so an action is a combination of "past", "present" and "future".

From the above discussion, we can deduce that temporality is the foundation of internal self-discovery and external innovation in the cognition of self-liberation.

The reason why we need to have the motivation and review of prior knowledge is that the evocation of past experiences helps to achieve the act of "cognition" and "identification" of a subject in order to know it with certainty.

And that these understandings have the potential to be used for specific purposes. This means that through experience or association, one is able to become familiar with the cognitive outcome of an event, which is based on the continuity of consciousness, i.e., time.

On the other hand, an individual's comprehensive awareness and knowledge of his or her physical and mental state and changes in people, events, and objects in the environment by using sensory, perceptual, thinking, and memory activities should be connected with past experiences.

This is not a discontinuity. This shows that both the past learning experience and the upcoming knowledge content should take into account the temporality or consciousness as a whole in understanding the theory and practice of a certain topic. But the study of it should start from two aspects

(A), static temporality: from the [before - after] arrangement principle, determined by the [before - after], these human beings in the process of understanding, from perceptual knowledge to rational knowledge, the perceived things, the

common intrinsic characteristics of abstraction, to generalize, is a study of the expression of self-cognitive consciousness, can be strictly in accordance with their order, and completely aside from the so-called changes (mundane world)

(B), dynamic temporality: time is not only in a fixed order, but from after before, which we have in Heidegger, in the process of knowing by means of concepts, judgments, reasoning reflecting the process of reality.

It is different from figurative thinking, which is the process of using scientific abstract concepts and domains to reveal the nature of things, expressing the results of knowing reality by inference; as well as in the sasa of natural and social phenomena, in accordance with existing empirical knowledge, experience, facts, laws, cognition

The logical conclusion of natural and social phenomena by means of generalization and deductive reasoning also proves that the present changes into the past and the future into the future.

The objectively existing [before-after] thing of all objects or phenomena in nature is a kind of irreversibility, but people examine the items non-continuously by "before and after".

Without a precise and brief description of the essential features of a thing or the connotation and extension of a concept before or after, we would immediately become "I".

Then we immediately become what I was before and after, and the distance between me and my "before and after" is gone. The separation between action and dream "before and after" is also gone.

What novelists and poets emphasize is precisely this "before and after" separateness, a notion within the scope of dynamics, where the prize passes, where time disappears, where there is a "before and after" separateness, where it is separating. Escape.

"Time will solve everything", that is to say, the separateness of "before and after" time, eliminating the process of acquiring knowledge or applying knowledge, or the process of processing information, which is the most basic mental process of human beings. It includes the separation of sensation, perception, memory, thought, imagination, and language.

It is the process of reflecting reality through concepts, judgments and reasoning in the process of knowing. It is different from figurative thinking in that it uses scientific abstract concepts and fields to reveal the nature of things and express the results of knowing reality.

In fact, the temporality in the eyes of philosophers is not so difficult to interpret, it is just a comprehensive awareness and knowledge of our own physical and mental state and the changes of people, events and things in the environment by using our senses, perceptions, thinking and memory and other mental activities.

Beyond the process that gives us the knowledge of external things, or rather the scope of information processing of external things that act on the human sense organs, we are only accustomed to it as a function of the instrumental state that we are in time.

And when we act or think about the things that are the target of our actions, when we use them for interactions of a certain nature between people and people or between people and things, we have to use again those mundane temporal concepts. On this level, we are not pure materialists.

(A) Static temporality

Criticism of Descartes as a dualist and a rationalist. Descartes believed that humans could use mathematical methods - that is, reason - to think philosophically. He believed that reason was more reliable than feeling.

For example, he believed that numbers and laws of physics, which are beyond the senses, are known through rational thought and cannot be known through the senses.

The book expresses Kant's exploration, criticism, and supplementation of the theory of knowledge. Kant argues that knowledge is possible because the human senses have an innate and universal intuition of space-time, and through the a priori form of space-time, the human senses can know things.

Secondly, reason is a synthesis of "concepts" through enlightenment of the substance that is unified by the human senses according to the "quantity," "quality," "relationship," and "state" of thought.

The criticism of Kant follows the Critique of Pure Reason and proposes the Critique of Practical Reason. If [the Critique of Pure Reason] deals with the world of natural objects, [the Critique of Practical Reason] explores the world of morality with human beings as the subject.

In the Critique of Practical Reason, Kant sets up the necessary existence of free will, God, and the immortality of the soul. Neither God nor the immortality of the soul can be proved in the realm of physical experience, except for the free choice of free will to comply or not with the Categorical Imperative.

However, God and soul immortality are related to the

intrinsic purpose of moral judgment, or moral motive, which has no real substantive content in moral philosophy, but is only a formal moral decree.

In this regard, the distinction between free will and the immortality of God or the soul highlights the possibility of conflict or inconsistency between reason and practice. Contrary to Kant, Leibniz, Descartes, and Bergson conclude that

Temporality is a dissolving force, but within a unifying activity it is not yet sufficient to be a real diversity, it cannot subsequently accept any unity.

Therefore, it cannot even exist as a "before and after" separateness of diversity, which makes the "before and after" separateness of time more quasi-diverse, one within unity, the beginning of the dismantling of the whole into fragmentary disintegration.

We cannot give priority to the continuous sameness of time, nor to the continuity of "before and after" separateness. Time should be seen as a unity in diversity.

Temporality is only a relation of existence within a single existence. Time does not exist in the first place, and it is both fractured and continuous. Only the self can exist in the state of the self as an individual using mental activities such as feeling, perception, thinking, memory, etc., to be aware of and recognize the changes in one's physical and mental state and in the environment of people, things and objects.

We can use various rhetorical techniques to visualize and interpret things: human existence is time-bound, and my mind that can produce differentiated effects on all phenomena is time.

1. (Past) means any moment or period of time before the present moment we are in, it can be a moment, but mostly refers to a period of time where the past precedes the self.

Because the Self precedes the Self, it is not what it is (and does not agree that the Self is correct). The processing of the Self/past is not what it is (not in favor of assuming that it is correct).

Self-existence is not what it is by transcending an irrevocable past that it is (in favor of self-righteousness), while in existence, it emerges.

2. (Future), the future, the time that has not yet arrived, self-realization itself is also a lack, self-realization

needs to become what it is.

Then self-realization itself is non-existence, that is, the lack of a self-realization. Instead of this non-existence, that is, the self needs to become what it is.

For example, I drink water because I am not drinking water, and I want to be the one who is drinking water. I have imagined through time [nothingness/consciousness] an uncompleted objective, uncompleted whole of all objects or phenomena that exist in nature.

In the first dimension of nothingness, which used to reside in the direct line of the ego, reality (Latin: Realitas) in its everyday application means "something that exists objectively" or "a condition that fits into an objective situation".

Reality is the sum of all actual or existent things, as opposed to the wholly imaginary, which resides after the ego.

As we have said before about ego-nature, it is important to note that ego-nature is not the same as temporality; they do not occupy the same space. Only in the stillness of ego-ness, temporality coincides with it as non-temporality.

3. (Now), this time refers to the time of speaking, sometimes including before and after speaking, or now and at present.

Objectivity is a central concept in philosophy, which refers to the reasonableness of thinking or judging something from different viewpoints or perspectives, the nature of a thing that exists independently of subjective thought or consciousness, corresponding to "subjectivity.

Objective facts are not influenced by subjective means such as human thoughts, feelings, tools, and calculations, but can maintain their truthfulness.

In the infinite game of all objects or phenomenal reactions and reactions in the natural world, it hides everywhere and may reside everywhere and nowhere.

(In fact, at this level of the process of acquiring knowledge through the formation of concepts, perceptions, judgments, or mental activities such as imagination, i.e., the mental function of the mind in processing information, Schart is not, either, a completely radical liberal)

Heidegger: The future of the past introduces the present, so he embarked on the life of death, only through the emptiness of death, to the environment and the ability to recognize the self

Sartre: The future of the past introduces the present indeed, but the present can likewise introduce the future of the past.

So he took the path of intervention, shouting the slogan that existence is prior to essence, that Dasein can be what it is and at the same time not what it is.

Sartre's approach to natural and social phenomena is based on existing empirical knowledge, experience, facts, laws, cognition, and tested hypotheses, through generalization and deductive reasoning, etc.

The temporality, which is logically concluded by inference, is the time of the actual universal law of man, and temporality is not imposed on existence from outside, a law of development. It is not existence, but the internal structure of existence that constitutes its own nihilization, that is, the self-existence.

(B) The Dynamics of Temporality

This chapter is in fact a summary of the previous ontology of temporality, mainly correcting Kant and Leibniz's view of change as containing the constancy of "self-reference".

If a static constancy is set to exist through time, then temporality can only be attributed to the scale and order of change in these sequences. The regular relationship between things, as suggested by the laws of kinetics, is a one-to-one correspondence of certainty.

It indicates that the existence of one thing must lead to the occurrence of another definite thing, and that chance phenomena are negligible.

The law of temporality reveals not a simple one-to-one correspondence between things, but a necessity, and a regular relationship between many random phenomena.

Most social events are random in their occurrence. The laws of nature are more often expressed as temporal dynamics laws, and the laws of social development are mainly expressed as temporal laws.

Therefore, natural science can accurately foresee the occurrence of natural events, while social science can only foresee the trend of social development, but not the occurrence of social events.

This is a characteristic that has been repeatedly identified in this book, and it is as "suspended" and deficient as the Self.

The so-called unity is only the unity of existence, and unity is first of all an emergence. To sum up: the temporalization of the Self causes this change as a temporal overall change.

The "present" becomes the "past", the "future" becomes the "past" of the "present", and the future becomes the future of the past. Individuals use mental activities such as sensation, perception, thinking, and memory to

The time when an individual uses his or her senses, perceptions, thoughts, memories, and other mental activities to become aware of and recognize the changes in his or her physical and mental state and in people, events, and objects in the environment is the objective reality of human temporalization.

(C), primitive temporality and psychological temporality: reflection

The premise is that when inferring, one first affirms some statements, assuming them to be true, and then proves that others must also be true; the premise is that these previously affirmed statements.

Schatte begins with a vague reference to the time of mental continuity. Then, on the next page, he starts to talk about the inference between the reflection and the reflected.

This is a logical form of analysis, which can be defined as the complex mental process of seeking or establishing rules and evidence to support or determine a belief, decision, or action, such as the notion of Spinoza's concept, and these inferences were already mentioned in the first chapter of the first volume.

It is only here that Sartre reintroduces them, adding the notion of self-conscious temporality. In other words, a premise is a ground or reason for accepting a conclusion in an argument.

And the conclusion is the result based on the premises (usually two). Premises and conclusions are relative and cannot be viewed in isolation.

It is necessary to decide which statements are premises and which statements are conclusions in a particular argument. The same statement that is a premise in argument A may be a conclusion in argument B. The same statement that is a premise in argument A may be a conclusion in argument B.

It is in the context of the argument that the original temporal premise comes to the fore; an argument must have at least one or more original temporal premises, but only one conclusion.

A logical triple argument has two premises and a conclusion; the two original temporal premises are called the major and minor premises respectively.

The distinction between major and minor premises has nothing to do with their position in the argument; a major premise is one that contains a "major word" (the object of the concluding statement), while a minor premise is one that contains a "minor word" (the subject of the concluding statement).

Examples are as follows

1. Anyone who is fair and upright and does not bend the rules is a good person...major premise

2. Jack is a fair and upright person who does not bend the rules ...... minor premise

---------- -----

3. Jack is a good person ....

Conclusion

In the above example, the sentence "Anyone who is fair and upright and does not pander to others is a good person" already contains the word "is a good person" in the concluding sentence, that is, it contains the big word, so it is the big premise.

The phrase "Jack is a fair and upright man who does not pander" is a minor premise because it contains the minor word "Jack".

Therefore, it is a very important step to identify those phrases that are the premises of the argument. In some arguments, the original temporal premise comes first and the conclusion comes second.

But not every argument is in this form. Sometimes the conclusion comes first and the premise comes second; sometimes the conclusion is sandwiched between two premises.

It is possible to tell which is the original temporal premise by using specific words such as "because", "since", "as", or "by reason".

However, not every argument necessarily contains these Premise-Indicators, but also Conclusion-Indicators such as "therefore", "thus", "therefore", or "consequently" to distinguish the conclusion.

Later, Schart proposes a rational and active process of reflection on the real world or any object, in the form of "inward dialogue", which is not purely conscious of turning back to the past.

The so-called inward dialogue is like talking to oneself, without the condition of the listener, and learning from the concept of experience. Then he proposes the temporality of temporalization and the temporality of psychology.

He also pointed out what they refer to. Primordial temporality: it is the temporality we mentioned in the previous chapter (past, future, present), which serves as the premise of the unreflected existence, the primordial temporality.

And psychological temporality: the consciousness of continuity, as the French philosopher Bergson pointed out, people are used to understand time in terms of the change in the number of objects presented in space.

But this not only misunderstands the characteristics of time, but also prevents us from truly understanding all the objects or phenomena that exist objectively in the natural world.

He proposes the concept of duration, and through it, further explores the relationship between life and time, that kind of temporality, the psychological product of an impure reflection on primordial temporality.

The difference between the two (primitive temporality and psychological temporality-reflection) is that, first, primitive temporality is self-temporalized and psychological temporality exists.

Later, Sartre mentions how impure reflection is a kind of reflection, the merging of the reflector and the reflected, and their movement process.

It is pointed out that impure reflection is self-deceptive [even though reflection itself is self-deceptive]. Second, reflection is not a sincere event, and third, the third movement of reflection is for others.

It is a rational and active process of reflection on the real world or on any object, not purely in the form of "inward dialogue".

The so-called inward dialogue is like talking to oneself without the condition of the listener, drawing lessons from the past, and becoming the other while maintaining the self.

These are the shadows of existence. It is this shadow of existence that psychologists study under the name of psychological behavior.

And then they explain what is the a priori basis of the psychological kind of time, of the kind of self (e.g., various personalities, various traits). Here, in fact, Schart is talking about how psychology is structured in terms of "existentialism".

(1) Different views on temporality

There are now two different views of time among philosophers: one believes that time is the basic structure of the universe, a dimension that appears in a sequential manner.

Therefore, it is also called "Newtonian time". The other school of thought believes that time is not a dimension that already exists, nor is it a "fluid" reality.

In conjunction with space and calculation, the word or symbol for all objects or phenomena that exist objectively in nature allows humans to prioritize and compare events.

In the tradition of Gottfried Leibniz and Immanuel Kant, the second school of thought believes that space and time "do not exist in themselves, but are the product of our individual mental activities such as sensation, perception, thought, memory, etc., our combined awareness and knowledge of our physical and mental states and the changes of people, events and objects in our environment, and the way we express things.

It is the study of the fundamental questions of life, knowledge, and values, and the deeper inquiry into time (the nature of things, laws, etc.) by analyzing and reflecting on them.

Ancient ontology is the branch of philosophy that studies concepts such as existence, being, becoming, and reality. It includes the question of how to classify entities into basic categories and which entities exist on the most basic level.

Adopting the tone of subjective solipsism and declaring the nature and properties of time, recent epistemology is a branch of philosophy that explores the nature, origin, and scope of knowledge.

The relationship between the theory of knowledge and

epistemology is currently controversial, with some arguing that they are one and the same concept, and others arguing that they are in fact two different concepts that have some close connections.

Adopting a subject-object dichotomy, an either/or attitude, and judging the nature and properties of time, modern "phenomenology" is a discipline that explores phenomena.

The phenomenology is the investigation of phenomena, whether they are imaginary or existent, as long as they can manifest themselves in human consciousness.

The theory of "existence" is the theory of what existence is and how it exists, that is, the dialectic of "the relationship between thought and existence". Therefore, Schart believes that time is an organized structure, "past", "present" and "future". As the three elements of time, they do not necessarily have to be brought together. As the three elements of time

They do not necessarily have to be brought together. "Self-action" is a transcendence towards its own possibility, which leads to the question of temporality. In order to clarify the nature of time, time is divided into three dimensions so that it can be discussed separately.

(2). On the question of the "past", there are two views.

One is Bergson and Husserl's philosophy of life, which refers to the philosophers' attempt to generalize a law of life, which organizes all the developments of the universe, such as human beings, creatures, etc., and presents philosophical insights.

Schopenhauer's theory of the will to live (the essence of the universe is the desire to live), Nietzsche's "will to power" (Der Wille zur Macht), and Darwin's theory of biological evolution are all commonly regarded as philosophies of life, and the theory of the existence of the "past" of time, that is, although space-time has turned to the past and merely ceased to exist, it still exists.

The second is Descartes' view that the "past" no longer exists: that doubt is the starting point, that the knowledge of sense perception can be doubted, that we cannot trust our senses.

Therefore, he does not say "I see therefore I am" or "I hear therefore I am". From this he realized the truth that what we cannot doubt is "our doubt". Meaning: what we cannot doubt is what we are "doubting.

Only then can we be sure that our "doubts" are real and not the product of falsehood.

Schart rejects both views, arguing that the study of the 'past' is inseparable from the study of the 'present', that is, that the present exists as the basis of its own past, and that the past is meaningful only in its connection with the present.

Schart also highlights that the word 'was' is a mode of being, indicating that the act of action has already been performed. For example, déjà vu.

"As an intermediary between the present and the past, 'was' is neither fully present nor fully past in itself, and in fact it cannot be either present or past, but is an indication of the temporal content of existence.

Critics of Descartes, empiricists, Leibniz, Descartes, Bergson, who criticized Kant and Kant in contrast, conclude that temporality is a dissolving force, but within a unified activity, it is not yet sufficient to be a real plurality.

It is not yet sufficient to be a real diversity - it cannot subsequently accept any unity so it cannot even exist as a diversity - it is more a quasi-diversity, the beginning of a disintegration within unity.

We cannot give priority to the continuous sameness of time, nor to the continuity of separateness. Time should be seen as a unity in diversity.

Temporality is only a relation of existence within a single existence. (Time first does not exist, and second is both fractured and continuous; only the self can exist in a state of ego-emergence (blossoming). We can elaborate it literally: human existence is of time, I am of time.

Therefore, it is now "for oneself". "For oneself" is the existence of the whole "being in itself". The same "for oneself" faces all of existence at the same time. "For oneself" is to be free from oneself and to be in existence, but not to be this "existence".

"For oneself" is the manifestation of existence in a way that escapes from the hidden. It is impossible to grasp the present in an instantaneous way. The present must be grasped in the connection of the past, the present and the future.

The present itself cannot be grasped; once it can be grasped, it becomes the past, and when it is not grasped, it remains the future. This is "what is not and what is not".

According to Schart, the so-called "temporality" is the continuous self-denying way of being "for oneself". Therefore, the real starting point of time is the future, not the past.

The past cannot be the starting point of time because time is not 'being in itself', it can only be 'for oneself' as a holistic, indivisible, and constantly self-denying way of being.

Sartre also explains the problems of static temporality, the dynamics of temporality, primordial and psychological temporality, and the time of the world.

But then Sartre makes a few more points: because the past is self, consciousness is "for oneself". "The past is an entity that I can no longer experience," and therefore Descartes' expression should be "I think, therefore I was.

There is an absolute heterogeneity between the past and the present, and the reason I cannot enter it is because the past exists. The only way I can become the past is to become Freedom myself, so that I can disappear in the past in the same way.

But this is unacceptable. The shame of yesterday was

the shame of self, and the shame described now is the describable "for oneself", the "for oneself" of Freedom.

It is the "for oneself" of Freedom.

For example, the pain of the Nazi concentration camp massacre was always the pain of others. And the pain of the past is expressed in the sense of "for oneself", Freedom's fixation on the past, like the pain of others, is silently fixed.

Finally, Schart thinks that the past is the opposite of value. Existence seeks the certainty of "freedom" in the form of "for oneself" towards value, and it wants to be sure of this constantly.

The past, on the other hand, is the "for oneself" of the past that is reasserted by the Self and fixed by Freedom. So in a sense the past is also a value used to escape anxiety.

As listed above, several famous existential philosophers in the West have discussed, defined and interpreted temporality from different perspectives.

When we analyze their views, we will find that their views have one thing in common: subjectivity. In other words, when they elaborate and understand the nature of time, they cannot leave the issue of human autonomy as the subject.

Specifically, their understanding of temporality is always closely linked to the awareness, consciousness, psychology, and life experience of human autonomy.

This has a certain contemporary significance in the existentialist thinking, because time is most closely connected with human experience, and time is connected with human behavior and experience.

Their view of one-sided truth interpretation must have its own meaning in the framework of the time.

Therefore, we cannot dismiss their temporal viewpoint as materialism, because it provides a way to think about the real meaning of life, "what is time?

Of course, their view of temporality sometimes overemphasizes the subjectivity of time, but neglects the objectivity of time, and thus leads to a one-sided interpretation of a limited truth.

Thus, after criticizing Descartes, the empiricists, Kant and Kant's opposite Leibniz, Descartes, and Bergson, Schartes concluded that

Temporality is a dissolving force, but within a unifying activity it is not yet sufficient to become a real diversity, "it cannot subsequently accept any unity, so it cannot even exist as a diversity"

It is more the beginning of a disintegration within each quasi-diversity, unity. We cannot give priority to the continuous sameness of time, nor to the continuity of separateness. Time should be seen as a unity in diversity. Temporality is only a relation of being within the same being.

The question of existence "for oneself" is discussed extensively in Being and Nothingness. In the view of Schart's view of temporality, only by understanding the meaning of "for oneself" existence can one truly comprehend

only by understanding the meaning of the existence of "for oneself" can one truly understand the true existence of "being in itself" existence.

In other words, to truly understand "being-for-itself" is to truly comprehend "being-for-itself", which, according to Scharthe, must be grasped in terms of the temporal and holistic nature of the existence of "being in itself" "for oneself".

"Freedom, as the totality of existence, has two basic mechanisms as life and death, so that while "for oneself" still exists, there is a "pending" Freedom.

Therefore, we must start from the basic mechanism of "for oneself": the whole life, the pending, and the termination to explain the existence of self.

This is the problem of "studying existence through rational reasoning and logic, which cannot be answered directly through perception".

That is, there are things that we cannot see with our eyes, such as objects in the dark, but we can perceive their existence with our hands and bodies; there are things that we cannot touch with our hands and bodies, such as distant objects or landscapes, but we can perceive their existence with our eyes.

There are things that we cannot see with our eyes and cannot touch with our hands and bodies, such as songs, music, words, etc., but we can use our ears to perceive their existence; there are also things that our senses cannot directly feel.

There are also things that our senses cannot feel

directly, such as ultraviolet rays, infrared rays, cells, particles and electromagnetic waves, etc. We can create various instruments and tools to perceive their existence.

It is said that "man accepts life without choice, then lives it under conditions of inevitability, and finally surrenders it with an irresistible struggle", which illustrates the finiteness of "being-for-itself". In Tillich's theology, "finitude" is one of the definitions of the nature of man and his condition of existence.

The "finite" is the limited nature of the "freedom" that man possesses. In Tillich's view, overcoming finitude requires a return to "being itself," and a return to "being itself" is the realization of the "new being" revealed by Jesus Christ.

Finitude is a Christian philosophical-theological term. Finitude is the term used in Christian philosophical theology to refer to the "infinity" of God. In Tillich theology, "finitude" is one of the definitions of the nature of man and his condition of existence.

The "finite" is the limited nature of the "freedom" that man possesses. That is, man has the freedom to choose, but cannot avoid the consequences of that choice. The "finite" is the "limited" reality of man's existence, which is limited by his environment, by others, by himself, and whose existence gives rise to an incomparable anxiety, because he is alienated from the infinite, while yearning for it.

In Tillich's view, overcoming finitude requires a return to 'being-itself', and a return to 'being-itself', the 'new being' revealed by Jesus Christ, makes the original non-existence a reality.

Death is the "being-for-itself"

death is the highest limit of being-for-itself. It is clear that being-for-itself

is inseparable from temporality, and temporality is the objective existence of being-for-itself

It is the condition of being-for-itself, which is the objective existence and development of all objects or phenomena in the natural world. Life in the world is also the "existence" of life.

Volume 3: the relationship with "others", the idea of "co-presence".

Section 1: Relationship with Others

1. The existence of others

We are "being-for-itself"-whose characteristic is to be known by others: what I "being-for-itself" know is the body of others' "self-existence", and the main thing I know about my body comes from the way others know it.

My body pushes me into the being of others and my being-for-itself. This is the structure of "being-for-itself" that cannot be studied without noticing the relationship between "being-for-itself" and existence.

2. Problems that are not easy to solve

According to Schart's philosophical viewpoint, existentialism emphasizes the existence, dignity and value of human beings, and fully represents the spirit of humanistic education. It is concerned with the "subject person", the "concrete and living person".

Existentialism does not explain the world, but rather the meaning of "existence". It emphasizes freedom of will and recognizes the human value of each individual's autonomy.

When Schartre calls "existentialism is humanism," he means self-transcendence and self-projection. Self-transcendence focuses on the sincere self-expression of the individual, while self-projection refers to the relationship with others, the relationship between others and others.

However, existentialism is not an easy problem to solve in terms of the relationship with "others", the idea of "co-presence".

Existentialism does not have a systematic approach to social activities that directly affects the physical and mental development of human beings, and its study of the universal, fundamental problems of the disciplines of

It is not a philosophy that systematically analyzes and reflects on the fundamental questions of life, knowledge, and values.

It is not a philosophy with a serious and prudent attitude, and most of the discourses are detached from social practice, but only through abstract analysis, synthesis, judgment, reasoning, generalization and other thinking activities.

Argumentation is only the process of forming reasons, justifying beliefs, and drawing conclusions in order to influence the thoughts and/or actions of others.

The conclusions reached are philosophical, prejudiced, influenced by emotions or preferences, or even distorted by interests or pressures, or unrealistic and imaginative, and the process of deducing conclusions from less analytical premises.

Rather than being a holistic, fundamental, and critical inquiry into the real world and human beings, existentialism is closer to a metaphysical idea.

2. Existentialism overly promotes the tendency to see things in terms of the subject's own needs, which are the attributes of viewpoints, experiences, consciousnesses, spirits, feelings, desires, or beliefs that an individual can have.

Its fundamental characteristic is that it exists only within the subject and belongs to the subject's state of mind. It can influence human judgment and the factors of truth, and human emotions consciously determine the purpose and regulate their actions according to the purpose.

It is a psychological process that regulates one's own actions according to the purpose and overcomes difficulties to achieve the predetermined goal. The will is the centralized manifestation of the dynamics of human consciousness and is a uniquely human psychological phenomenon.

The freedom to initiate, persist, stop, and change the control and regulation of mental states and external behavior is expressed in the action of human initiative to change reality, making the study of universal, fundamental issues of the discipline, including existence, knowledge, values, reasoning.

This makes the discipline of studying universal, fundamental problems, including the fields of existence, knowledge, values, reason, mind, language, etc., in danger of becoming irrational, or even anti-rational, as opposed to all mental elements of rational thought.

It is difficult for modern society and culture to accept that the relationship between the individual and society is interactive, and that the relationship with others is absolutely free.

3. Existentialism is the ability of human beings to

follow the existing empirical knowledge, experience, facts, laws, cognition, and generalization and deductive reasoning of natural and social phenomena in a normal state of mind.

In order to obtain the expected results, have the confidence and courage to calmly face the status quo, and quickly and comprehensively understand the reality to analyze a variety of feasible solutions, and then judge the best solution, and the ability to effectively implement it, logical inferential summary.

If one pays too much attention to, or puts too much emphasis on, the existence of a human being as a deposit towards death, then there is an excessive use of subjective imagination to exaggerate the thing to be described, in order to achieve a certain expressive effect.

In order to achieve a certain effect of expression, the rhetoric of exaggerating or reducing the image, characteristics, effects, and degree of things, which exceeds the reality, is suspected of enhancing the pessimistic effect of life that is intended to be expressed, and may have an adverse effect on the development of the immature mind and body.

Therefore, when faced with conversations that are not easy to resolve in relation to "others" and "co-presence", we must first not run away and not be afraid.

When faced with a potentially conflicting conversation, most of the people involved will probably make peace with it and avoid it if they can, but avoidance and patience will not solve the problem.

In addition, the process of learning about the outside world, or the process of processing information in the face of the outside world, is slowly losing confidence in our sensory organs.

For example, in our daily life, we often experience things that are not so easy to decide, such as a relationship or a job, and at the beginning, we are full of confidence and hope in life or the development of everything.

But with the progress of all the things and objects of this objective existence, slowly we will find that problems appear one by one, and then we should not continue, and at first only give up part of each other's views, principles, etc., in order to avoid disputes or conflicts.

To avoid disputes or conflicts of compromise, but in

the end is still inexperienced, do not know how to properly handle the situation, choose to break up or resign.

This may cast a shadow on your atmosphere in the face of the same situation in the future, so confidence will be more or less negatively affected. The reason for this is that you have chosen an immature escape.

We often try to solve problems by talking and communicating, but always learn to control everything, know how to stop when appropriate, and not to do everything to excess.

The social matrix of "leave a line, we will meet in the future", taking into account the feelings of others, so when expressing their own ideas, fear, and even anxiety, is the fear of blocking your desire to talk about the possible, you use the habitual thinking of guessing instead of the facts, so it will make the conversation to stop, or not solve the problem.

Therefore, only their normal state of mind in order to obtain the expected results, have the confidence and courage to calmly face the situation, and quickly and fully understand the reality of the analysis of a variety of feasible options, and then judge the best solution, and the ability to effectively implement the rational program.

And the ability to effectively implement the rational thinking, the face of the problem is not easy to solve, do not run away, not afraid, then to be able to face the correct conflict, and ultimately solve the problem.

For example, in the relationship with "others", the idea of "co-existence", in some conversations are not so easy to solve, often because of different positions, rather than the problem of people as we think.

The root cause of this is that we are too accustomed to speculating about other people's intentions, and we tend to take such opinions as the truth.

In order to avoid such a situation, we must first look at the root of the problem and find the source.

If you realize that you hate or are angry with someone, or are unhappy with someone you don't like, you should stop yourself in time and separate that person from your position, so that you don't keep making wrong inferences and prevent further harm.

Many times, when we are faced with problems that are not easy to solve, when we have a relationship with "other people", when we have a conversation that is not easy to solve, we are at our wits' end because we cannot distinguish between small talk and communication.

It is true that chatting can help you solve your problems because the atmosphere of chatting is more relaxed, but chatting cannot help us face problems that are not easy to solve, and chatting and communication are two different things.

In fact, facing problems that are not easy to solve is a process of "problem solving".

It is easy to think that the methods and techniques in different fields must be different, but a lot of truths may be "applicable" to the same problem, and sometimes thinking about the problem too complicated may make a simple problem more difficult to solve.

"The first step to solve a problem is not to collect a bunch of solutions or techniques immediately, but to "think" first and identify what the real core problem is, so that there is no problem to solve at all. Facing problems that are not easy to solve

The most important first step is also to "think", to think about "why we need to organize", to think about "what is the real problem to be solved", to think about "our purpose to face the problem that is not easy to solve".

If we do not first clarify our relationship with "others" and the purpose of "co-presence" thinking, it is likely that our efforts to face problems that are not easy to solve will be in vain for half a day.

Section 2: Objective reality and formal reality.

As I have already stated, man exists concretely in the immediate reality and condition of existence, but the structure of reality - the structure of "being-for-itself" - this being-for-itself reveals to me a being that is not "being in itself", a being that is my being. (The existence of being-for-itself). (Being in itself), shame reveals 'being-for-itself' to others (no shame alone).

And "reality" is the sum of the actual state of existence, the thing itself, all actual being, as opposed to appearance, and the ability to recognize the environment and the self, as well as the clarity of cognition.

The consciousness, which is not related to the cognitive ability of a person to recognize the environment and the self, and the clarity of cognition, is all the real things. Actuality means "the whole of reality", "the whole of reality". There are two types of "reality": Objective

Reality and Formal Reality.

First, objective reality.

It is a special social phenomenon, consisting of a certain system of speech, vocabulary and grammar, and the external existence related to perception.

Perception is the perception of objects in the external world through the senses, specifically when faced with the concrete existence of facts in front of the eyes and the knowledge of objects in their condition, as opposed to imagination.

In early psychology, the study of universal, fundamental issues, including existence, knowledge, values, reason, mind, and language, was often associated with sensation, or sense-perception.

Sometimes it also refers to "inner perception", which contains the tendency to see things in terms of the subject's own needs; it is the awareness of the attributes of perspective, experience, consciousness, spirit, feelings, desires, or beliefs that an individual can have within.

In psychology, the terms perception and sensation are either used interchangeably without distinction, or perception is considered as the process of knowing something external to the person who perceives it, adding to the process of knowing something external to the psyche.

It is the process of processing information about external things that act on human sensory organs. It includes sensation, perception, memory, thought, imagination, and speech, and refers to the process of cognitive activity.

It is the process of receiving, detecting, converting, simplifying, synthesizing, coding, storing, extracting, reconstructing, forming concepts, making judgments and solving problems.

In psychology, it is the process of acquiring knowledge through mental activities such as concept formation, perception, judgment, or imagination, while the image of an object stimulates the retina and produces a sensation, which is the image of an object obtained by the pure senses.

In fact, the sensory effect occurs only when the sensory organs are connected to the central nerve, such as "seeing a teacup", and when the teacup is looked at more closely.

It becomes the overall view and understanding of the external world when external stimuli are applied to the sensory organs, and it organizes and interprets our perception of the external sensory information.

If we further describe this teacup and add our own opinions to it, we can still say that the objective things act directly on human sensory organs and the human brain reflects the objective things as a whole.

In fact, more mental activities are added, and these activities are connected very quickly and are not easy to cut off, so it is difficult to make a strict distinction in the interpretation of nouns. Perceptual activities have the following four general characteristics.

First, the objective environment in which the objective object does not exist.

In the objective environment, there is something that can be distinguished and exists independently within itself. But it does not need to be a physical existence. In particular, abstraction is also usually regarded as physical existence, or the existence of explicit rather than abstract, generalized objects, whereas in the mental activity of imagination, objective objects do not exist.

Second, the generation and completion of perception, based on the activity of reasoning.

The generation and completion of perception is the result of the moment of contact between human and external objects, during which the conclusion is deduced from known or assumed premises, or the reasoning (reasoning) based on the known answer is rarely involved.

Third, the degree of precision and quality of perception is unequal.

It refers to the mental state of concentration, seriousness, which is an attitude that does not distract and concentrate energy on something.

Concentration is deeper than seriousness, and the subjective initiative of concentration is stronger. Concentration on learning is self-initiated, while serious learning may have other factors. For example, one can hear very clearly, or one can turn a deaf ear.

Fourth, the nature of the perceived object can be understood through perception.

The result is that the act of "knowing" and "recognizing" a subject is known with certainty, and that this knowledge has the potential to be used for a specific purpose. It means the ability to become familiar with something through experience or association.

02. Two issues have received the most attention in discussions of philosophical inquiry.

In the discussion of holistic, fundamental, and critical inquiry into the real world and human beings, two questions are most important: first, is knowledge obtained through perception reliable and how certain is it? Second, is perception itself an obstacle in the process of knowing things in the external world? The summary is as follows

First, is the knowledge gained through perception reliable and how certain is it? First, is the knowledge gained through perception reliable and how certain is it?

Is the knowledge obtained through perception reliable and what is its certainty? On this issue, it is often criticized that knowledge based on illusion or illusion has no solid foundation and cannot be the source of knowledge.

However, there are also people who believe that knowledge based on perception may be erroneous and less certain than mathematical knowledge, but its possibility of error is still within the acceptable range.

Moreover, the certainty of learning and experience acquired by all people in the process of learning and practice need not be the same as that of mathematics, so that perception is still reliable and the act of "cognition" and "identification" of a subject is still within acceptable limits.

The act of "cognition" and "identification" of a subject is a way of knowing with certainty, and these knowings have the potential to be used for specific purposes. It means that through experience or association, one is able to become familiar with a source of further understanding.

Second, is perception itself an obstacle in the process of knowing external things?

One knows things in the external world through perception, but is perception itself an obstacle in the process of knowing external things? On this question, naive realism holds that human perception can grasp the properties of objects in the external world, and therefore perception does not become an obstacle to the knowledge of external things.

However, the representative theory of perception and the causal theory of perception argue that things in the external world are stimuli, and that only through perception can we have sense data that is exclusive to our senses.

We cannot know external objects directly, because perception lies between the perceiver and the external world. According to this statement, some philosophers argue that since we cannot directly know objects in the external world, we cannot understand them at all.

Even if they exist, it is difficult to be sure that our knowledge of them is correct, because perception itself may be wrong.

In addition to the above two topics, philosophers study universal, fundamental subjects, including the fields of being, knowledge, value, reason, mind, and language. The fundamental nature of perceptual activity (perceving) is also often studied

These two topics are also of interest to psychologists, as well as the effects of perception on human cognitive abilities and the clarity of cognition.

Although perception is based on sensation, sensation does not necessarily lead to perception. Perception can certainly provide some basic information, but the embellishment and meaning of such information depends on the individual's past experience to give some kind of meaning and to recognize the relationship between things.

Therefore, cognition can be described as a mental process of subjectively organizing objective sensory materials; it can also be described as the intermediate process from sensation to thought.

The cognitive processes that are generally referred to are the mental processes from sensation, perception, and thought. Thought, also known as "concept", is the result of its activity and belongs to cognition.

People's social existence determines their thoughts. All ideas based on and in line with objective facts are correct ideas, which promote the development of objective things.

On the contrary, it is the wrong thought, which hinders the development of objective things. Thought is also related to the way a person's behavior and emotional approach to the important manifestation.

03. Specific meaning and characteristics of each stage of the perceptual cognitive process.

Individuals use sensory, perceptual, thinking,

memory and other mental activities to perceive and understand their own physical and mental states and the changes of people, events and objects in the environment, representing the symbols and relationships of various things.

The meaning and relationship of these symbols are the products of perceptual experience. The time from "sensation," "perception," and "thought" is often very short, and it is difficult to draw the boundaries of each stage.

1. Sensation and perception.

Perception is based on the characteristics of objective things, through sensory stimulation to the nerves, and the brain to produce a physiological response to identify the reaction or psychological feelings based on the sensation, between the two, although there is no clear demarcation limit, but generally can be distinguished as follows.

(1) sensation is only some simple factual impressions, while perception is beyond the characteristics of various objective things, through sensory stimulation to the nerves, and the brain to produce the sum of physiological reactions or psychological feelings of recognition

reactions, and a new product of the psyche.

(2) The process of selection is necessary to obtain awareness through sensation; that is, from the characteristics of objective things, through sensory stimuli to the nerves, and in the brain to produce the recognition of physiological reactions or psychological feelings of data, to select a part of the data to be sorted and selected.

In other words, from the characteristics of objective things to the nerves through sensory stimuli, and among the physiological reactions or psychological feelings of the brain that produce recognition reactions, a part is selected for sorting and selection. As for the choice of data, it depends on the individual's needs, orientation, motivation and other changes in the situation.

(3) The characteristics of objective things are transmitted to the nerves through sensory stimuli, and the physiological reactions or psychological feelings of attention in the brain that produce a recognition response vary less individually.

(3) the characteristics of objective things are transmitted to the nerves through sensory stimuli, and

the brain produces the physiological response or mental perception to recognize the response.

2. Attention and perception.

Perception and attention are quite similar and difficult to distinguish from each other, which has led many psychologists to treat these two processes as one without differentiating them. Although there is no obvious boundary between the two, they can be broadly distinguished as follows.

(1) Intentional attention occurs before perception and is a common psychological feature that accompanies the mental processes of perception, memory, thought, and imagination.

Attention has two basic characteristics, one is directionality, which means that mental activity selectively reflects some phenomena and leaves the rest of the objects. The second is concentration, which refers to the intensity or tension of mental activity staying on the selected object. Directionality is expressed as the selection of many stimuli that appear at the same time.

Concentration is expressed as the inhibition of interfering stimuli. Its generation, scope and duration depend on the characteristics of the external stimulus and the subjective factors of the person. Therefore, attention is also referred to as preperceptive attitude.

(2) Attention itself does not determine the meaning and organization of the perceptual experience; the meaning of the overall response perception of the various properties of the stimuli directly acting on the sensory organs differs for many people in the same situation.

(3) Attention does not necessarily produce perception; the orientation and concentration of mental activity on a certain object is accompanied by mental processes such as perception, memory, thought, and imagination.

A common psychological characteristic of attention is only the preparatory stage of perception, so that the observer is close to the facts, with the role of exploration and selection, while perception allows the observer to recognize the facts, with the role of identification and discrimination.

In summary, perception not only consists of certain stimuli from the material world, but also acts directly on certain sensory organs of the organism, such as light causes vision, sound waves cause hearing; stimulation in the senses, nerve impulses caused by the senses.

The nerve impulses caused by the sensory nerves are transmitted to certain parts of the cerebral cortex to produce sensations based on selective attention, i.e., attention is the tendency to selectively process certain stimuli and ignore others.

It is the tendency to selectively process certain stimuli to the exclusion of others. It is the tendency of the senses (vision, hearing, taste, etc.) and the perception (consciousness, thought, etc.) to selectively focus on certain objects.

When a person is paying attention to something, he or she is always perceiving, remembering, thinking, imagining or experiencing something. A person cannot perceive many people, things and objects that are touched by all human activities at the same time, but only a few things that are the target of action or thought in the environment.

To obtain a clear, deep and complete reflection of things, it is necessary to make mental activity selectively directed to the people, things and objects that are the target of all human activity when acting or thinking, and to put the objective sensory material

The process of mental activity in which two or more concepts, which seem different but related, are added to things or phenomena to form a meaningful whole and are explained. Therefore, psychology has always attached importance to the study of perceptual phenomena

Second, the reality of form.

Form (English: Form) is an important concept in ancient Greek philosophy. It corresponds to material and describes the essence of things, referring to the "first entity".

Form and substance, as part of Aristotle's "theory of entities," have been important to later studies of universal, fundamental issues, including existence, knowledge, value, and reason.

It has had a major impact on the holistic, fundamental, and critical inquiry of the real world and of man, including the fields of being, knowledge, value, reason, mind, and language.

The term "understanding" is used to refer to the ability to explain known facts and principles and principles in one's own words, writing or other symbols, and is called "understanding the real", a common phrase, more often used in oral language, and is a partial term.

The word 'in' has the meaning of language assistance, which makes the meaning more complete. The word is mostly used as adjectives and adverbs, and has various meanings. The specific meaning expressed depends on the "context" of the way of thinking, and involves the relationship between logic and concepts.

01. Formal definition

A "thing" is an object related to human consciousness, which may have a specific image or may exist only in awareness; that is, the general name "thing" in common parlance.

Therefore, "things" refer to a wide variety of objects: either a single thing, such as the externally visible; or a category, i.e., the sum of several identical things; or an intellectual one, which is abstracted or generalized and not directly referred to by name; or divisible or indivisible, depending on the object referred to.

In short, a thing is an object relative to the human psyche that can be perceived and constructed as a mental object as a whole, in a strict sense, something that is distinguishable and exists independently within itself. But it does not need to be a physically existing entity, it is the essence of things.

02. Formal properties

The word "abstract" originally means "to extract a higher concept from a solid. The so-called abstract painting is to extract the feeling and spirit of the painting

It means to extract the feeling and spirit of the painting from the entity. In the Ministry of Education's Mandarin dictionary, abstraction means "philosophically, it refers to the activity of analyzing the commonality of individual and accidental things. As opposed to concrete."

2. Expressive, revealing thoughts, feelings, life experiences and other internal forms

3. Idealism is a beautiful imagination and hope for the future, and it is also a metaphor for the concept that a certain form is the most perfect.

It is people's aspiration and pursuit of future society and their own development, which is formed in the process of practice and has the possibility of realization, and is the centralized manifestation of people's world view, life view and struggle goals.

Satisfy the immediate material and spiritual needs, but also look forward to the future life goals, expect to meet higher material and spiritual needs

4. A priori, in the usual sense, is understood as opposed to "experience", meaning that which precedes experience but is indispensable for constituting experience.

However, this formal concept involves the exploration of fundamental issues in modern Western philosophy, and there are subtle differences in the meaning of the term in different contexts.

5. Perceptibility. Something that has a perceptible distinction and exists independently of itself (English: Entity). But it need not be a physical being.

In particular, abstraction is usually perceived as a formal physical entity. It can be used to refer to something that may be a non-thinking being, inanimate object, or belief, such as a human, animal, plant, or fungus.

In this respect, formal solids can be considered as an all-inclusive term. Sometimes, the formal entity is taken as an ontological broad meaning, whether or not it refers to a material being, as is often the case with the immaterial form of entity-language. Moreover, formal entities sometimes also refer to existence or essence itself.

03. Etymology and development of formal reality

Formal reality is a discipline in which philosophy studies universal and fundamental issues, including the fields of being, knowledge, value, reason, mind, and language. Philosophy differs from other disciplines in that it has a unique way of thinking.

It is derived from the ancient root idein, which can be translated into English as to see, and is synonymous with morphe, literally the perceptual appearance or shape of a thing.

In Western thought, there have always been two forms of practical concepts, one of which is used to denote the appearance of all objects or phenomena that people see visually with their eyes, that have physical existence, or that exist objectively in nature in an explicit rather than abstract and general way, that is, "phenomena", and the position on which this study analyzes or criticizes problems and issues has the same meaning as the etymology.

The other kind, proposed by Plato, is the formal

reality, which means the inner structure of things that people see with the "eyes" of their souls, or the ability to interpret known facts and principles and principles into formal reality with their own words, texts or other symbols, that is, "ontology".

From Plato's philosophical point of view, true existence can only be found in the upper world, where everything is a purely formal reality, which is abstract and internal and must be grasped through reason (the soul's eye).

The two formal realities of ancient Greece, which speak of the origin and nature of the universe, constitute the two pillars of the systematic and organized laws or arguments of Western form.

Aristotle further analyzed the formal reality as one of the four causes (material, dynamic, formal, and purposive) that constitute the entity in his "theory of entity" and considered the formal reality as the "essential cause" of the four causes.

The result of modern thought or the mental process of cognition. It refers specifically to the human point of view or social consciousness, which has transformed the ancient Greek view of formal reality and subjectivized it.

This begins with Kant's view that the formal reality of things is unknowable, and that one can only know things within the limits of one's own cognition, so that the formal reality becomes a product of innate cognitive ability, rather than the inherent nature of things themselves.

Later, in Hegel's complex psychological process of seeking or establishing rules and evidence to support or determine a belief, decision, or action, he concludes that "the formal reality of concepts is the living spirit of real things.

Real things are true only by virtue of, through, and within these formal realities." In effect, Hegel reverts to Plato's position.

04. Analysis of Aristotle's formal reality

The formal reality itself consists of two parts. One is the external formal reality of things; the other is the internal formal reality of structures and combinations. In fact, these two are unified.

How the content is concealed in the inner substance of a thing, it naturally has a certain shape of expression to the outside. This is common knowledge.

In terms of the nature of things themselves, there are perfect formal realities and distorted formal realities. This human-related formal reality is the concept of the content and law of the entity.

First of all, the relationship between content and formal reality is discussed. The content is the entity, which is all the objects or phenomena that exist objectively in nature.

Formal reality is the abstract expression of content. There is no one-to-one correspondence between the two, but multiple contents correspond to a certain form. By examining all kinds of objective objects or phenomena in nature, we can summarize a formal reality.

For example, "the sky is not the earth" and "the sun is not the moon" can be collectively summarized as "p is not q". Content must be expressed through some kind of formal reality. The formal reality exists in dependence on the content.

Next, let us examine the relationship between formal reality and law. Bacon thinks that formal reality is the structural law of things, the substance of which is hidden within things. Formal reality determines all objects or phenomena that exist objectively in nature.

For Bacon thought that the formal reality is the law. But we know that the cup is cylindrical. How can you think that it is this cylinder shape that determines all objects or phenomena that exist objectively in nature? So we think that the form is really a result of the law of knowledge, acting on the content.

If we follow the inevitable connection between things and determine the inevitable tendency of their development, we will get a beautiful form.

On the contrary, there will be a distorted form of reality. The pursuit of beauty is a human instinct. The form is based on the content and originates from the law. Therefore, it precedes our daily life experiences and ideas.

As to where the concept of cup comes from, we believe that it is based on the inherent and necessary connection between objective and inherent things, which determines the inevitable tendency of things to develop.

The individual is the comprehensive awareness and knowledge of his physical and mental state and the changes of people, things and objects in the environment by using mental activities such as sensation, perception, thinking and memory.

It is the thing that recognizes the information processing activity of the objective world. The form actually connects our cognitive organs to things. If an objectively existing object or phenomenon in nature cannot be analyzed as a formal reality, then we have no way to start and develop it.

We would not be able to begin to study it. We cannot study God, for example, because there is no sensible part of God at all.

05. Formal reality and materiality

Aristotle states in the Metaphysics that a particular objective entity of all objects, or phenomena, in nature is derived from the combination of both [formal reality] and [material], and with that can be logical.

It is also logical to explain the occurrence of an event, and to prove the truth of a thing by one or more objective existences in a reasonable way, i.e., to show or conclude it by persons or facts.

The concepts of "potentiality" and "reality" are proposed. It is also pointed out that "material" is the material of which things are composed, and "form" is the individual characteristic of each thing.

For example, all the tomatoes in the world are hidden in the flesh, which is the material of the composition of things.

But from the outside, each tomato has a different color, size, and shape, which is unique to each tomato.

Just like "you can't find two identical leaves in the world", this is the individual characteristic of each thing. It can be said that the material is the thing itself, and the form is indeed the definition of the essence of the statement.

The relationship between reality and possibility is the relationship between formal reality and material. The material is the possibility that has not yet been realized, and the formal reality is the reality that has been realized.

For example, if you bury a seed in the earth, it exists objectively as a possibility in all objects or phenomena in nature, and you know that it will grow naturally, this is a potential ability that may come into play, or a possibility.

When it really takes root and grows leaves, it will look more or less the same as what you imagine, that is, it is roughly the same as its possibility, which is the reality.

But possibilities can be divided into those that come from birth and those that come through action or learning. For example, if you learn to swim, play a musical instrument, play the piano, play basketball, ride a bicycle, etc., these are all possibilities acquired through learning and stimulated; for example, a child with normal five senses can hear, see, and smell after birth. The possibility exists in order to achieve the desired goal.

Possibility refers to the potential trend of development within an objective thing; reality refers to the possibility that has already been realized.

Possibility and reality are the two inevitable stages in the development of things or phenomena. Any transformation of a thing from one quality to another is a transformation of possibility to reality.

Thus reality itself is the result, for example, seeing, hearing, smelling, playing the flute, or playing the piano as described above ...... They exist as results.

Section 3: Influence on future generations

Influence in Kant's Philosophy

Aristotle believed that "entities" are the origin of

things. In explaining how entities are constituted, Aristotle introduced the famous "Four Causes" in the Physics to explain the causes of things, i.e., he believed that everything exists with a material cause, a formal cause, a dynamic cause, and an end cause.

That is to say, something that is distinguishable and within itself, and exists independently. But it does not need to be a physical existence. In particular, abstraction is usually regarded as a solid, consisting of formal reality and material. This theory deeply influenced Kant's metaphysics.

In Kant's theory of innate sensibility, he argues that human innate sensibility, which intuitively observes formal reality in space and time, arranges and organizes material to form facts that can be observed and observed.

It is usually used for all objects or phenomena that exist objectively in nature in particular. Mathematics, on the other hand, is composed of purely intuitive abstract, universal ideas that serve to specify the domain or class of entities, events, or relationships, and this material is combined with comprehensive judgment, a form of reality.

Among innate rationalism, Rationalism, European

Rationalism, is a philosophical approach based on the recognition of human reason as a source of knowledge, a theory that holds that reason is superior to sensory perception, also known as theory alone.

Rationalism, from the point of view of formal reality, is a methodology or theory that recognizes that truth cannot depend on the senses, but on reason and deductive reasoning. Kantian science is the study of the structure and development of the celestial bodies in space and the universe.

It includes the structure, nature, and laws of motion of celestial bodies. It is composed of concepts abstracted intuitively as material sensibility, and innate rational concepts as formal reality.

The unified objective material world is an objective reality that exists outside of the consciousness and does not depend on the consciousness. It includes both the known and the unknown material world, from the smallest atomic quark to the largest celestial body of the universe.

It is in perpetual motion, change and development, systematic, complex and infinite diversity, constantly known to human consciousness, and transformed by man to say, the sum of various phenomena is the material.

This innate reason defines the sum of phenomena as the material, and thus the form of nature.

From the above two aspects, it can be seen that in Kant's holistic, fundamental and critical inquiry into the real world and man, he clearly inherited Aristotle's systematic and organized laws or arguments of the material and formal reality

It is a logical way of analysis in which human beings analyze natural and social phenomena in accordance with existing empirical knowledge, experience, facts, laws, cognition, and tested hypotheses, through generalization and deductive reasoning, and so on.

It is also a logical and inferential summary of the complex mental process of seeking or establishing rules and evidence to support or determine a belief, decision, or action, and can be briefly described as the exact nature of a thing, or the connotation and extension of a concept.

At the same time, Kant also studied the position on which Aristotle based his analysis or criticism of problems and issues. From a certain standpoint or perspective, the view of things or issues is further developed.

Kant's logical inferential summary of natural and social phenomena in the form of real and material human beings, in accordance with existing empirical knowledge, experience, facts, laws, cognition, and tested hypotheses, through generalization and deductive reasoning, etc., has developed into a process of cognition in which people

The field of cognition, in which people reflect reality by means of concepts, judgments, and reasoning. Before that, Aristotle's study of existence and the nature of things. Metaphysics is a branch or area of philosophy that is regarded as the primary philosophy and the "fundamental problem of philosophy".

For questions that cannot be answered directly by perception, it deduces the answer by rational logical reasoning under a priori conditions, and cannot contradict empirical evidence.

It is a discipline in which human reason explores the most general aspects of things and their ultimate causes, and the discussion of formal material is limited to the question of physical existence.

It is clear that Kant's theory of innate sensibility and

his theory of innate reason extend this structured and coherent account of the relationship between certain observed phenomena, a systematic combination of concepts, variables, definitions, and propositions, into the realm of logical cognition.

In addition, Kant's "formal reality" is also clearer than Aristotle's, which draws out one or several characteristics of complex objects and pays attention only to the actions or processes of other characteristics (e.g., the mind thinks only of the shape of the tree itself, or considers only the color of the leaves, unconstrained by their size and shape).

Aristotle's use of "essence" to explain form is too vague. Kant describes "formal reality" as the perceptual intuitive formal reality of space and time, as well as the areas of causality and possibility. This makes the relationship between formal reality and materiality clearer and more understandable.

2. Influence in Hegel's Philosophy

A similar view continues in Hegel's view of the formal reality as a fundamental question of life, knowledge, and value, which is explored and reflected upon through analytical thinking.

Hegel believed that the concepts and regulations we obtain are not arbitrary, but exist objectively in nature, and that all objects or phenomena, by themselves, are determined by the nature of things.

Similar to Hegel's theory of cognitive form, what is the definition, explanation, and distinction of the Kantian "object-identity"?

The so-called "thing-self" is a fundamental concept of philosophy proposed by the classical German philosopher, Kant, and is also translated as "thing-self" or "self-existent thing".

Kant regarded the Self as something that can be perceived but not known by human cognitive ability, but which exists. It also shows the unknowability of his philosophy, dualism, and the fact that when your will is the subject

Assuming that all phenomena (objects) are necessarily like your body (direct object) pointing to your subject, the self-substance acts on the human senses, which are the source of the human senses.

The innate spatio-temporal form of the sensory material cannot be the knowledge of the Self at all, once it has been organized by the innate spatio-temporal form of the subject.

The innate spatio-temporal form of the sensory material is actually organized, and then processed by the a priori scope of the subject of knowledge to form a scientific knowledge with universality and necessity.

This kind of knowledge is not the knowledge of self-contained things, but the universality of which is a widely shared characteristic of the human subject, not limited by time and space. It can also refer to the commonness and inevitability of the occurrence of events.

In this way, there will always be an insurmountable gap between knowledge and the "self-existent thing". In Kant's view, in order to push one's knowledge forward, one must eventually move more and more towards the moral purpose in practice.

Therefore, one must strive for the knowledge of the Self or the Self of things, which is then the abstract "idea" that one sets up out of one's rational nature, namely, the soul, the universe, and God. Although they cannot be known, they can be believed.

And what those phenomena point to is the object-self,

that is all. Of course, there is a hidden contradiction in the phrase "with your will as the subject", that is, the subject itself cannot be known, that is, it cannot be described by "will" or "object self".

In other words, it cannot be described by "will" or "object", but under the assumption of the great superconsciousness, it actually means that every phenomenon

The higher the degree of self-consciousness, the more it can express the substance of the subject.

Most philosophers accept that "man" has the highest degree of self-consciousness, and therefore the phenomenon manifested by man is the most complex and individualized. Even in Schopenhauer's interpretation, this degree of self-consciousness can be so high as to negate the will, which he considers as the source of all divinity.

Subjects are free from time, space, and causality; not so with concepts, which are an abstract form of phenomena and must, to some extent, comply with time, space, and causality.

Quote from "newsunny0912 (The Sun in the West)": I

was recently reading Kant's philosophical theory, and I came across a term called "object-autonomy".

In the book, there is a description of the object-self, "Kant thinks that we can only know the phenomenon, but not the thing itself (object-self), and the phenomenon is not equal to the object-self.

This is the limitation of reason and science, because reason can only deal with the phenomenal world." Therefore, in terms of human beings, we can only know the external forms, observable facts or events that are manifested in the development and changes of things, but not the object.

Therefore, in the case of man, we can only know a man's visible body, but not himself, because his exterior is not equal to his substance that is hidden within the self, and natural science can only know and deal with his exterior.

The concept that arises from the observation of his external appearance, such as the "beautiful man", can also only be used for the eyes of those female "hangers-on" who are not rich and ordinary looking.

The "rich and beautiful" and "handsome" have

gradually become synonymous with the best men and women in today's society, but they are unattainable figures, and of course, it cannot be said that this is what he is inside. It is evident that the concept and the object-object are actually quite different.

However, he believes that an a priori world cannot be assumed, because this would deprive both the subject and the object of their freedom, and the process of knowing is the process of knowing the formation of the object.

Moreover, content and form are indeed unified with each other, rather than divided into parts from the whole.

We can get the purest and most realistic truth, that is, the knowledge that objective things and their laws are correctly reflected in the human mind, that is, we can know the essence.

In contrast to Kant's view that the process of knowing external things, or rather, the process of processing information about external things that act on human sense organs.

It is the existence of something that cannot reach beyond knowledge and is absolutely unknowable. It is the basis of phenomena, and people admit that they can know phenomena.

Unlike the existence of the "object-self", which is the basis of phenomena, Hegel expands the possibilities of formal knowledge and gives it infinite dynamism.

Object-identity is a very common yet profound term. Most people take for granted what they are used to in their lives, and it becomes a reflex action in their daily lives.

This is why not many people can be philosophers, and most of those who do become philosophers are like madmen, writing and mumbling. In order to clarify what knowledge is, Kant wrote a thick book discussing which is knowledge, "the house is white" or "the white house"?

What is the "thing-self"? When someone says to you, "Look, there is a daffodil," the sound vibrates your eardrums and strikes your hearing cells, causing the thoughts in your mind to appear white, with images of trumpets, petals, marigolds, and so on.

Then you search for similar images in space with your eyes, and you find "there is a daffodil there. But are the daffodils that you see in your eyes and the daffodils that you see in your mouth the daffodils themselves?

In fact, there are three completely different things here: "the daffodil that comes out of your mouth", "the daffodil that you see with your eyes", and "the daffodil itself".

Daffodils are just the name we give them, and white is just the color we see with our eyes. The object of a daffodil may not be white, and the object of a daffodil may look like a trumpet, a heavy petal, a marquise, and so on.

The beautiful appearance of the daffodil is only produced by our sense organs, and the external world produced by our sense organs is not necessarily the real appearance of the external world itself.

But does this external body really exist? If all we know depends on our own sense organs, then we can never know whether we are deceived by our sense organs.

Like Zhuang Zhou's dream (butterfly), Zhuang Zi's fable is the most literary imagination. The purpose of the fables is to depict the spiritual state of "forgetting oneself" and "materializing" after one has realized the way of nature and has no distinction between right and wrong.

This fable is recorded in the "Theory of Qiwu": "One day Zhuang Zhou dreamed that he became a butterfly, fluttering up and down, vividly alive, and quite happy with himself.

When I woke up, I felt that I was lying in bed, or Zhuang Zhou. But at this time still do not know, whether Zhuang Zhou dreamed of himself as a butterfly, or the butterfly dreamed of itself as Zhuang Zhou? Although Zhuang Zhou and the butterfly are different, they change from each other in dream and consciousness, which is the so-called "materialization".

Materialization means the unification of matter and self: matter can be transformed into me, and I can be transformed into a crop, and there is no difference between matter and self, so there is natural unity. However, in order to achieve the state of "unification of the object and the self", one must first "forget the self", as it is said here, "I do not know who Zhuang Zhou is at this time".

Only by letting go of the ego's prejudice of subjective self-love and clinging to it, and by freeing oneself from the physical body, the emotional consciousness, and the cognitive-sensual self, can one be free of the spirit and transcend the ego to reach the state of "forgetting the self".

Once we can forget the self, we will not be forced to distinguish between the material and spiritual selves from nothingness, and whether they can become one with each other.

Einstein, the founder of relativity theory, believed that there is no absolute time, that time is fast and slow, and that the future and the past are only human perceptions, and that there is no corresponding existence of a physical body. However, Einstein could not accept that our universe has no physical body.

He believes that even if you don't look at the moon now, and you were not born in 1957, you can predict that the moon will still be there in 2050, and that the moon is not something that you personally imagine.

Ball doesn't think so. If your brain can form an image of the moon because of some telecommunication signal, how can we be sure that this signal is not spontaneously generated by our own body, like a dream?

There does not need to be a moon entity there in order for us to "see" the moon. The world may not have a physical body, but it exists because of you, and when you disappear, your physical world disappears.

People often feel happy and sad because of certain things, but who determines these happiness and sadness? When you are watching a horror movie, there are many vampires in the movie, chasing the female lead, the sound effects create a very scary atmosphere?

Should the director's approach be to make the heroine feel scary, or to make you, the audience, feel scary inside? When someone says you are very beautiful and handsome, "you are happy to hear it", you feel very happy all day long, is this happiness given to you by the person who said it?

Or, one day, you read a short sentence or a short story that makes the speaker and the listener feel very funny or generate a sense of humor, or another action (action) type of joke.

Is it because the joke is very funny, or is it the funny action that makes you feel very funny?

If your answer is yes, then you belong to the Einstein fan school, that is, you believe that the Self should exist, that there are really people there to praise you, so that you can feel happy, and that there are really jokes, so that you can laugh.

If your answer is no, then you belong to the Poe school, where it's up to you to be happy or not, to have someone praise you, to be happy or not.

Someone praises you, you can choose to be happy or unhappy, someone insults you, you can choose to be angry or not angry. Your emotions grow on you, who can blame you for being sad? Who can stop you from choosing to be happy?

The worldview that believes in the existence of things in their own right, that is, classical physical philosophy, is like Laozi's "Tao", where everything has a source called "Tao".

If it can be applied as a specific reason, it is not a reason that can be applied permanently. The philosophy of quantum mechanics, the world view that things do not exist on their own, is like the Buddhist philosophy of Sakyamuni, which says that everything is empty and only the mind is responsible.

Therefore, there are three types of cognitive ability and clarity of cognition in the environment and the self by using the mental activities such as sensation, perception, thinking, and memory to conduct a holistic, fundamental, and critical inquiry into the real world and people, as well as a comprehensive awareness and understanding of one's physical and mental state and changes in people, events, and things in the environment.

The first one: seeing the mountain is a mountain: it means that people only see all the objects or phenomena that exist objectively in nature, and the part that is in contact with the outside world, and when they see an apple falling from a tree, they only see the apple falling from the tree.

The second type: Seeing a mountain is not a mountain: It refers to a learned person who sees something and can deduce the consequences from the previous reason. When he sees an apple fall from a tree, he knows that it is because of gravity, and gravity, based on facts or premises, can predict the moon's gain or loss on the first and fifteenth days of the month, and the ebb and flow of the tide between sea and night.

The third type: to see the mountain or the mountain: refers to the wise person who knows that all things must follow the rules of their own accord, in accordance with the implementation and development of instinct, without having to go through the human system to establish it, but can treat it with a normal mind.

He knows that everything has a gravitational force, which is the fundamental law inherent in the existence and movement of the universe, and that it does not depend on the will of man, and the mystery of nature.

Section 4: The whole universe is reality

Parmenides (540-470 B.C.), the founder of the Eleatic School of ancient Greek philosophy, argued that thought and reality are inseparable. The whole universe is reality, which is eternal, immortal, single, continuous, immovable, and limited.

Only the real can be thought of and described, and only the real has a real name. Parmenides also proposed the proposition that "thought and reality are the same", which is the first doctrine in the history of Western philosophy that thought and reality are the same.

Georg Wilhelm Friedrich Hegel (1770~1831), the master of German conceptualism in the 19th century, believed that the whole universe is a manifestation of The Absolute Spirit.

"The Absolute Spirit is the only reality, and the reality

is the result of thought, or the mental process of cognition.

The Absolute Spirit is the only reality, and reality is the result of thinking, or the mental process of cognition. The real and the spiritual are the same, and thus the real is the same as the mind and thought.

Hegel argues that objective existence is reflected in the fact that the result of thought activity in human consciousness, or the viewpoint formed, is the same as the real, that thought is the real, and the real is thought.

We cannot seek reality apart from thought; it is not outside of thought. The English philosopher Francis Herbert Bradley (1846-1924) wrote the book Appearance and Reality.

He believed that reality is absolute experience, which includes thought, feeling, will, emotion, truth, goodness, and beauty, etc. There is not and cannot be any reality outside of the spirit.

Bradley believed that the ultimate reality is the overall perception and understanding of the external world by the human brain when external stimuli are applied to the senses.

Perception and cognition have a wide range of

influences on life, from the basic survival needs of infants to the schoolwork of school-age children.

For example, auditory perception helps babies distinguish between the voice of their parents and the voice of a stranger; tactile perception helps us identify the difference between what we touch or eat; smell and taste perception help us avoid eating bad food.

The sense of smell and taste help us to avoid eating bad food, the sense of proprioception helps the child to realize how much force is needed to play ball or hold a pencil, the sense of sight helps the child to distinguish misspelled words, and so on. This absolute reality has three symbols and signs that can be used as characteristics of things.

The first is non-contradiction: the process of analysis, synthesis, judgment, reasoning and other cognitive activities based on images and concepts, whether the object exists, whether it has a certain attribute, and whether things have a certain relationship between them, the absolute standard of affirmation or denial of truth.

The second is the single unity.

By unity, we mean the property, tendency, or

sameness of mutual attraction and connection between two contradictory parties. It has two situations.

1. Both sides of the contradiction are interdependent under certain conditions, and the existence of one side is predicated on the existence of the other side, and both sides coexist in a unity. For example: tall and short, fat and thin, happy and sad, etc..

2. The contradictory parties are mutually transformed according to certain conditions. That is to say, all contradictory parties are always under certain conditions, like their own opposing contradictory parties after a struggle.

Under certain conditions, each side changes to the opposite side of itself, to the position of the opposing side. For example, joy and sorrow, difficulty and ease, victory and defeat, etc., can be transformed into super opposites under certain conditions.

Without certain conditions, it is impossible for both sides of the conflict to be transformed. It is also the whole collective or the whole thing that is complete, not bad at all, with proper coordination.

Thirdly, in terms of absolute nature: the recognition of

the phenomena and external connections of the objective things, which people acquire through the sensory organs in the process of direct contact with the objective things, is equivalent to the reality.

In short, perception is the brain's overall view and understanding of the external world when external stimuli act on the senses.

It is the organization and thinking based on observation that rationalizes the reasons for changes in things, the connections between things, or the laws of development of things. Experience is reality, and anything other than perceptual experience is not reality.

The twentieth-century philosopher Alfred North Whitehead (1861-1947), who wrote Process and Reality, proposed an all-encompassing view of reality, arguing that every element of reality, i.e., every "true reality," is essentially a self-developing or "real" reality. is a process of self-development, or self-creation.

This process is realized through the selection and rearrangement of the materials provided by the context of such objects. The object perceived in his philosophical system is the "real" past, which is conveyed to the "real" present.

The object of perception in his philosophical system is a complex of many sensations from the past "reality" to the present "reality". These sensations are the whole nature of actual entities or actual occasions, and it is in the whole nature that "reality" arises and disappears.

Section 5: Obstacles to Egoism

The realist only has a fixed view of other people's concept of the objective existence of all objects or phenomena in the natural world.

1. The "other" is treated as a purely perceptual image formed on the basis of perception, a form of perceptual understanding.

2. In physics, it refers to the process by which two systems interact with each other and generate energy exchange.

3. "Other" means the perceptual image formed on the basis of perception, a form of perceptual awareness, the existence of other perceived by others.

Actualism, also translated as actualism, is a view of Western philosophical ontology, which holds that ontological reality is independent of human senses, beliefs, concepts and ideas.

It is a view of Western philosophical ontology that holds that ontological reality is independent of human senses, beliefs, concepts and ideas. Unlike some teleologists, reality is derived from or influenced by consciousness. What the world believes to be reality today is not the same as what ancient Greek or medieval philosophers said.

In content, it is not the same as, or sometimes even the opposite of, what the realists believe, which has become teleology or egoism. This is clearly a reference to Husserl.

Thus, Schart again takes his thinking from Kant's philosophical logic: Kant's approach is based on the idea that a mere subject is philosophically a being with a unique consciousness and a unique personal experience, or another being external to itself.

The universal law of subjectivity is established from the viewpoint of another entity external to itself and related to it. Subjectivity, also translated as subjectivity, is literally closely related to subject, subjective, and subjectivism.

In practical application, any idea that mostly adopts or defends a subjective or subjective viewpoint or approach has the quality of subjectivity.

The word subject is derived from the Latin sub and jacere, meaning "to throw under the surface" or "to place under the surface as a foundation", and the word substance (substance, substratum) is something that is distinguishable and exists independently within itself.

But it does not need to be a physical existence. In particular, abstraction is usually regarded as a subject in a similar sense; it refers to a real being that is the basis of a load.

It follows that all objectively existing objects or phenomena in nature which are loaded are qualified by the existence of the loader and must be established by the loader.

The original meaning of subject is found in the logic that Aristotle valued, i.e., the logical subject, also called subject or subject; the opposite of it is the predicate.

The main or crucial part of the subject of all objects or phenomena that exist objectively in nature is the actual object that exists in the world, or all objects or phenomena that exist objectively in nature as a whole with actual content.

Therefore, in the general statement, [the existence of things] is used as the subject, and a certain qualification is used as the description.

Moreover, in the Latin original, the term "to carry, to bear" is used in the discipline of analytic reflection on the fundamental questions of life, knowledge, and values.

It means that the ego is the subject of its actions, that is, the ego is clear that the individual uses mental activities such as sensation, perception, thinking, memory, etc., and is aware of his or her own physical and mental state, and of the changes in people, events, and things in the environment, and that certain situations belong to and are in the ego.

In addition, the subject as opposed to the object means that the individual of the ego uses mental activities such as sensation, perception, thinking, memory, etc., to become aware of and recognize changes in his or her own physical and mental state and in people, things, and objects in the environment.

In this case, the subject may refer to the mental subject, or the whole collective composed of [body] and [soul], or the whole objective existence of all things and objects.

After the emergence of the meaning of the subject, the discussion of the whole, fundamental and critical inquiry into the real world and human beings is not based on the object or objective facts, but on the subjective or subjective feelings, which are called subjective or subjective.

Thus, subjectivism considers that the situation or tendency of the subject of knowledge is to determine the act of "cognition" and "identification" of a subject in order to know with certainty, and that these knowings have the potential ability to use the principle of validity for specific purposes, which is the opposite of objectivism.

In the discussion of the theory of knowledge, the individual's comprehensive awareness of his or her own physical and mental state and the changes of people, events, and objects in the environment by using mental activities such as sensation, perception, thinking, and memory, as well as the subject of knowledge and the object to be known, can be said to be the focus of discussion.

From the point of view of the subject, the act of "cognition" and "recognition" of a subject is the standard by which to know with certainty, and these knowings have the potential ability to be used for specific purposes.

It may be based on the individual's comprehensive awareness and understanding of his or her physical and mental state and the changes of people, events, and objects in the environment through mental activities such as sensation, perception, thinking, and memory; or it may be based on cultural background or ethnicity.

Others believe that certain forms of human nature or thought should prevail. Rationalism and empiricism differ in the subjectivity they advocate.

The rational subjectivism believes that the object is determined by the subject; if there is no one who uses mental activities such as sensation, perception, thinking, memory, etc., and is aware of the state of his own body and mind and the changes of people, things and objects in the environment, and the subject who can recognize the cognition, then there is no cognizable object.

Then there is no cognizable object at all. Because

knowledge is knowledge, it is necessarily some operations or an activity of the main body of thinking, this kind of knowledge out of the characteristics of the body (i.e. subjectivity), and may not hinder the objectivity of knowledge.

On the contrary, the subjectivity of knowledge makes it possible and meaningful for the act of "cognition" and "identification" of a subject to be known with certainty, and these knowledges have the potential ability to be used objectively (objectivity) for specific purposes.

In Kant's theory of knowledge, the term 'subjectivity' refers to the result of thinking or the mental process of cognition. In particular, it refers to the subject of human viewpoint or social consciousness, that is, the authority of transcendental reason.

Thus, Kantian subjectivism and empiricism advocate the idea that knowledge is the act of "cognition" and "identification" of a subject in order to know with certainty, and that such knowledge has the potential to be used for specific purposes.

This fact or state is called knowledge, which includes knowledge or understanding of a science, art, or technique.

In addition, it also refers to a set of learning, experience, or a series of information acquired in the process of learning and practice through research, investigation, observation, or experience, which is very different from the doctrine that is composed of subjective abstract impressions and concepts.

Some scholars believe that subjectivism more or less means that the standard is based on changing feelings and inclinations. Therefore, they do not consider subjectivity as a purely existential theory of knowledge.

However, in the inquiry of value issues, it is advocated that the individual's feelings or experience of lessons learned in the process of contacting external things is the criterion.

In this view, the ultimate standard of moral values is a kind of social ideology. It is the sum of behavioral norms that regulate the relationship between people and individuals, individuals and society.

The ultimate standard of beauty is personal taste; this school of thought is called value subjectivism.

It is not about the individual. The subject is only the

individual's comprehensive awareness and recognition of his or her own physical and mental state and the changes of people, events, and things in the environment by using mental activities such as sensation, perception, thinking, and memory.

Those common qualities that belong to individual cognition cannot determine the diversity of individuals, just as Spinoza believes that human beings have an everlasting attribute, which makes an entity or a substance into an entity.

They make an entity or substance what it is, and what it necessarily is, without which it loses its identity.

Kant here specifies not only general objects, conditions of possibility, but physical objects mathematical objects, objects of beauty and ugliness, and objects of various spheres

as well as objects that exhibit teleological properties (conditions of possibility), but as objects to which others are given in our experience.

In this way, in Kant's view of philosophical thought, it is necessary to ask how the knowledge of others is potentially possible, that is, to determine the conditions of potential possibility for the experience of others.

Schart, who considers Kantian logic as a process of analysis, synthesis, judgment, reasoning, and other cognitive activities on the basis of images and concepts, studies the universal, fundamental problems of the discipline

Including the field of existence, knowledge, values, reason, mind, language and other fields of study and analysis or criticism of problems, issues, etc. based on the position that the results of thinking about others or the mental history of cognition. Specifically, the deficiencies of the human point of view or social consciousness are evident: 1.

1.The cognitive object of Conde is not a "thing-self".

This is because cognition is the activity of thinking in which the human brain reflects the characteristics and connections of objective things and reveals their meanings and effects on people.

In a broad sense, cognition includes all cognitive activities of human beings, i.e., the phenomenon of perception, memory, thought, imagination, understanding and production of language and other mental phenomena collectively known as representations.

2. Causality unifies temporality.

Causality (English: causality or causation), also known as causality, is the relationship between an event (i.e., "cause") and a second event (i.e., "effect"), where the latter event is considered to be the result of the former event.

Generally speaking, cause and effect can also refer to the relationship between a series of factors (cause) and a phenomenon (effect), the characteristics that make something effective, meaningful or useful in a certain period of time. In a certain period of time the most valuable and effective part of the whole, the divergence into a consistent. It is a characteristic that is not valued enough when it is out of time. 3.

3. the concept of coordination does not apply to others.

The concept of harmonization, proper coordination, proper handling of various relationships within and outside the organization, creating good conditions and environment for the normal operation of the organization, and facilitating the achievement of organizational goals, is an abstract, universal idea that serves to specify the scope or class of entities, events or relationships that do not meet the requirements of objective conditions and are suitable for application to others.

Shat's I-other relationship: it is not my personal use of sensation, perception, thinking, memory and other mental activities.

We treat negation as something that is distinguishable and exists independently within itself. But it does not need to be a physical existence.

In particular, the abstraction of the other as a physical structure is not the I and the person I am not. This does not mean a particular component of nothingness [that separates the other] from [myself].

He also concludes that the error of the realist is to believe that the other is grasped through the body of the other, whereas idealism reduces my body and the body of the other to thinking from different points of view or perspectives, or to judging the rationality of something.

It is a system of representations of the nature of a thing that exists independently of subjective thought or consciousness.

Both of these are, in fact, a formal and physical space that separates me from others. In fact, it is a tacit acknowledgement that my relationship with others is an externally separated negative relationship.

And this kind of connection between people and people or people and things, of a certain nature. It is a generalized meaning of things or phenomenal conditions that can produce other things, or phenomena, and in the end, one has to turn to God.

Therefore, the state of interaction and mutual influence between me and the primordial things of others must be concealed within all the objects and phenomena of objective existence, with a two-way negation of substance, where others define me and I define others.

Section 6: Shame and Shame

As we grow older. When an individual compares his or her behavior with existing moral concepts, such as the realization that his or her behavior does not meet the standards of social moral behavior, or harm the interests of others around him or her, it will also arise.

It is a manifestation of self-moral consciousness, and is inseparable from the formation of individual moral concepts, the establishment of moral relationships between people, and respect for those around them.

It can stop or correct the immoral behavior of the individual, and maintain or promote the healthy development of the individual's moral character. The lack of shame can cause serious obstacles to the moral development of an individual.

From a psychoanalytic perspective, shame has a self-deprecating (self-attacking) component, and appropriate shame is beneficial to the moral development of an individual.

It is a social ideology. It is the sum of behavioral norms that regulate the relationship between people and individuals, and between individuals and society. Such as honesty and hypocrisy, good and evil, righteousness and unrighteousness, justice and partiality, etc. are in the field of morality.

However, excessive shame will harm the individual's self-evaluation based on the formation of a self-esteem, self-love, self-respect, and require the emotional experience of respect by others, the collective and social, is not conducive to mental health.

1. Self-perception of existence

First of all, most people understand that anger hurts the body. When people are angry, it is easy to cause many physiological and psychological problems in the body and improper operation, so it is natural to malfunction.

When we let out our anger, we have a window to vent it, and often the negative emotions will be relieved. Physical and psychological damage will also be reduced to a certain extent.

On the contrary, the feeling of shame (shame, guilt) tends to hide the problems that are hidden in the substance of things, because the feeling of shame refers to the emotional experience of self-condemnation when an individual realizes that he or she, or the group to which he or she belongs, has violated social norms and moral codes of behavior.

This is evident in the early stages of an individual's moral development, such as when one's behavior is condemned by the people around him or her, or when one is aware of how the people around him or her will treat this behavior, the more shame one feels, the more one doubts oneself, and the more shame one feels.

Second, shame is also a kind of denial of our (self-existence), distrust. The more shame we feel in our hearts, the more we distrust and doubt ourselves, and the more we lose our self-confidence.

Subconsciously, we think that we are not good enough to do everything, and then we do not concentrate on our work, and then we really do not do well in all the people, things, and objects that we touch in all human activities. The vicious cycle results in a deeper level of negativity towards oneself. This is the constant influence of each other.

In fact, the most negative human emotion is: shame. It is the source of almost all the unexplained emotional discomfort we can perceive. What is shame?

It is an emotion brought on by a strong sense of self, a judgment of "selfhood" against "self-existence", a devaluation of self-worth. This is a memory that you and I share. Think of a time when you were a student and had such a moment.

When you were criticized and humiliated by your teacher or other people in public, you felt so bad, so humiliated, that you wanted to find a hole in the ground and go in immediately? This is the sense of shame.

Shame blocks positive emotions, such as the pleasure of being comfortable or childish curiosity. Shame also has the ability to "control" other emotions.

Whenever shame is present, we become wary and our expression of other emotions is suppressed. Unlike most other emotions

Shame does not disappear over time, it lingers in our subconscious. It is also the emotion that is least recognized and least released.

It is the most secretive emotion in our lives: the urge to express our emotions when we are upset, sad or angry.

However, when we feel shame in our hearts, we will reduce the opportunity to express ourselves because of the social matrix of shame, fearing that we will inadvertently expose what we are hiding to others.

In fact, it is normal to feel a certain amount of shame in daily life, but if shame has affected a person's questioning of who he or she is and how much he or she is worth, then it is a very dangerous situation for the body and mind to be in danger.

Every time you are accused or a small failure occurs, this pathological self-deprecating shame will be repeatedly evoked and experienced. And sometimes pathological shame lingers in the relationship.

People with pathological shame perceive themselves as inadequate, such as constantly feeling self-deprecating and dependent on others, and because of the subconscious presence of "not being good enough" in this self-conscious perception of existence, they secretly and continuously feel shame.

2.There are three main differences between guilt and shame.

Guilt, also known as guilt as opposed to innocence, is a state of mind in which a person feels remorse or shame when he or she discovers that he or she has broken a moral or normative wrong or crime.

People choose their own course of action based on reason and will, and react in a social context, but when they violate moral norms without being detected by others, they feel guilty and blame themselves.

Guilt is seen as the result of innate, original sin, and that is why guilt compensation is seen as dependent on the will of God.

Sigmund Freud (1856-1939) thought that guilt was the result of desire, of delusions, of contradictions arising from guilt, either by pressing the delusions into the subconscious and becoming a symptom of future psychological pathology, or by applying psychological mechanisms to find liberation.

Guilt is mainly focused on a certain behavior, while shame is focused on how one looks to others and whether one's existence is meaningful.

01. Guilt is about behavior, shame is about oneself.

Shame is a painful emotion that arises from a sense of fault, deficiency, or incompetence, or inferiority, essentially in the context of the relationship between the person and the self.

It is triggered when the self realizes that one's appearance, speech, behavior, or origin is inferior to others or to existing norms, or when one's personal performance is not ideal, or when one feels a loss of self-control or confidence.

02. Guilt has more power than shame.

Guilt is a human action, a way of acting, and a reaction to the environment and other organisms or objects that have crossed the line, feeling remorseful but still retaining a sense of control over their own existence and feeling empowered to hurt others.

The shamer is a methodological approach to self-being, a disciplined, systematic analysis of written or spoken discourse, a fear of losing one's connection to those important others, and a sense of the weakness of self-being.

03. Guilt is expressed, shame is concealed.

In The Primacy of Perception, M. Merleau Ponty analyzes that guilt begins with an objective self-awareness of the self, i.e., guilt begins to emerge when a person becomes conscious that he or she has hurt another person.

"People who feel guilty about "being-for-itself" long for forgiveness and understanding, and sometimes do compensatory actions to reduce the guilt of "being-for-itself". Thus, being-for-itself guilt inspires action and is easier to [express] and to [identify].

Aristotle (384-324 B.C.) was the first to speak of

"shame" in the Western philosophical tradition. In Book V of Nichomachean Ethics, Aristotle speaks briefly of shame.

Shame is not a state of virtue or character, but a fear of disgrace and blame. When a person feels embarrassed and humiliated, he or she often blushes and is in a state of shame.

In Shame and Guilt (1953), C. Piere & M. Singer argue that shame is a reflection of the tension between the ego and the ego ideal.

He argues that shame is different from guilt in that the former arises from failure and the latter from evil. Shame, on the other hand, is directed toward the self as "being-for-itself". Blame is always accompanied by misconduct and is therefore not a virtue.

But guilt negatively keeps one from doing evil, and can be considered conditionally a good thing. Shame inhibits people, it gives them a sense of inferiority, and it deprives them of the power and confidence of their "being-ness".

People want to hide the shame of "being-for-itself" and not let others see who they really are. "What are some of the behaviors that people exhibit when they have a strong sense of shame about being-for-itself?

Section 7: Relationships with "Others

Often, our relationship with others is actually a mirror that reflects each other's inner needs and reactions. All smiles, tears, anger, jealousy, gratitude, blessings, and joy ...... reach the source of the heart directly.

We see our own problems, conflicts, hurts and shortcomings in others. It is also easy to see the inner workings of a relationship. Therefore, when there are problems and conflicts in a relationship, the first person to review and awaken is oneself.

How to better accept and recognize others depends on whether you can better improve yourself and get training and nutrients from the relationship.

When you are jealous, it means you are not doing well enough; when you are demanding how others are, should you ask yourself first; when you want to be cared for by others, have you also given them tenderness?

When we look at others who are too arrogant, are we aware of our own low profile; when they have a violent temper, have you treated them with tenderness; and so on and so forth? The problem can be digested by giving up, not caring, and putting it here.

01. anger and violence

Shame is the first and most fundamental cause of all "being-for-itself" violence. Anger and violence are both strategies to deal with the overly painful feeling of being-for-itself shame. "Being-for-itself uses the seemingly powerful emotion of anger to cover up the powerlessness of shame.

In order to cope with shame, people externalize the harshness of being-for-itself to the outside world and to others, and then become angry and violent towards the outside world and others.

This feeling is more bearable than the shame of simply being-for-itself. In particular, the more narcissistic the "being-for-itself" is, the more vulnerable it is to shame, and the more likely it is to show violence in response.

In addition, it is worth noting that people who have a high sense of shame about "being-for-itself" will use violent means to hurt their partner's self-esteem in order to gain a sense of self-esteem about "being-for-itself" in their relationship with their partner.

So often if your partner is often angry or violent, it may be that the essence of "being-for-itself" is that his sense of shame is out of balance.

02. Fantasy of perfection and masochistic tendencies

Schart believes that the pathological "being-for-itself" shame makes people feel weak, and they easily have the illusion of perfection, such as "if I am strong, I will never experience misfortune again", to fight against the powerlessness of "being-for-itself".

This fantasy of "being-for-itself" is impossible to exist in the reality of daily life. But because this fantasy of "being-for-itself" cannot be realized, they will encounter more problems.

As a result, they experience more shame and guilt about being-for-itself, creating a vicious cycle of mutual influence.

But this unrealistic fantasy is important because it gives them a sense that "all the misfortunes I have suffered are due to my 'being-for-itself' not being my ideal self.

They can also feel a sense of being-for-itself control over fate and a sense of basic justice in the world. This feeling of "being-for-itself" is better than pure shame, although it is also negative.

Some "being-for-itself" patients develop masochistic tendencies under the illusion of "what they think they should be".

With this being-for-itself tendency, passive suffering becomes pleasure, anxiety becomes excitement, hatred becomes love, separation becomes integration, helplessness becomes strength, and shame becomes ultimate victory.

All the "being-for-itself" passivity becomes active, and one begins to enjoy the illusion of being-for-itself masochism.

Therefore, being-for-itself shame is a very complex issue, and it is a time-consuming battle of being-for-itself to live with the shame in oneself. The goal is not, of course, to eliminate shame altogether (that's impossible), but rather to put the 'being-for-itself' into perspective.

Rather, it is to keep being-for-itself to a certain degree and within a certain frequency, so that being-for-itself does not become the undercurrent of our sense of self-worth. Therefore, being-for-itself shame is the most damaging emotion among all emotions.

And it is a normal emotion. You are not the only one who feels this way; there are millions of people in the world who have experienced or are experiencing similar feelings - shame: such as

Others explain to me that I am "Freedom" and that they are "being-for-itself" for me.

Shame in front of others, shame of self (I am 'being-for-itself' for me)

The shame of being-for-itself in front of others: I have lost the transcendence, the constant possibility of being-for-itself. I would like to deny this freedom, but the shame of 'being-for-itself' proves that I have an objective self-knowledge and therefore feel ashamed.

Jean Paul Sartre also argues that shame begins in interpersonal encounters, when an individual feels that he or she has a poor image in the minds of others and that the "self" has become an "object" (object) for others to denigrate, shame will be triggered. This leads to the self-deception of "being-for-itself".

Shame proves that my "being-for-itself" exists for him, so that I am in the freedom of others, and through the freedom of others, I express the existence of my "being-for-itself".

Everything "being-for-itself" happens because I have a complete, inscrutable "nothingness" - the freedom of others. My "being-for-itself" begins to have an appearance of a nature, and the original fall of my "being-for-itself" is the "being in itself" of others.

Therefore, "being-for-itself" is ashamed - it is "for oneself" → he thinks that the possibility of my "being-for-itself" becomes a contingency outside of me. The possibility of my 'being-for-itself' dies, so that I experience the freedom of others 'being in itself'.

This freedom of being-for-itself death is reborn in the freedom of being in itself. That is, others have all the rights to choose, name, and explain my being-for-itself, while my being-for-itself desires the death of my constant possibility to be appropriate.

My "being-for-itself" is confronted with the instrumental totality of the instrument, of the "self-existence" that is transcended and organized by others as something of the world. Because of the gaze of others, my "being-for-itself" is no longer the totality of the situation.

I can no longer use my instrumentality as it was intended, because I know that someone is always watching and considering me, and so I exist for others "being in itself".

Section 8: The idea of "co-existence".

Man is a social animal, and how to co-exist with others in a harmonious way is a subject that cannot be ignored.

The so-called "co-existence" is the result of thinking or the psychological process of cognition. It refers to the human point of view or social consciousness, and the meaning derived from it is the human understanding of natural or social things, the meaning given to the object, and the spiritual content transmitted and communicated by humans in the form of symbols.

All the spiritual contents that human beings communicate in the communication activities, including intention, meaning, intention, understanding, knowledge, value, concept, etc., are included in the field of meaning. In other words, is it possible to interact with others in a meaningful way?

The interpersonal relationship with others refers to the interaction between people. The harmony of interpersonal relationship depends on one's attitude and ability to deal with and treat people.

The difference in co-existence of interpersonal relationships gives people different emotional experiences. For example, if the co-existing relationship is harmonious and intimate, there will be pleasant feelings for each other, while if the co-existing relationship is not harmonious or distant, there will be unpleasant, strange or hateful feelings.

In the experience of life's journey, we may be mere passers-by, or we may have stopped somewhere, but we inevitably co-exist with some people, and these co-existing relationships may only be a momentary encounter.

These relationships may be just a momentary encounter, but they are woven into the fabric of our lives. In fact, the establishment of co-existing interpersonal relationships is a basic need of individuals. If the result of mutual communication and interaction can satisfy the needs of both parties, the relationship between them will definitely tend to be harmonious and intimate.

However, if the opposite is true, then the relationship between the two parties will definitely tend to be incompatible, distant, or even antagonistic. Good co-existence of interpersonal relationship makes people have the feeling of openness, mutual trust and understanding, and then sincerely cooperate and harmoniously get along with each other.

However, alienated or poor co-existence interpersonal relationships lead to feelings of resentment or distrust, which naturally make co-existence impossible and may even lead to confrontation and hatred.

1. Husserl, Hegel, Hedgerl - Single World

Husserl: Husserl, like Kant, treats pure consciousness as if it were an innate form, and "my experiential self and the experiential self of others appear simultaneously, and my non-priority".

And my non-priority" is to show me the object to be observed, the thing that is always referred to, existence as meaning, interiority and interiority, my existence and other existence are only intellectual relations. It is commonly called the single world.

According to Leibniz, the modern German

philosopher, the monad is a dynamic, indivisible mental entity, the basis and final unit of all objects or phenomena that constitute the objective existence in nature.

The monads are independent and closed (no "windows" for access), yet they interact with each other through God, and each of them reflects and represents the whole world. Leibniz's monad theory reveals the nature, function and development of human consciousness.

The singleton has a spiritual, dynamic nature, and is a dynamic spiritual entity; the singleton has the principle of continuity; inorganic objects and plants have "microsenses", animals have "souls", or clearer perceptions, and human beings have an integrated I, or a comprehensive I.

Man has an integrated I, or "I-thought," also known as the "unity" or clearer perception of self-consciousness, and sees the physical body as the intermediary of the soul in understanding the universe; the soul and the physical body are inseparable, each following its own laws and in harmony with each other. Monadic idealism solves the important epistemological problems of matter and consciousness, soul and form

Leibniz's monism is an objective idealist system that tends to compromise with religious theology, but it also contains some reasonable dialectical elements, such as the idea that everything moves by itself.

It also contains some reasonable dialectical elements, such as the idea that everything moves on its own. This book systematically discusses the author's monism. It is argued that the monad is the final element that constitutes all things, and that it has neither extension nor parts.

Since there is no part, it cannot be naturally created and destroyed by the combination and separation of the parts, and each monad has no window of access to things.

There are no two monads in the world that are qualitatively identical, and there are no two things that are identical. The monad is similar to the soul in that it has perceptions and desires.

The lowest is the monad that constitutes inorganic matter, while the highest monad is God.

Although the monads are independent of each other, all the objects or phenomena that exist objectively in nature are mutually influencing and interacting with each other, thus forming a harmonious whole.

This rationality of thinking or judging something from different viewpoints or perspectives, the nature of a thing that exists independently of subjective thoughts or consciousness, corresponds to "subjectivity".

The objective truth, which is not influenced by subjective means such as human thoughts, feelings, tools, calculations, etc., but can maintain its truthfulness and harmony, is predetermined by God when He created each single thing.

Hegel: Narrow consciousness is the initial stage of the development of consciousness, and its three stages are

Self-consciousness is a comprehensive awareness and knowledge of one's own physical and mental state and the changes of people, things and objects in the environment by using mental activities such as feeling, perception, thinking and memory.

The truth of the objects that a person knows by the cognitive ability of the environment and the self, and the clarity of the cognition, is attributed to the self-consciousness (subject, self).

Self-consciousness = absolute spirit. Hegel indeed

pointed out that self-consciousness is initially empty, and is a purely self-referential existence. Self-consciousness is the mutual recognition of the opponent's use of sensation, perception, thinking, memory and other mental activities, and the comprehensive awareness and recognition of one's physical and mental state and the changes of people, things and objects in the environment.

The mutual recognition of the master-slave relationship. Schart is indeed borrowing from Hegel's relationship of the interactions and interconnections between master and slave (Hegel treats the other as another self, a self that is my object and reflects me in turn).

Schart considers Hegel to be more progressive than Husserl in that Hegel tells us that the negation of the other is direct, immanent, and mutual, in a direct thought, feeling, belief, or preferred intuition that quickly emerges without much thought process.

The presence of the other makes I-thinking possible. Schartre argues that Hegel breaks with the solipsism that asserts that the ego is the only being, and that the things of the external world and the states of mind of others are merely the contents of self-consciousness, dependent on the mind of the ego, and do not really exist in themselves.

However, Hegel's theory is imperfect, although his theory is concerned with [self-existence] and [other-existence], and he believes that any individual who uses mental activities such as feeling, perception, thinking, memory, etc.

Although his theory is concerned with [self existence] and [other existence], and he believes that any individual who uses mental activities such as sensation, perception, thinking, and memory to become aware of his own state of mind and body and the changes of people, things, and objects in the environment includes the existence of others.

However, Hegel's theory of logical thinking, which is a holistic, fundamental and critical inquiry into the real world and human beings, is still an epistemological terminology.

The great driving force of interconscious competition is that each individual uses mental activities such as sensation, perception, thinking, and memory to

The great motivation of inter-consciousness competition is each individual's effort to transform his or her self-reliance into truthfulness by using mental activities such as sensation, perception, thinking, and memory, and by integrating awareness and knowledge of his or her physical and mental state and changes in people, events, and things in the environment.

This kind of truthfulness is in essence a kind of self-love-obsession for the self, that is, Hegel is still epistemological, and can only be the arrogant scale of existential epistemology.

Secondly, Hegel is an optimist in his epistemology. The ego uses mental activities such as sensation, perception, thinking, and memory to perceive and recognize the truth of his own state of mind and body and the changes of people, events, and things in the environment.

It is able to reveal an objective unity among all kinds of consciousness through the knowledge of others about me, and through the name of my knowledge about others.

This is completely impossible in the eyes of Shat.

First, there is no common scale between others, objects, and the subject, me, at all.

Secondly, I cannot grasp him in his existence/his

subjectivity, because any universal human brain reflects the characteristics and connections of people, things and objects touched by all activities of objective human beings, and reveals the objective existence of nature.

It is because any universal human brain reflects the identity and connection of people, things, and objects touched by all objective human activity, and reveals the meaning and effect of all objects or phenomena objectively existing in nature on human thinking activity, which cannot be derived from the relationship of all consciousness.

From the study of the sources of knowledge, the process of its development, the methods of knowing, and the relationship between knowledge and practice, epistemological optimism, Hegel moves on to the study of conceptual philosophical branches such as existence, being, becoming, and reality, ontological optimism.

It includes: how to classify entities into basic categories and which entities exist on the most basic level. A certain cosmic spirit that exists objectively and independently, which is actually a logical thought, detached from man and separated from the objective world.

The idea of absolute spirituality, expressed only in conceptual form, holds that diversity is capable of and should be transcended towards the whole.

The critique of Husserl and Hegel explains that Sartre's existence of others is in fact such an existence.

In order to abandon egoism, Husserl, when he reached this level, measured existence in terms of knowing, and Hegel in terms of the identity of existence and knowing.

There are two major schools of thought on the nature of knowledge, i.e., what knowledge is. The conceptualists believe that the content of knowledge is formed by the concepts in people's consciousness.

Extending and expanding it, objective knowledge need not be consistent with external things or existing objects.

Actualists affirm that actual objects do not change as a result of our knowledge, so that the act of "knowing" and "identifying" a subject is based on confident knowledge, and that such knowledge has the potential to be used for specific purposes.

It means that through experience or association, we are able to become familiar with the content of something, i.e., we construct it from our feelings and experiences of what exists; by extension, what corresponds to the object becomes objective knowledge.

Heidegger: Co-presence is still abstract. Schart: Co-presence must be predicated of being. Thus the logical thinking of Scharthe and Heidegger's philosophy.

The main difference between Scharthe's and Heidegger's philosophical logical thinking on this subject is that Scharthe argues that although Heidegger emphasizes the existence of a certain nature of connection between me and others, between persons and persons or between persons and things, the "common" in his "common" is not the most fundamental.

"Common" still implies abstraction, and Schart thinks that "common" must be predicated on "for the sake of...". The other is not the thing that is the object of action or thought.

He remains in the form of a person in his connection with my person and person or person and thing, in a certain nature, which he uses to determine my being in being, that is, the mere being of others as if they were being in the world.

To be in existence is to be entangled in the world, not to be attached to it forever, and it is in my being in existence that the social matrix makes its constraint about the way, the method or quantity, the quality, etc. that holds me.

Our connection of a certain nature between man and man or man and thing is not a face-to-face opposition, but rather a shoulder-to-shoulder interdependence.

Since I make a world to exist as an instrumental complex in the service of the fact and condition of my concrete existence in front of me, I am in the position of being able, in my own words, texts or other symbols, to

I am prescribed by a being who makes known facts and principles and principles into an explanation of my existence, and this being makes the same world, as a complex of for it form real, instrumental."

Only this "for ......" is a prerequisite to make "coexistence" possible. From a Scharttian philosophical point of view, the ontological basis on which Heidegger's "coexistence" rests is actually similar to the generalized, inconceivable view of the Kantian master body.

This is because the structure of the "being in the

world" that appears as me, and the co-existence of the unknowable "self-existence" that opposes the phenomenon, cannot be a basis for "co-existence" with the ontology at all.

For example, "I am with Jack" or "I am with Anthony" is in fact a structure in which I have a physical existence, or an explicit rather than an abstract, general existence, and a concrete existence of facts and conditions in front of me.

But according to Heidegger's philosophical view of "co-presence", this is impossible. In Heidegger's "co-presence," the other becomes an object that depends on assistance and cannot stand on its own, and has no capacity to become this other.

2. The Body

The body is an existential 'gaze'. Unlike Descartes' mind-body dualism, the mind of the body is not an intellectual but an existential 'gaze'.

For example, when I am under local anesthesia, the doctor feels my foot the same way I feel my foot. A human being or an animal's physiological organization as a whole, the trunk and limbs are in charge of themselves, unrestricted and unbounded.

The ability to recognize the environment and the self, and the condition of the clarity of cognition are not physiological conditions, but the individual's ability to use sensory, perceptual, thinking, memory and other mental activities, the comprehensive awareness and knowledge of his physical and mental state and changes in people, things and objects in the environment, is the condition of his existence.

The individual's comprehensive awareness and knowledge of his or her own physical and mental state and the changes of people, events and things in the environment by using mental activities such as sensation, perception, thinking and memory

The mental process of knowing and understanding things through the activity of consciousness of the body becomes the target of action or thought only when "otherness" is present.

I never unify my ability to perceive the environment and the self, and the clarity of cognition in the body, but I try to unify it in the bodies of others.

The body in Schatt is closely related to the idea of

"otherness", which emerges from existentialism. The point of departure of "self-existence" is the body and the mind, neither the act of "cognition" nor "identification" of a subject in order to know with certainty.

And these knowings have the potential to be used for a specific purpose, not as a function of rational abstraction. The "autonomous being" can never have an object knowledge of my body.

It can only become an "autonomous being" when it is "for others", when it becomes a person, thing, or object that is touched by all human activities.

The existential theory of the whole physiological organization of the human being, sometimes referring specifically to the trunk and limbs, is that 1. I bring my body into existence; 2. my body is known and used by others; 3. others behave toward me as if I were the subject of objects to them.

Such a three-dimensional description of the body establishes the interaction and interconnection between the "other" and the "I" as the basic primordial thing. So Schartre considers the "self-existent" body: it is the contingent necessity of my contingency, the contingent form of reality obtained.

This necessity is the manifestation of contingency (between two).

One, I am not the basis of my being, dasein under necessity, existence is contingent.

Second, my intervention in "this being" is necessary, then I am here and not in other, which is contingent. It is either something else or "self-existence".

The body is, after all, the contingent form of my situation. The idealist, Descartes, is thus struck by the fact that the body manifests itself as the individuation of my intervention in the world. The soul is the body, just as the "self-existence" is a personalization of its own ability to perceive the environment and the self, and of the clarity of that perception.

3. The specific relationship with others

Schart has explained the basic relationship between the Self-being and others, and the three-dimensional structure of the body as its own. Next he looks for some cases or applications. These should be the most exciting, applied and easy to understand parts of the book.

When I first encountered it, I thought it was a vulgar relational science with "sex" as its theme. Schart uses the basic model of human "sexual" attitudes toward others as a way to specify the types of relationships I have with others.

This is because Schart considers the attitude of "sex" as the basic primitive act of relating to others, which contains the primitive contingency of the "existence of others" and the primitive contingency of the existence of the self (personhood). Many of the complexities of interpersonal relationships are a multiplication of these primitive behaviors.

The first attitudes toward others are: love, language, and masochistic eroticism. Schart believes that love is an integrated awareness and knowledge of the individual's state of mind and body and the changes in people, events, and objects in the environment, using mental activities such as sensation, perception, thought, and memory.

Because love is much more than mere physical possession of lust, it is difficult to succeed and satisfy.

The extreme attitude of masochistic pornographers is to plan to be subsumed by others and to disappear in subjectivity, so that I can get rid of myself, so that I can be free of myself.

It is the tendency to look at things from the perspective, experience, consciousness, spirit, feelings, desires, or beliefs that an individual can have, based on the subject's own needs.

Its fundamental characteristic is that it exists only within the subject and belongs to the subject's state of mind. It can influence the factors of human judgment and truth. This distorted attitude ultimately leads to failure.

4. The concept of masochistic eroticism was introduced in

Masochism: I find a wonderful temptation in pain, and nothing can arouse me more than an authoritarian, brutal, and unfaithful woman. But masochism itself is supposed to be a failure: why?

Why, masochism makes the perpetrator the perpetrator and deprives him of his freedom. At the same time, instead of giving freedom to the perpetrator, I treat it as an object. "Thus, the masochist ends up treating others as objects and moves beyond them to his own objectivity."

One does not only have to dominate the Other as a

kind of object, but one has to dominate the Other as a free being (self-being).

In other words, man wants to possess the Other as a free being. This is a great desire of man.

Our desire is not only for the body of the other (the body as self-existent, Body), but also for the fidelity of the other's consciousness to himself.

For example, in the act of caress, what we "get" is not like "getting" after eating an apple, but an objective reality that does not depend on one's subjective consciousness.

It is an objective reality that does not depend on one's subjective consciousness. But we still want to caress, and we still expect to gain something from the act of caressing (with a sense of belonging, we begin to learn intimacy).

According to Schart, caress is an act in which we expect ourselves to identify with others.

Does Schartre propose an immoral, purely free ontological world? Existentialism, in fact, is the placement of nothingness in the situation of being, turning the whole of nothingness into a kind of being.

He tries to delineate the most hypocritical side of

human interaction. Nietzsche was the first to suggest that Saudi Arabia was in fact following in his footsteps. Under the inscrutable "nothingness" of Europe, no value is worth mentioning, and the freedom of the individual's autonomous will is re-explored.

The second attitude towards others: indifference, lust, hatred, sexual sadism. The extreme sadism of this attitude demands a non-reciprocal master-slave relationship, to act as a free possessor of the right to exist.

The object of his action or thought, or specifically of his lover, is treated as an instrument to manifest the flesh of the object with pain. This attitude, like the first one, is a failure.

However, all these aims of "possessing others" will ultimately fail. For these aims are contradictory in themselves. In human relationships, each person wants to dominate the other, to possess the other.

But when each person wants to do this, the person cannot be different and become the object of other people's actions or thoughts, or the object (object) of love for the other specifically, and be possessed by others. This is why the purpose of "possession of others" will ultimately fail.

This is an example of a specific attitude toward others, and the point of the exposition is to show that there is an irreconcilable contradiction between "self-control" and "self-control". When one 'self-control' is in relation to another 'self-control'

One is to become the object of the other, or the other is to turn around and watch the other's gaze and resist.

But whichever attitude of "self-control" we have, we cannot truly integrate the two "self-control" into one.

The relationship between me and others is not a republican fusion, but a consciousness of "our 'self' consciousness" versus the consciousness of "I exist for others - to be 'self' with others". In addition, there are two other forms of experience of our 'self' existence, the 'self' corresponding to the existence of the gaze and the existence of the gaze of the 'self'.

5. Object-us

Example: The teacher improvised a class hygiene inspection, and together we saw two young lovers who were deserting in shame (my classmates and I - for his presence → us, because the teacher intervened) and a third party (snitching, asking for strong support, different from the confrontation between us and you) intervened.

A small couple is together or a third party is looking at one party, and I am looking at her, and I am together with him; the different "self-realizations" are gathered in the name of us-objects through a main purpose of being together.

It means that people who are in the same class suddenly have to behave as a third party (snitching, asking for help, different from the confrontation between us and you) so that the reality of the oppressed class can exist.

They bring this reality into being by their gaze. This part is extended by Schart in the Critique of Dialectical Reason, where we develop into a vowed group (a kind of being that assembles in order to find the truth). Under the gaze of a third party, we are united in the name of the object.

6. Co-subject - we

Collective labor - the "I" individual → the impersonal "people". For example, "I intervene in the crowd of passengers". But the subject-we is not ontological, it is psychological, co-presence is like I intervene in the same rhythm as others (stepping).

I do not use collective co-existence as an instrument, nor am I held hostage by the rhythm - experiencing the motive of our-subject co-existence, their rhythm = my rhythm, non-primitive. The subject-us assumes a prior dual recognition of the "existence of others" in order to realize itself.

Heidegger: Co-presence is primordial. Schart: We - the subject is not primordial (I exist because of the prior presentation of others, I am below others as I am. (But in fact there is no other.)

Heidegger: Co-presence is still abstract. Sartre: Co-presence must be predicated on the assumption of a prior dual recognition of the "existence of others". Sartre and Heidegger are making this "co-presence" possible.

From Sartre's point of view, the ontological basis on which Heidegger's 'co-presence' rests is actually similar to Kant's abstract view of the subject.

This is because the co-existence of the ontology that appears as the structure of my "being in the world" cannot be the basis of an ontological 'co-existence' at all.

For example, "I am with Pierre" or "I am with Anne" are

in fact the constituent structures of my concrete existence.

But according to Heidegger's view, this is impossible. In "co-presence", the other becomes a completely non-self-sustaining object, without the ability to become this other at all.

In order to complete the theory of the relationship between me and others, Scharthe further raises the question of the co-existence of "we". The "we" is a particular kind of experience in my relationship with others, which arises in a particular situation on the basis of "being for others," and being for others precedes and establishes co-presence with others.

Sartre argues that there are two quite different experiences of 'we': the 'object-we' that is co-present when I relate to others under the gaze of a third party, and the 'subject-we' that is co-present in some individual collective activity or collective labor.

There is no symmetry between these two experiences; the former reveals that the actual one-dimensional space of human existence has only length, not breadth or depth. It is a primitive experience of "existence for others" and a mere diversification.

The latter is a psychological experience realized by the individual in the world of social history, a purely subjective experience. Therefore, the experience of "our" common existence, although real, cannot change the conclusion that the relationship between me and others is a "conflict".

There is no way out of the dilemma of either surpassing or being surpassed by others. The nature of the relationship of consciousness is not 'co-presence' but 'conflict', which is the destiny of the self.

7. The Other is Hell

Schart is the first to criticize the actual theory of "autonomous existence" and the psychological history of the thinking or cognition of Husserl, Hegel and Heidegger about the "autonomous existence" of others. It refers specifically to the individual's point of view or social consciousness.

Actualism, also translated as actualism, is a branch of Western philosophy that considers the study of concepts such as existence, being, becoming, and reality.

It includes the questions of how to classify entities

into basic categories and which entities exist on the most basic level. Reality in ontology is independent of human senses, beliefs, concepts and ideas.

Unlike what some idealists believe, reality originates from the individual's mental activity of sensing, perceiving, thinking, remembering, etc., and is either a combination of awareness and knowledge of the state of one's body and mind and the changes in people, events, and objects in the environment, or is influenced by consciousness.

What is considered realism in the world today is not the same as the realism of ancient Greek or medieval philosophy, and sometimes it is even the opposite

Actualism sees others as persons other than oneself, or as real circumstances, as entities whose objective existence is reflected in one's consciousness, the result of thought activity, or the formation of opinions and systems of concepts.

Husserl's a priori "self-existence" is not free from the egoism of a philosophical theory that proposes that only one's own mind is the existence of confirmation, and it is still not free from the relationship between "self-existence" and "self-existence".

Hegel was more advanced than Husserl (although he preceded him in his life), and his "master-slave relationship" already made it clear that the existence of "self-existence" is dependent on the "autonomous existence" of others.

He also concludes that the error of the actualists of "self-existence" lies in the belief that the "autonomous existence" of others is grasped through the bodies of others, while idealism reduces my body, and the bodies of others, to an objective system of representations.

Both of them are in fact using a formal space to separate my "self-existence" from the "autonomous existence" of others.

In fact, it is a tacit recognition that my relationship with others is an externally separated negative relationship. And the meaning of this connection of a certain nature between man and man or man and things, as already mentioned, is that in the end, one has to turn to God.

Therefore, my original relationship with others must be a two-way negation of the substance hidden within things, in which others evaluate and assess me, while I also evaluate others from a third party's point of view, from a macroscopic viewpoint.

However, Scharthe argues that Hedger still makes the mistake of 'being-for-itself' epistemology and formal 'being in itself' optimism, failing to see the transcendence of other people's immanent being itself.

Heidegger, in Schartre's view, although proposing a relationship of "being", sees the relationship between people as "co-being", which is still an abstract relationship, or similar to Kant's "being-for-itself" subject as an abstract "being in itself" ontological foundation.

Schart considers the mental history of the result of the thinking or cognition of his others. It refers specifically to the human point of view or social consciousness, which starts from "being".

He sees the relationship between others and 'being-for-itself' as a connection of some nature between being and existence of persons and persons or persons and things, rather than being able to identify a person or thing as this person or thing, rather than another relationship.

The existence of others creates a split in the world centered on being-for-itself, so that the "diversity of consciousness" creates conflict and confusion.

The state of interaction and interconnection between others' being-for-itself and me occurs through the "gaze", where I, being-for-itself, feel that the two things that naturally belong or are in harmony with each other are separated or even opposed to each other.

This also means the alienation of the nature of the species to which a person belongs by the combination of many identical or similar human things.

Therefore, the conflict of interests between human beings and each other, the I-love-obsession, is always there. If "being-for-itself" is too attached to the attention of others, too much attention is paid to others' "being in itself" comments.

Then "being-for-itself" becomes a victim of others, and others are hell. This shows how important it is to change the rational cognitive consciousness of "being-for-itself" and to liberate it

Volume 4: Possession, Acts and Existence, the relationship between 'acts of being' and 'freedom'.

The last volume focuses on what is the value of 'being-for-itself'? And how do we "for oneself"? The last volume focuses on the value of being-for-itself and how we can be for oneself.

And the "for oneself" of "for oneself", etc.... Finally, Schart is going to give a concise explanation and confirmation of existentialism, the concept of "existence before essence", and the final reflection of the thought form of "being-for-itself".

Section 1: Possession, Acts and Being

What does Scharthe mean by "being prior to essence"? Sartre's philosophy of Existentialism is one of the great philosophies of the twentieth world, and its influence is so far-reaching that it is still widespread in modern Western society.

Sartre's ideas were informed by a series of great literary works, for which he was awarded the Nobel Prize in Literature. "Existence before essence" is a central idea in Sartre's philosophy, so what is "existence before essence"? What does it mean for our life in reality?

(1). Existence precedes essence:

In Sartre's view, man is like a seed that drifts into the world by chance, without any essence to speak of, but only existence, and to establish his essence must be proved by his own actions. Man is not something else, but simply the result of his own action.

To say that "existence precedes essence" means that man first exists, rises in the world, and then determines what he ultimately is. Simply put, there is no innate plan of what is good or bad, but rather, man is born, and through his own choices, he determines what he is.

It is man's nature that is determined by his actions after birth and through his choices. The nature of man is not given by God, nor is it determined by the environment, but is acquired in the process of man's "free choice" and self-creation.

There is a famous quote by Jean-Paul Sartre: "Act, and in the process of acting you form yourself; man is the result of his own action and nothing else. This is different from traditional Western thought, where traditional

Christianity led people to believe that

Man was created by God, with an idea, and then created according to a certain intention, and man, like the hammer, has an innate nature and then exists.

Shat did the opposite, saying that there is no innate nature of man, but that man's nature is determined by his actions. So he said, "Cowards make cowards themselves, and heroes make heroes themselves.

Since there are no innate rules, people have to be responsible for their own actions, and this is the main expression of the distinction between human beings and material things, because the nature of all things is predetermined, and its characteristics are predetermined.

For example, a stone is the characteristic of a stone, and all the characteristics are predetermined, and it cannot change its nature because of its movement, which is common to all things whose nature precedes their existence.

Since human existence precedes nature, we have to think about our own existence, what kind of person do we want to be, what is our ultimate nature?

Is it good or bad? Is it good or evil? In every free choice

of action in life, at every turning point in life, be careful to take responsibility for your own actions!

(2). Why Act?

Previous existential theories of freedom of will have offered the autonomy "for oneself" approach to human beings. For example, Stoicism's "live according to your nature" and Spinoza's "exist to the nature of nature", but they are both condescending, in the sense that one does what one is told to do according to one's expected norms.

"If one's "being-for-itself" dissolves in the activity of others, and the purpose of morality is to cease to exalt one to the supreme dignity of "freely" ontology, then the spiral of self-life, too, is caught in the vortex of another "for oneself" matrix.

The greatness of Kant is to give the autonomous subject to man, replacing "being" with "doing" as the supreme value of action. "That non-empirical, purely rational absolute command" is the highest principle of the "absolute command". "Man 'being in itself' is the morality of teleology.

(3). To be and to possess

Schart is to take the subjectivity of mere possession,

i.e., "for oneself", as the basic existence of human action; to take it as the point of departure of philosophy, and to derive from it the existence of the external objective world.

Therefore, the "existence of human action" or "mere subjectivity" is regarded as the basis and origin of all things in the universe that I love and possess. Human existence is considered to be the first.

Based on this, Sartre, in contrast to the traditional concept of "essence before existence", puts forward the view that "being as existence precedes essence" in regard to human existence and essence. The so-called "existence" refers to "the actual act of man" or "the act of the self".

This means the subjectivity of an individual. The so-called "essence" refers to the fact that man acts according to his own will in order to create his own constancy. The basic meaning of the proposition that existence precedes essence is that man exists initially only as a purely subjective possession, the essence of man's action.

The essence of human behavior, the various characteristics of human beings are later chosen and created by subjectivity itself. Therefore, there is no human nature in the world, because there is no God who set human nature in the world. Man is nothing but something created by the free will itself. Unlike things, they cannot choose and make the nature of their own actions.

The nature of things is given by man's freedom of will, and is made to have a certain meaning and value by man through the action of consciousness, according to his own needs and purposes.

In other words, it has the nature of an act, and then it appears in the world as something. Obviously, before something appears, the essence of its action already exists in the possession of human consciousness.

Therefore, the essence of a thing is prior to its existence. In this way, man and thing are distinctly distinguished, thus affirming the value and dignity of man, emphasizing the subjectivity of man's action, emphasizing the status and role of man, and inspiring people to take positive action to create the essence of their own action. Thus Sartre's existentialism is in fact a philosophy of action.

(4). Existential Psychoanalysis, Existential Psychoanalysis

Two major errors are to be guarded against

The error of empirical psychology - asserting that action, desire, desire is the content of self-love-obsession consciousness, desire meaning in the desire for self-love-obsession itself.

Ans: It avoids the concept of transcendence. Desire is the consciousness of the object of the desire for the object of my love. Desire is the consciousness of I-love-obsession itself in the original structure of seeking and transcendence.

2. Psychology is the completion of the concrete things that people want to achieve the total empirical desire, that is, the intention established by empirical observation, giving a definition of the person (sexual perversion is due to the empirical nature of its parricide complex)

Schart: We should use a descriptive psychology or a phenomenological psychology or an existential psychoanalysis. That is, instead of listing behaviors, intentions, preferences, etc., we should describe them, identify them, and ask questions.

Existential psychoanalysis is based on the principle that the person is a whole and not a collection. Therefore it is expressed in its actions in the least meaningful and most superficial things.

Objective: To expose the empirical behavior of man, that is to say, to clarify completely the revelations contained in any of them and to establish them conceptually, that is, to understand the revealed free choice.

The point of departure: it is experience - the ontological and fundamental understanding of the individual before he or she has a "self-existence". It is not a rational activity, nor is it the same as cognition or reflection, but any action towards a purpose, a conception of the self itself, can be understood.

It is by comparing these actions that we can express in a different way the only revelation realized.

(5) Acting and possessing: possession

The ontological analysis of action and desire is the principle of the existential psychoanalysis. The specific ontological study of desire possession is structured as follows

Desire - the lack of being for oneself - because of the existence of the lack of being for oneself ("freely" - "for oneself"), the contingent ideal self (God - man), the pure effort of man to become God.

For this "being-for-itself" effort, there is no established specimen in which everything other than the analytic object is composed, and nothing is like this self (being and possessing). And I love to cling to the effort that expresses this desire.

One kind of desire: the desire to be, the means of possessing for "being-for-itself" (e.g. chopping wood for an axe) and another kind of hidden desire: writing, scientific exploration, sports, painting, etc. My works are a continuum of my "being-for-itself" manifestation, but a creation frozen in freedom, with my infinite mark, my infinite thought.

I pursue the synthesis of my "being-for-itself" and non-self "being in itself" in the act of free planning of my will, and I take possession of the literary work by this method (self-being-self). To chop wood and make an axe is also to belong to me twice (God and man).

To know = "being-for-itself" into oneself = the ontology of existence in oneself: "for oneself" into oneself in relation to this self-relation of desire. Therefore, the double rule of desire. I love to hold on to desire as the lack of existence.

(1) The desire on the one hand has to become some kind of "freely" - "for oneself", its ideal desire for existence. (2) In most cases the desire is prescribed as a contingent, concrete, self-contained relation that it plans to assign to itself.

Conclusion, as revealing the nature of existence

According to Schart, what ontology can teach psychoanalysis is, in fact, first of all the true origin of the meanings of things and their true relation to human reality. In fact, only ontology can be placed on the level of transcendence and grasp the world from this point of view alone.

In fact, only ontology can be placed on a transcendental level and grasp the existence of the world from this point of view alone, together with its two ends, because it is the only one that starts from the point of view of the Self. It is even the concept of personhood and the concept of situation that enables us to understand the symbols of the existence of things.

Sartre's psychoanalysis is, in short, an ethical call to understand phenomena: there is no irreducible mental material.

The person as the consciousness of existence in oneself is the center of intentional belonging. But this "being in itself" is the unification of the intention of "being in itself", which in turn means a spontaneous choice of "being-for-itself" existence by the subject.

This choice of "being in itself" is absurd and tends to be a kind of meaningful transcendence. This brings us back to the ontological mediator of being-in-itself: to clarify the relationship between human reality and existence.

Existential psychoanalysis describes the act chosen by the person who chooses freely of his own will, and at the same time makes this act of choice intelligible. Therefore, there is only ethical self-knowledge.

Only 'being-for-itself' can be self-knowledge; only self-knowledge can be truly 'for oneself'. To know is to act. Such active, purifying reflection and action on the unreflected world must always be in mutual flow.

Each person represents a certain ideal, and this ideal represents a certain type of lived experience, and the identification and presentation of this type of lived experience is only possible in its action on the world.

Only by going back to a choice made in a certain way is it possible to truly understand "being-for-itself".

For others, it is only by assuming full responsibility for their chosen actions that they can give full meaning to their chosen actions. Ontology only enables us to specify the final purpose, the basic possibility and the value of being in itself.

At the same time, it is also a plan of being-for-itself towards "self-activity", the division of the world into its own. All 'being-for-itself' is a 'passion'. Religiously, "being in itself" is "God".

The passion of man is the opposite of the passion of Christianity. Man was born as man's "self-depreciation" to bring God into being, but man's "self-depreciation" is useless. Man is just a bunch of "being-for-itself" useless passions.

Section 2: The relationship between "being-for-itself" and "freedom

The first condition for action is freedom of will, action (practice): to transcend the simple and stable determinism of the world through action in order to transform the world in its materiality.

This is the most important part of the book Being and Nothingness. Schart believes that the four previous sections of the lengthy discourse are in fact all meant to boil down to the formulation of this central issue, which is the question of 'human freedom'.

Since the experience of oneself, others, and "for oneself" itself are determined by action, the eternal possibility of action should be regarded as the essential feature of "for oneself.

I. Schart's Absolute Freedom

01. Relative freedom

When we talk about freedom in general, we usually do not hesitate to discuss freedom together with the power of autonomy.

In other words, whether a person enjoys freedom or not, we will judge from the perspective of whether he is able to make complete self-determination. For example, the typical unfree person is considered to be a slave and a prisoner.

A slave is the opposite of a free person in our

conception, i.e., a slave is not free. A person in prison, on the other hand, is said to have lost his freedom.

Why do we regard these two types of people as typically not free? The reason why we consider prison inmates to be unfree is that they are subject to many restrictions and fetters, such as prison windows, locks, and guards.

These objectively existing things are all restraints for him, which prevent him from doing what he wants to do. His space and habits of life were under close supervision.

Moreover, when we say that a slave is not free, it is also because he is under considerable restraint, even more so than the former.

He cannot have his own power and property, he cannot decide what he wants to do, but he is completely determined and controlled by someone else.

Compared to ordinary people (including non-slaves), they enjoy a much greater degree of freedom and autonomy than the above two. Therefore, freedom is a concept of degree in the eyes of the general public, and we can say that the degree of freedom of a certain person is higher than that of a certain person.

In other words, person A is less bound than person B and has more autonomy. If a person is bound to a certain degree, we say that he is not free, or has little freedom.

If a person's sire bond is small to a certain degree, we say that he is free. This concept of a degree of freedom pervades every society and every era.

In other words, not everyone in any age or society enjoys the same freedom.

Some groups and societies always have more freedom, while others have less freedom, or even none at all. If we ask a society formed by capitalism or a society formed by communism, which society is free?

We usually answer that the former is freer than the latter, because we can see from their different marketing markets that one has a complete system, while the other has a queue for food stamps in order to get the food they need.

Therefore, the main question of how much freedom is at stake is: "How much of my movement is restricted? How much room do I have for unrestricted movement?

Generally, we say that I am free to the extent that no one else interferes with my activities, and I am not free if someone or some people interfere with my activities.

If he or they interfere to a certain extent, we say that I am under compulsion and cannot do what I want to do. When we are in this situation, we lose our freedom.

This interference or coercion may come from someone else (an individual or a group of individuals) who imposes physical interference and restrictions on us; or it may be a state or legal restriction on our movement.

Or it may be the pressure of public opinion on us. However, no matter what kind of coercion or interference, it is a restriction on what one may desire or what one may want to do.

Such discourses of freedom are mainly political or sociological, that is, they deal only with the range of valid choices that one can make to describe one's freedom.

However, in the eyes of philosophers, one's freedom is not determined by a valid range of choices. In fact, freedom involves the question of whether or not the theory of decision is valid. That is, the philosopher is concerned with the question of whether freedom is possible.

If the thesis of determinism is valid, there is no such thing as freedom, but merely the product of being determined.

In terms of determinism, we often find that there is a natural process, an action, in our bodies. For example, when I drink water to quench my thirst. Both of these are subject to a biological and environmental influence.

So how can we see here that the physiological response of trees is different from that of humans (especially the part that we usually think of as animal) and not compare them to each other?

Moreover, we cannot deny in any way that we are born with tastes, beliefs, and reactions to things around us that are already influenced by our surroundings (social or family) and genes.

So how can we make our own decisions? How can we enjoy the freedom to make decisions in a genetically predetermined environment?

Therefore, some determinists argue that a person's behavior can therefore be analyzed. In other words, a person's behavior is strictly determined, it is the convergence of a series of external causes.

A person's behavior has an external cause as a precondition. Therefore, as long as the series of external relations of the person is clarified

Then the behavior of an actor can be strictly foreseen at all times, foreseeing his future. Thus, in the eyes of the determinist, the so-called future is in fact a reappearance of the past, a consolidation of the past.

The behavior of an actor becomes a kind of repetition - a constant manifestation of the past, always appearing in the present, in the future, with a face containing the nature of the past.

All the conditions of his actions are given, the past dominates everything. In this way, man cannot have freedom and is bound to act under the cycle of cause and effect.

However, this only leads to the question: if man is already determined, on what grounds can we morally evaluate him in a determined situation?

For this reason, many philosophers have tried to solve this problem. In order to be morally responsible, they introduced the concept of will to affirm the freedom of choice.

For them, this will depends entirely on the knowledge of the good by the intellect. Reason does not only know the concept of goodness in general, but also in each specific case.

It is able, in each specific case, to identify what the good is, and thus to determine the will, which necessarily seeks what is considered good. A person's freedom is independent of social, genetic, biological, and family factors.

A person's freedom comes from the fact that he "knows" what he is doing. Therefore, he must be responsible for his actions.

The will is dependent on reason, freedom is a necessity that exists because of knowledge, and the decision of the will depends only on the pure activity of knowledge.

However, for the person who does something wrong, he "does not know" what he is doing; in this sense, his actions are not free.

Moreover, although some philosophers have argued from the point of view of destiny determination, man is like other animals in all his internal and external structures, in all his actions, and in all his internal and external structures.

In all that he does, he is determined by the cosmic forces (that give all life); like everything else that happens, his actions appear to be nothing more than God's.

His behavior appears to be nothing more than a naturally predetermined and inevitable activity of God. Yet, man remains free. For, man is still the cause of his own actions.

Although, the decisions of an actor actually come necessarily from the union of man with his environment and his faith.

But, right here, the external environment or beliefs are only subsidiary causes. The real determining factor comes from the person himself.

Therefore, man is still responsible for his own actions. From the above discussion of freedom, we can see that human freedom is affirmed either politically, or ethically or religiously.

We can find that when talking about freedom, strictly speaking, they do not consider "freedom" to be universal.

There are some people who can achieve freedom, and some who cannot. Or perhaps, universally speaking, people are free, but this freedom is subordinate to the laws of religion and the universe.

Therefore, no matter how much they claim that man can have freedom of choice, this freedom is not absolute, but only limited.

02. Absolute freedom

This relative, limited freedom, in Saudi Arabia's view, is the same as no freedom for man. If we want to talk about a freedom or not, it must be holistic, universal, absolute, and instinctive.

That is, one must speak of human freedom in terms of human existence. This freedom must be a human instinct, which every human being necessarily possesses. In this way, the question of whether or not one should be "responsible" can be truly addressed.

In Saudi Arabia's view, human freedom, whether political, social, intellectual, or religious, is not strictly speaking a freedom.

In terms of political and social freedom, there is no such thing as freedom because everyone is exercising their limited right to choose under some degree of restraint and interference. Some people have the right to choose between apples and pineapples.

Some people may have so much power of choice that they can decide the fate of others in addition to their own ownership. But no matter how much self-determination one has, and how little stifling and interference one has, stifling will always exist.

As long as there is a sire bond, freedom is not possible. For the religious or rationalist, Schartes believes that freedom is also impossible. For freedom is bound to reason, to God. In Schart's view, these fetters indicate that man is determined and is not free.

To be free, freedom must be free from all fetters and interference, outside of determinism. However, if man is free, then man's existence cannot be free.

If man is free, then man's existence cannot be within determinism, subject to fetters and interference. For man to be free, he must exist outside of determinism, free from all fetters, coercion, or interference.

In other words, freedom is possible for a person only if the person himself is a free being. This free human existence

For Schart, it is human reality - that is, self-existence, consciousness. How, then, is man free? Do not external political circumstances, social customs, physiological cycles, religious beliefs, etc. already determine human behavior in advance?

How does Saudi Arabia prove the freedom of human beings? In other words, human freedom is demonstrated through the ability to "deny". This "negation", for Saud, is not only a "negative" judgment between human existence and the existence of things, but also the negation of all determinants.

For example, we generally think that the commands, values, and moral judgments of a class, a society, or the common sense of a whole community will necessarily determine our behavior.

However, in the case of the rule "Thou shalt not lie," we find a gap and a possibility between moral commands and human behavior.

In the case of the phrase "Thou shalt not lie," everyone recognizes it as a rule or a command. However, this does not mean that everyone acts in accordance with this rule.

In fact, we often find many people telling lies in our daily lives, even though they are aware of the value of "Thou shalt not lie". Why is there such a contradiction?

Do they just want to impose on others a value that they do not want to follow? No. We can find that they also impose this command on themselves, and they also think that "lying" is a wrong action.

They would rather blame themselves for lying than to have the prohibition of lying lifted and made a reality by unconditional permission.

In this way, the truth is no different from a lie. Therefore, they also impose the prohibition of "lying" on themselves.

Thus, we can see the possibility of a "negative" judgment in the connection between objective forms (institutions and orders) and actors - I do not "always" not lie, I am often "dishonest".

This "negative" judgment distinguishes human existence from the world that is built up by norms, by the ways of behavior and values of social and ethical life.

Social constraints, political interference, and religious coercion are all excluded from the actor because of the "negative" judgment. There is no connection between command and action - the command does not serve as a cause or precondition for an actor's action, and the actor can only act on its own.

Schart thinks that if a command, a norm, necessarily determines the actor, then there should be no so-called possibility ("negative" judgment) between the command and the action, but rather a complete connection between the command and the action in a necessary and certain way.

However, in the actual example, we can find that "negation" appears between the command and the action. Therefore, the determinist's claim that an actor's behavior must be predicated on an external cause is not valid.

In contrast, for an actor, all norms, social and ethical ways of living, and values exist only as an objective fact, external to the subject.

Similarly, in the case of "do or die", although the actor carries out an order under a decisive condition, it is also carried out by way of "death", free from these decisive conditions.

For example, honor is the unconditional value and command of the feudal family, and the actor must maintain the family's honor at all times. If necessary, in a dangerous situation, the actor must give up his life to save the honor.

At first glance, it seems that the actor is dictated by the family's supreme command - honor - and can even sacrifice his or her own life.

Yet, says Schatt, it is not. "Death" is in fact a "negation" of the inevitable decision of external conditions. When the actor proposes "do or die," the actor is, in fact, "dying.

In fact, the actor sees "death" as a way of absolute rejection. By "death", he detaches himself from all external decisions and recognizes that there is a power within the actor that is higher than the power of external causes.

This power determines his own behavior, making possible any premise, external condition, moral requirement.

Therefore, the external conditions such as my class, nationality, family, history of my group, heredity, personal situation, habits, etc., do not necessarily determine my behavior. I am a free being, a self-determined actor who always transcends this objective decision (human reality).

For no value, no norm can determine or bind my actions, everything is mine - I am completely free.

Freedom is the same as human reality.

Freedom, in Schart's view, is not a limited choice that unfolds in a fettered environment. That is, it is not just a choice between apples and mangoes, not just a political system in which one is free to buy and sell, not just a choice based on reason, and not a choice based on the law of God.

On the contrary, a free man is completely free from these fetters - political, religious, customary, ethical, and even from man's own reason - beyond these fetters.

Only then can a person call himself "I am free" and "a free man" can be established. In short, what is called freedom (a free man) must be completely free from determinism, in order to be free to choose, to be fully self-determined.

Schart's account of freedom is not on a political or metaphysical level. ········...he tends to equate freedom with the human instinct (or, really, the inevitable) to choose.

Freedom, for Chartres, is the primary condition for action, or an autonomy of choice. But what is a free person? Can one be free simply by being free of external determinants? Schart doesn't think so.

He believes that to be free, a person must also be free from himself (ego, psychological state) at all times in order to be fully described.

For example, if I hate Peter, my hatred of Peter is a state that I perceive by reflection - by reflective consciousness. This state is "there" under the gaze of reflective consciousness and is real. However, this "hatred" does not exist as an inherent nature of consciousness.

On the contrary, it exists as the object of the reflective consciousness. In other words, this "hatred" is the existence of a real object, which is outside of the consciousness, like the existence of an ink bottle.

For example, when I saw Peter, I felt a feeling of extreme agitation and anger due to disgust. But is this disgust hatred?

No, says Schart. For the feeling of "disgust for Peter" is limited to the moment of seeing Peter. It does not imply the past, nor the future.

And precisely because there is no past or future involved, it is highly likely that I will cease to hate. However, hatred is not derived in this way. This hatred is the hatred of the "I" and is inseparable from the ego.

It is given together with the ego in an eternal state, and it (my hatred) was already there when the momentary feeling of "hatred of Peter" appeared, and will appear again tomorrow.

Therefore, a momentary feeling of disgust (awareness of something) does not become "my hatred". "My hatred is not in the consciousness, it is an object that exists beyond the transient nature of consciousness and presents itself in each moment of the experience of revulsion, disgust, and anger.

For reflective consciousness, the state is given and is a concrete and intuitive object. If I hate Peter, my hatred of Peter is the state that I perceive by reflection.

Since the psyche has been distinguished from consciousness, the psyche is the object of reflective consciousness. Then the "ego" also exists as an object for the reflective consciousness. The ego, says Schart, belongs to the psychological aspect.

This hatred cannot be separated from the ego. If the ego could be separated from hatred, then hatred would disappear behind the ego and hatred would not exist;

however, it is not the nature of hatred that comes first, and then this hatred accepts the ego as if it were a light.

There is no gap between the ego and the hatred. The ego and the hatred are a concrete and absolutely full event, which is "my hatred".

Schart said that the ego is to the mental state as the world is to various natural objects, and it is the law of the existence of the mental state. In contrast, mental states are to the ego what objects are to the world, enabling the ego to exist. The ego and the mental state are a real existence, inseparable.

Each new state is directly tied to the ego, just as it is tied to its root.

In this way, consciousness is also detached from itself (mental state and ego) and becomes a free being completely detached from determinism - nothingness. This existence of nothingness is for-itself existence

(for-itself), that is, consciousness, the reality of man.

In Schartre's view, freedom and the real existence of man must be identical (for-itself existence, consciousness) in order to fully describe man's freedom.

A free man (the reality of man) is an existence of nothingness. When we think of man as consciousness, or as self-existence, we are pointing out that man is a being of nothingness, creating nothingness.

Freedom is this non-being as consciousness, as self-existence - the human reality - of the human being. (The freedom is the form of human reality.

Simply put, the way of existence of human reality (non-existence, consciousness, self-existence) is freedom. Freedom is derived from and identical with nothingness.

Nothingness, in its literal sense, is not a thing at all. Therefore, man, as a kind of nothingness, does not exist as something that can be defined or determined.

Man's existence is not some kind of pre-existent nature or essence. He is completely free and undecidable. This freedom, in the view of Chartres, is the true and absolute freedom.

Therefore, man is nothing but free, for he cannot even be called "I". To be a free man means that man's reality, his consciousness, is in a state of emptiness, which makes him free in himself.

Even the "ego" (nature, essence), which philosophers

once thought to be the most representative of man's existence, subjectivity, and freedom, was excluded from consciousness by Shatt and could not represent man's freedom.

Therefore, man has no essence, and his nature is an empty existence. Only in this way can man truly escape from his predetermined destiny.

For if nature is always prior to man's existence, it necessarily determines what man's existence should be, just as a paper cutter, before it is made, already has some concept that makes its making and use possible.

But what a person should be should be decided and chosen by the person himself, not by something that is called a person in advance.

Therefore, in order for man to truly make choices and self-determination, man's existence must be free from all metaphysical values and reach a state of pure emptiness and transparency. This is the first principle of the existentialists.

Human existence must be prior to essence. This human existence does not mean the existence of the self as if it were a natural thing, but the existence of the self as if it were nothingness and lack of existence. This self existence

For Saud, it is the human reality. Man's existence, first of all, is nothing, it cannot be defined, and only then can it be what he thinks it is.

If man is not definable in the eyes of existentialism, it is because at the beginning man cannot say anything.

It is only afterwards that he can say what he thinks he is, and then he will be what he thinks he is. And because of this, the concept of man is impossible. Man is free, and there is no humanity that can be considered fundamental.

(1) Being and doing: freedom

Sartre believes that self-realized being is absolutely free, not bound by anything, including itself, but constantly denying, creating, and developing itself, precisely because man is completely free to make himself, and man is fundamentally free.

At the same time, man's freedom is prior to his nature. Man does not first exist in order to become free, but there is no difference between man's existence and his "being free.

There is no difference between man's existence and his "being free": "free to choose". God is dead, man is free in this world, and man's choice of action is free. This is because man's choice has no innate pattern, no guidance from God, and cannot be judged by others; man is the sole director of his own actions, but he is responsible for them.

Thus, freedom in Sartre's sense does not mean the achievement of ends and success, but only the autonomy of choice.

Freedom is freedom as long as one can choose, and even not choosing is a choice, that is, choosing not to choose is also freedom. In Sartre's view, man's absolute freedom only means that man is thrown into the dusty world in isolation.

For nothing can determine him. He has to decide for himself, to choose for himself, to make himself.

Man's life is a project of constant choice, of constant freedom to choose, to create his own nature, to make himself in the future.

(2) Human Existence and Freedom

In the history of philosophy, there are many philosophers who have presented their positions on which to analyze or criticize problems and issues.

Aristotle, for example, believed that freedom is a voluntary choice of behavior. Kant believed that freedom is the freedom of man, and that self-discipline of the will is the most important part of his view of freedom.

In Hegel's logical thinking, freedom is directly linked to necessity, and freedom is the knowledge of necessity. Sartre's existentialist view of freedom: there is no escape from man's free choice.

In many cases, freedom refers to the freedom of the psychological state in which man himself consciously strives for the realization of an ideal or the achievement of an end, and in the existential philosophers who followed, freedom itself has a unique meaning.

Existentialism, also known as existentialism, was first founded by the philosopher Friedrich Heidegger, and it was during the time of Sartre that existentialism had its great influence.

Sartre's idea of freedom occupies an important place in his philosophy of holistic, fundamental, and critical inquiry into the real world and human beings.

But it also had some related negative effects.

Therefore, in this paper, we try to analyze the connotation and, more importantly, the extension of Sartre's concept of freedom.

In the medieval period, the essence of man was considered to exist as a religious-philosophical aspect, and thus man emerged accordingly.

In modern times, we still hold on to the ideas, culture, morals, customs, arts, institutions, and ways of behavior that have been passed down through history from generation to generation, that "nature is prior to existence. It is a traditional concept that has an invisible influence and control on people's social behavior.

Existentialist philosophers, on the other hand, have abandoned this traditional concept of ideas, cultures, morals, customs, arts, institutions, and ways of behaving that have been handed down through history from generation to generation.

The traditional notion of invisible influence and control over social behavior is abandoned. Shat believes that "existence is prior to essence" and that it is in the ability to put known facts and principles and principles in one's own mouth, words or other symbols.

The earliest way to explain Schart's concept of freedom is to analyze it in detail, to understand it from a certain cognitive point of view, and to understand what he means by human freedom and essence, and to interpret the important idea of "existence before essence".

001. Scharthe's nihilistic view of freedom

Scharthe's philosophical view is that the existence of human ego life itself exists as an impractical and inscrutable nothingness, and that human freedom, as a human specification, coexists with human existence itself.

The declaration of the existence of human ego-being itself is a clear expression and declaration of its free nature, and existence is the freedom of human ego-being itself.

Therefore, the freedom of human ego-being itself extends and expands outward in width, size, and scope, which is the initial starting point of Schart's view of freedom, "the independent nothingness arising from human reality".

This possibility is proposed in the explanation of "how universal knowledge is possible" and "how moral practice is possible".

This is why it is called freedom. Freedom, in Schartan philosophy, is the condition required for the creation of nothingness, which is to be present within the human mind and is an inerasable nothingness.

For man's objective material world, which is independent of his consciousness, existence itself is self-chosen, and all the objects or phenomena that he can accept as objective exist in nature.

All the objects or phenomena that he can accept as objective exist in nature are formed from the state of being thrown, and therefore present the decisive connection or trend prescribed by the inner nature of things.

In the process of development of things, necessity and contingency are linked to each other, interacting and transforming each other under certain conditions.

Freedom is not an objective material world independent of human consciousness, that is to say, the emptiness of human existence, arises out of it, and therefore on this basis

Therefore, on this basis, Schart is of the opinion that human self-existence itself is the nothingness of existence, and that there is no fundamental, characteristic specification of things, and that the existing human being can, in an acquired situation, transform itself.

The human being can transform himself in the future, and Schatt also believes that the human ego is absolutely free, and that freedom determines the nature of all human beings and their changes in the future.

But behind Schart's absolute freedom, in fact, there is also a profound decision by the nature of things, and to know the inevitability of things is to know the nature of things, where the innate problem lies.

Schart's view is that man is born free, and that the objective material world, which is independent of man's consciousness, exists as freedom, but existence itself is an impractical, inscrutable nothingness, and does not have the fundamental, characteristic specification of all objects or phenomena that exist objectively in the natural world.

Therefore, man is free to change his own nature, as long as he knows what he is going to do and what he is going to be, and it is up to man to determine the final outcome of his own existence.

Therefore, here we can see the rational and active

process of reflection on the reality of the world or on any object by such a free mind as Sartre's, in the form of "inward dialogue".

The so-called inward dialogue is like talking to oneself, which has an important contingent element, ignoring the external influences of the living world, the ego that each person has.

It also has ideas, culture, morals, customs, arts, institutions, and ways of behavior that have been passed down from generation to generation and from history to history. People's social behavior has an invisible influence and control of the role of the self.

The unique historical situation, the background of its own life world, these decisive linkages or trends prescribed by the nature of things.

In the process of development of things, necessity and contingency are linked to each other, interact, and transform each other under certain conditions, which are not intrinsic contents, but affect the determinacy of the human self.

002. There is no escape from human freedom

The ideas, cultures, morals, customs, arts,

institutions and behaviors that have been passed down from generation to generation and from history. They have an invisible influence and control on people's social behavior.

The basic social relations of the self, which are determined before each person is born, are full of many objects or phenomena that exist objectively in nature and are hidden within the substance of things.

The decisive connection or tendency stipulated by the nature of things. In the process of development of things, necessity and chance are linked to each other, interacting with each other and transforming each other under certain conditions.

Although in this by the nature of things, the decisive link or trend. In the process of development of things, Shat emphasizes the importance of freedom of choice and action.

However, what is presented behind it is a more profound judgment and assertion of things, and in introducing the assertion that man is born free in the very existence of his own life, Scharthe cannot escape from this freedom.

In introducing the assertion that human life itself is born free, it is impossible to get rid of the unfreedom behind this freedom, the fact and the situation that really exists in front of us. That is, in everyday application, it means "something that exists objectively" or "a condition that fits the objective situation".

It has been said that "man accepts life without choice, then spends it under conditions of inevitability, and finally surrenders it under an irresistible struggle.

The idea of predestination, that all encounters or changes of an individual or a group are already determined by God, and that freedom, which is not humanly possible, exists precisely as non-freedom, on the other hand.

Therefore, Scharthe proposes that the existence of human life itself is thrown into the world, and in the process, it is impossible to get rid of this state of non-freedom.

Therefore, in Scharthe's thought of freedom, this aspect is presented as an important state of non-freedom, and although the banner is set as true absolute freedom, the final end or destination of all things and objects in objective existence becomes the throwing in of non-freedom.

Moreover, Schartre considers the freedom he speaks of as a state that transcends experience and can be self-evident in any situation. Man as a being precedes essence, such a being.

In philosophy, essence (English: essence) is an eternal attribute or set of attributes that make an entity or substance its very essence, and which necessarily exists without it.

Without it, it loses its identity, is chosen by itself, and assumes itself, without any innate stipulation.

However, the ideas, culture, morals, customs, arts, institutions and ways of behavior that have been passed down from generation to generation and from history.

The invisible influence and control of people's social behavior, in the actual life, which kind of development of things or for people to follow in the world, is still each person's own choice to decide, necessarily there is no such state, only a variety of objective things within the potential for a variety of development trends exist.

Therefore, freedom becomes the free choice of man

when the existence of self-life itself exists as a non-fixed form, as a potential trend of development within this objective thing.

It exists only in the activity of man's self-selection, not through the choice to obtain freedom, but freedom itself is the act of choice, and all the values of the whole world are presented in this choice of man.

In philosophy, essence (English: essence) is an eternal attribute or set of attributes that make an entity or substance what it is.

They make an entity or a substance fundamental to it, and it necessarily exists, without which it loses its identity, and in this process it becomes the freedom of each individual to interpret this state of itself.

This freedom to use mental activities such as sensation, perception, thinking, and memory, to be aware of and recognize changes in one's own state of mind and body and in the environment of people, things, and objects, is above the innate state of one's own life existence itself.

But in life there are many contradictions, for example, the inability to choose one's own environmental and social factors, which exist as states that have been determined since one was born, and therefore cannot be resolved.

Freedom in the philosophical viewpoint of Scharthe is limited to the individual's choice of his own mental and physical state and the integrated awareness and knowledge of changes in people, events and things in the environment by using mental activities such as sensation, perception, thinking and memory, which makes the situation of freedom, which does not exist, a reality.

Since freedom in life means choice, even if one abandons choice, it is still a choice. Therefore, Schart thinks that the greatest non-freedom is the freedom to choose without conditions.

003. Freedom leads to the non-freedom of responsibility

No one can escape from this acquired choice, which exists from birth, which is inherent in oneself, which is not inherent in oneself, which refuses to exist, and which cannot exist because of freedom, which is itself a choice.

Therefore human life itself exists as part of choice, and therefore human justice, in its everyday application, means the reality of "objectively existing things" or "conditions appropriate to objective circumstances", adding to the contingency of existence in general.

Here, Schart's view of freedom enters the realm of the belief that any event, including a decision made by human beings of their own free will, has external conditions that determine the occurrence of that event, and not some other event.

The freedom granted to man cannot be freed, and this freedom exists by necessity in man's self-life itself, so that man is destined to become unfree, to exist, and man cannot escape this freedom, and therefore cannot change it.

This freedom refers to a predetermined and inevitable (some objective law or destiny); a predetermined freedom that is beyond the will of man or human power. This is followed by the idea of responsibility, which was introduced by Scharthe.

It is after this free choice that man bears the result of his own choice, and the requirement of the nature of freedom is to bear the result of the choice, that is, the responsibility behind this free choice, and the individual needs to be responsible for what he does.

Therefore, when people enjoy this state of free choice, they inevitably enter the state of responsibility, and even in life, the responsibility of the whole world arises, the oppression of the individual.

This also makes people enter the state of responsibility from which they cannot free themselves. Freedom implies responsibility, implies even greater anxiety, and anxiety becomes another form of unfreedom.

In this respect, Sartre's view of freedom shows two inevitable aspects, both of which have a state of non-freedom.

The complex psychological process of seeking or establishing rules and evidence to support or determine a belief, decision, or action under the premise of Sartre's view of freedom as being prior to essence enters into the ideal situation of an illiberal society.

Although the place from which it starts, as a more important part, proclaims freedom, it is caught up in the immense unfreedom in its physical existence, or in its explicit rather than abstract, generalized discourse.

Therefore, the individual's choice to use mental activities such as sensation, perception, thinking, and memory, to become aware of and recognize changes in his or her physical and mental state and in people, events, and objects in the environment, is not only the choice of the individual himself or herself, but also of the whole society.

Each person has to be responsible for himself or herself, but also for others, and even for the freedom of the whole human race, which exists as a not so wonderful state.

Of course, the meaning of the concept of freedom as stated in the "philosophical view of being and nothingness" of Scharthe cannot be completely eliminated from the process of human freedom seeking, and its main spiritual content and substance is a more important aspect.

This inference, analysis, and then judgment of "existence before essence" is an important aspect in the development of a holistic, fundamental, and critical inquiry into the real world and human beings.

It also has an important significance for the future development of modern philosophy in that it is the spiritual content of the understanding of natural or social things, the meaning given by man to the object, and the symbolic form of transmission and communication.

2. Action and Freedom

01. Freedom and choice

Man's existence (consciousness) is a free existence, and the place of God is completely taken by the freedom of man. Man is destined to be free, and nothing can determine man's nature.

Nothing can determine man's nature, make choices for him, but man can make choices for himself, and he must make them. That is to say, for the Saudis, freedom is the primary condition for action, or the spontaneity of choice. The Saudis equate freedom with the ability to choose, and talk about freedom in terms of the possibility of choice.

This is in fact a statement that as a human being in the life of experience, "one must choose". Every second, every future that I choose, comes from my freedom.

That is why Schartrecht says that man must choose, and choose spontaneously, not be determined. I am choosing and deciding my future every second of every day.

For example, if man's existence should be righteous, this does not mean that righteousness innately determines man's existence, but that man chooses himself to be a righteous man.

Or there is no innate circumstance that determines whether a person is good or evil by nature; if someone is evil, then that person chooses to be evil.

For, before a person becomes something, he is non-existent, and in order to achieve existence, he must move toward that which he wants to become.

And in this activity of moving toward, in fact, man makes a choice, and this choice is to take all possible actions in order to make himself the kind of person he wishes to be.

This is the basis for Schart's claim that man is bound to choose. When a non-being wants to achieve existence, in the process of this orientation, man is making a choice, choosing between good and evil, between honesty and lying, etc., the kind of person he would like to be.

So, even in a poor economic environment, as a thief or as a person who still suffers from hunger, it is a self-choice to be what one wants to be.

Simply put, because there is no determinism, man is free and necessarily chooses. There is no action that does not simultaneously create an image of the person he thinks he should be.

Freedom is prior to truth, not in the sense that we are free because we are human and can choose, but in fact we are free and are therefore human and must choose.

What does this necessary, free, original choice tell us? It explains that man can create the life he wants, the kind of life he wants to be, his nature, his essence. The existence is preceded by the nature, says Schart.

People always have to exist first - free, nothingness for their own existence. Then there is the so-called essence of man, rather the essence of the individual (for there is no universal existence) - the form of the kind of man one wishes to be.

All actions taken by the subject are aimed at creating one's own life, shaping one's own "to be.

For example, a young man is faced with a dilemma, a difficult decision - to join the army or to stay with his mother. For him, joining the army was a desire for a better existence, a desire to be the hero who sacrificed for the nation.

With his mother, he was willing to be the filial son who helped her live. In either case, the young man is making a decision about what he should be (and what he wants to be).

In other words, the young man's choice is a decision about his own nature (to be a filial son or to be a hero).

Therefore, there is no universal moral standard (one should be filial to one's parents) that determines in advance what you should do, or a universal human nature (human nature is good) that determines in advance what you should be.

Because you still have to choose. I determine my own existence, I am the master of my own being. My existence depends entirely on the actions I choose.

Man's real existence is completely equal to his chosen action, and nothing else. For example, what makes Proust a genius depends entirely on his entire oeuvre.

The genius of Racine is expressed in his series of tragedies. These works represent their creative power, and this creative power represents them as geniuses.

A genius is a genius not because he is a genius by nature, but according to what he does, and apart from his works, without creativity, genius cannot be established.

Therefore, the nature of man is based on his actions. Of course, this does not mean that a writer is determined by his works (his creations) alone, but there are a thousand other things that help to explain what kind of person he is.

Therefore, the real existence of a human being is not explained by anything other than his actions (creation, choice, etc.). A person is no more and no less than a series of actions: he is the sum, the organization and the set of relations that constitute these actions.

02. On Motivation and Passion

Since every action represents a necessary choice, and is based entirely on the fact that the real existence of man is a kind of emptiness. However, perhaps we still have some doubts, because we always find that when an actor acts, does it not always involve a question of motivation or will?

For example, we most often ask people: Why do you do what you do? And we often hear people answer.

Is this what I want? Don't we often find that an action almost always involves some social, moral, or religious reason, or even the actor's own reasons or motives, which make them take certain actions?

What is the explanation for this Saudi Arabia? Although people are free, we still find that there is still a so-called motive or will in the consciousness? If this is true, is it possible that the Saudis say that man is free in his existence?

If there is a motive or a will, how can consciousness be made void and how can freedom be possible? Let us proceed in this direction.

First of all, let us see what Schart has to say: the human condition is a condition of free choice, without excuses or assistance, so that anyone who excuses himself by his passion or by inventing some determinist theory is self-deceived.

For the determinist theory, the general orientation is that the behavior of an actor is strictly determined in a motive, and this motive is determined in some transcendent, innate purpose.

Purpose-motive-activity is a continuous entity. There

is no nothingness between these three, they are connected like an unbroken chain, and all actions must have a cause that causes them to occur.

For these determinists, the motivation may be due to social, religious, or political circumstances.

These contexts, which are the fundamental constraints on man's place in the universe, include

Man is born a slave in a pagan society, a nobleman in a feudal society, a proletariat. However, the eternal limitation is death.

These limitations, for the determinist, limit our motivation and influence our actions. Conversely, my actions are only a means to achieve conformity to these situations, to these limitations.

My actions are not free, they are already established under these circumstances. This end-motivation-activity continuum is not only found in the discourse of the determinist.

In general, for example, in the research of scientists or in the questions and answers of ordinary people, we often hear the question "Why do you do this? What are your reasons for doing this?"

In asking these questions, we believe that there must be a reason or justification for every action, and that this reason is sufficient to explain why the actor acted as he did.

We usually refer to this original point as the motive, the purpose. In this way, it seems impossible for people to behave without a motive or purpose. The relationship between motivation and behavior is like the causal connection in nature.

Whatever purpose or motive I have set before, my subsequent behavior will follow. In other words, in order to realize my previous motive and purpose, I will actively take a series of behaviors that can achieve my previous purpose, or behaviors that satisfy my previous motive.

These two are the cause and the effect, which are closely linked together as a logical process of inference, one step after another, and there cannot be any other possibilities.

There can be no other possibility, no other result outside of this series, but a complete presentation of necessity. But is it really the case that the motive necessarily determines the behavior in a certain form?

For Schart, motive is only an external, objective fact, as an objective apprehension of a real situation, not as a subjective fact.

Moreover, this objective apprehension can be formed only under the guidance of a predetermined purpose and within the limits of a project for one's own existence toward this goal.

It is possible to form it. In short, a motive is a tool to achieve a desired goal of a subject. For example, some justice practitioners, even psychologists, when faced with a murder case, will always provide what they call a strong motive as the reason why the perpetrator killed the person.

The motive may be the circumstances of a relationship between a couple or someone's financial situation. For them, this motive becomes an irrefutable proof in the inner body.

Saudi Arabia, however, thinks otherwise. The motive merely represents an objective fact, but does not constitute the reason for the killing. For the appearance of this motive is possible only within the limits of the purpose proposed by the subject.

When the act of killing is carried out, the offender has already considered the objective consequences of the killing, that is, the purpose has already been set forth (possession of hundreds of billions of dollars of property, etc.).

It is within the limits of this purpose that the motive - someone's financial situation - becomes the motive. It is only because of the subject's plan (purpose) that the motive reveals itself and is revealed.

In contrast, for the main project, the motive becomes an excuse, a tool, and a means to achieve this purpose.

In general, motivation is an objective understanding and grasp of the situation, or we can say that the real situation is the motivation.

However, this motive can be revealed only under the purpose proposed by the subjective consciousness. Therefore, the appearance of the motive must be revealed only when the subject puts forward a goal.

The goal proposed by the subject is the subject's free choice and action, therefore, the motive must be realized through our action.

In other words, it is my choice that determines the

motivation, the reality of the situation, and without the choice of action, the motivation cannot appear. Therefore, the subject's choice and action are inevitable and therefore free, and all choices and actions represent the original choice of the subject.

It is certain that the motive is an objective state as an event, but it cannot be a cause of action, because the subject's planning is necessary for it to be recognized as the motive.

We therefore call the motive, an objective grasp of the prescribed situation, because this situation is revealed under the guidance of a certain purpose, indicated as an instrument for achieving this purpose.

What about passion? What about the will? Can passion or will determine choice, action? For example, a soldier on the battlefield, faced with many threats from the enemy, may panic and flee out of fear.

But he may also think that he should stay put, even though at first glance it seems more dangerous to resist than to flee. For most people, the act of flight is an act of passion; the act of resistance is an act of will.

The will to resist one's fearful state of mind is seen as

free, a free will, free from psychological causal decisions.

However, for the Saudis, these two states are still expressions of free choice (expression). That is, either the passion of fear or the will must be identical with the act of escape or the act of resistance, and we cannot distinguish choice or action from will or passion.

Schartre says that an action without fear is not fear, an act without resistance cannot be will, an act without love is not love, and an act without pleasure is not pleasure. In short, without action, performance does not exist. Both the will and the passions are expressions of the choice to act, the way to express oneself "freely".

They (will, passion) are the same as the necessary choice of action (the original choice). Fear, like courage, is free.

Fear intends a situation with non-reflective consciousness (i.e., I avoid facing a dangerous situation), while courage intends a situation with reflective consciousness (i.e., I face a dangerous situation with grace).

Either one (the act of voluntary introspection, the emotional expression of fear, other passions or anger) is my choice, out of my free choice.

It comes from my different choices (which may be fearful or courageous) in the face of the same dangerous situation. Emotions are not as some mental excitement, but as a response of non-reflective consciousness to a specific situation.

The will, like the emotions, is some subjective attitude by which we achieve our original freely proposed purpose. Of course, we should not understand original freedom as a freedom that precedes the activity of the will or the passions, but rather as a freedom that is completely simultaneous with the will or the passions, and in which the will and the emotions each manifest themselves in their own way.

As we can see, Schartes strives to state that all our actions are free and come from the person who is free.

The existence of this free man is not something metaphysical; it is his own action, his own choice. Freedom is the inevitable choice. Nothing can determine my choice, prevent me from being free, I am completely free, or I am forced to be free because of it.

In Saudi Arabia, the for-itself, the nothingness, the

(nothingness, human consciousness, freedom, and free choice are all of the same meaning, the same The same.

In the existentialist view, human existence is not a mere "being toward", a static existence of "being there".

Human existence is a tendency, a constant questioning by oneself, a constant intention to challenge oneself. It is active, not passive, as the common term "being" is.

What is commonly called "existence" and "being" is dead and dull, at the mercy of the outside world. However, the existence of human beings in the sense of existentialism is a trend, an intention not to be satisfied with the status quo, or not to be willing to fall.

In other words, life is basically a being toward. Human existence contains intentionality.

This intention is not a human (i.e., collective human) intention, nor is it a rational intention, but an intention that emerges from the reality of individual existence. The purpose of this intention is to bring one's existence to perfection, and ultimately to transcend one's finitude and achieve absolute freedom.

Therefore, one's existence cannot depend on external material or psychological forces, on the power of others, or on society, but on the intention of one's existence itself, which constantly breaks out of the limited situation. So it is with Saudi Arabia.

The real existence of all people is their own self-creation, self-shaping, self-building, and self-striving; it is the intention (tendency) of the real existence of people. What is intention? It is a choice of purpose, an act of choice.

For example, my intention is to have a nice dinner. I choose that intention, and the restaurant I go to expresses the particular meaning of my choice of that intention.

For every action, every choice, no action, no choice is possible without intention. Every free action is intentional, directed toward an end, a goal.

Therefore, intention is the action of a person, which is the same. When Schart speaks of freedom, action, original choice, consciousness, self-existence, and the real existence of human beings, he is at the same time stating that human beings are intentional.

This intentionality represents man's freedom of

action, choice, his ability to take himself where he wants to be (life, essence), to a purpose, beyond the limited environment, beyond himself.

Intentionality is the action towards a purpose. This action is autonomous, and there is no physical presence behind this action. The intentional action is the whole of the subject, which is freedom.

Therefore, what is freedom? For Saud, freedom is not just a nothingness, but an action, an action, a choice to take oneself to the place one desires, and this action, this choice, is inevitable for every human being.

Human freedom is the inevitable and absolute ability to determine one's own future and purpose. Freedom is not something that can be defined.

Rather, it is the intentional nature of the subject itself - the constant action of transcending one's own finitude and moving toward one's own future possibilities - that proves that man is free.

Schart thinks that no one can stop being free. Freedom is the existence of man. It is revealed in all of man's plans.

The actions of any man are in harmony with a larger

meaning - his style of life. His plan is - to choose the way of life. Freedom is not something that precedes free choice, but freedom is simply choice.

Man's existence is free to decide for himself, free to choose, free to move towards a goal, a project of the subject himself.

On the contrary, this goal exists only through man's choice. If it is my goal to be a filial son and I must be a filial parent, it is because I have chosen it; or if Mount Jade is regarded as a mountain I must climb, it is because I can want it, I can choose it, and it is my goal.

Thus, the existence of a goal is entirely determined by human choice. At first glance, Schart's view of human freedom seems to be quite positive and optimistic.

I am the whole of this action, I am the choice, the action itself.

There is no internal or external factor that can limit my freedom and determine my behavior. The existence of human beings seems to be perfect in all unexpected ways.

However, in reality, it is not. Although I have made a choice, I have a purpose to act. However, this purpose, this future, is only a possible existence.

That is, my past-present-future is not inextricably linked as a cause and effect. Perhaps, I decide what I should be, I decide what kind of environment I should live in, I decide my future life.

However, there is a gap between this choice and the goal. That is, there is a "negative" judgment between my past, my present, and my future.

A gambler makes a free determination to stop gambling, and as soon as he walks into the table, he finds that all that determination melts away.

The determination to "stop gambling" is always there, and in most cases, when faced with the table, the gambler always recalls the determination he made to get relief.

Because he doesn't want to gamble, or rather, he made the determination the day before, he still thinks he doesn't want to gamble anymore, and he believes the determination is valid.

However, as he stood at the table, he felt an extreme anxiety. This "anxiety" (anguish) is an indication that his previous determination was completely ineffective.

This "anxiety" is like a crack between the past and the present, separating the past from the present. Even if the gambler really did stop gambling the day before, and the determination still seems to be there today.

But that determination is stagnant and ineffective. For when the gambler faces the table, his anxiety tells him that he must make another choice, a choice to "stop gambling," that he must make another determination to "stop gambling.

001. Being and doing: freedom

This means that past actions, choices, are not the cause of present actions for the present subject consciousness.

It is as if it is a part of the world, existing in the world, that is, it is the in-itself. Its appearance is only possible through the action of the subject (for its own existence).

This anxiety shows that the person must act (recall this determination to "stop gambling" and "stop gambling") in order to make the determination to "stop gambling" of yesterday visible.

This means that I am free to determine my past and give it meaning. However, it is important to note that when Schatt sees the past as a present existence that has nothing to do with present action (self-existence, intentionality, choice of action)

This is also an implication that I must be constantly choosing. Whenever I choose the future I want, at the same time as this choice, this "choice" becomes the past, and I cannot make it "being what it is" (being what it is).

According to Schart, the past is a solidification of being for oneself. When I say "I was tired," I mean that the fatigue was as something that could not come back: my past became present.

I was facing it as if I were facing the presence of an external thing. I am not my past, but was it. I could not be my past, otherwise I would not be what I am, that is, a self-presence that is absolutely free and full of possibilities.

The "past" exists as being in the self, that is, as an object of consciousness (present consciousness - intentionality).

Past consciousness and present consciousness are not a longitudinal continuation (the past is the cause of the present behavior), but are opposed at a certain distance.

Therefore, the past and the present cannot be the

same, but there is an "internal negation" that explains the connection between the past and the present consciousness.

It is the present self that faces the past self while constantly escaping from the past self. The past choice is still there, but the present me is constantly running away from the past choice.

The present consciousness can only be the awareness of the past me, but it cannot be the past me. This "anxiety" (internal negation) explains everything.

Therefore, the present me is not as something fixed, but is only constantly facing (i.e., knowing, remembering) the past and escaping from the past choices.

Of course, there is a good side to this constant re-choice, because one can always break free from the baggage of the past and move toward a new future. If someone has broken the law when they were young.

It does not mean that he must live his life forever with the name of a criminal. According to Saudi Arabia, he can start over again anytime and anywhere. Therefore, one always leaves the past behind and moves forward to the future.

The present is the for-itself, that is, man's consciousness escapes from the past and enters the future. But what can the future say about the present?

In addition to the anxiety of the present consciousness in the face of the past, Scharthe also points out that there is another kind of anxiety: the anxiety of the future. I was walking on a narrow path without guardrails on the edge of the cliff.

For me, this cliff was something to avoid, it represented the danger of death. In order to pass safely and leave the threats of the world (the purpose), I thought of some possible actions.

Watching for rocks on the road, leaving the edge of the road as much as possible, etc. However, I still feel anxious (anguish). I clearly understand that for the present me (fear), these imagined behaviors (future me) are only a possibility.

The present me (fear) does not necessarily determine my future behavior, and the present me is not the future me. Because, when I think about all the protection measures in the future.

I am actually facing the future: I plan for the future, I

think about the future, I desire the future, I hope for the future.

However, I am not yet this future. This future exists only as a goal that I desire, as something that I am facing.

Therefore, I am anxious. I am anxious that my present situation is only a lacking being, because I am not yet that which I desire, I am not yet the state I want to be.

The future makes the present desire an action toward the future, not an existence that can be the future of so-and-so. In other words, the present me is only an action of intention, a lack of existence, a non-existence.

I feel anxious because my actions are only possibilities, which means that in rejecting all the motives of the situation, I simultaneously treat them as insufficiently valid.

At the moment when I think of myself as fearing the precipice, this fear is not decisive in terms of my possible behavior. ……...precisely because I do not see it as the cause of this subsequent development, but rather as: a demand, a call, a wait, etc.

What does this anxiety tell us? It explains my freedom. Because, whether it is anxiety in the face of the past or anxiety in the face of the future, everything tells us that human existence, as a being, is a constant necessity.

It is the existence of a person who is forced to be free, as a person who must constantly make choices for his own existence. Human freedom is a forced freedom. The fact that I am now constantly escaping from the past (in my own existence) means that I must be constantly starting over, choosing again.

The fact that I am now constantly moving towards the future means that I am in a state of lack, and in order to make up for what I lack, I am always planning my future and moving towards it.

But for Saudi Arabia, this whole of the "lacking-deficient" is impossible to achieve, although one always aspires to it. For as soon as I choose my future, that choice slips into the past, into a present what I am - never the what I was.

Then, I must begin again, choosing a new future, and so on and so forth. The conclusion is: I must be chosen, I must be free, and I am forced to keep choosing, forced to be free.

Nothing but freedom. Consciousness is the mere

negation of the given, as a detachment from a given that exists and an intervention in a purpose that does not yet exist.

But on the other hand, this inner negation can only be the fact of an existence that is always in a retreat from itself.

········... he can find no relief, no support, in the east or west of what he was.

······ ... freedom is freedom only because the choice is always unconditional.

········...Since freedom is an existence without support and without a springboard, it is planned to be constantly renewed in order to exist.

I am always making my own choices, and I can never exist as a chosen being, or I will fall back into mere self-existence.

········...Choice, to the extent that it is ongoing, generally indicates that other choices are possible.

········...Because I am free, I plan my full range of possibilities, but I therefore propose that I am free, and that I am always able to nullify and transmute this original plan.

Man is completely free, and nothing can limit his freedom. Therefore, man must be responsible for his free self-determination. For there is no longer any causal determinism that can justify man's actions.

Man must take full responsibility for his own actions. If we say that a person is a habitual thief, it is easy to attribute his behavior to poor circumstances and to feel compassion for him.

But in the case of Saudi Arabia, because people are free, his theft was not the result of a poor environment. The reason for the theft is that the man chose to steal.

Therefore, there is no more excuse, no more dependence, all decisions come from the free choice of man; man is free, and absolutely free.

002. Freedom and Responsibility

In the Western philosophical tradition, "being" has always been regarded as the eternal essence behind the phenomenon, not related to time.

Western philosophy, dominated by Descartes' doctrine, regards "being-for-itself" as the subject and nature outside of man as the object, and measures nature with a dichotomy of "subject/object", with manipulation and conquest of nature as the highest goal.

For example, "freely" - "for oneself", I put the weight of the world on me alone, and nothing and no one can reduce this weight.

I am responsible for everything, except my responsibility itself, because I am not the basis of my existence. Thus everything still seems to state that I am forced to be responsible.

Heidegger argues that this way of thinking implies a misunderstanding of "being": it sees "being" as an eternal state of being, an objective entity that is outside the body, present and observable.

This is like a child complaining about his situation. "I am not the one who has to live in this world. Rather than following the philosophical tradition of exploring the properties and uses of beings, Heidegger is more concerned with the fundamental question first posed by Aristotle.

What is "being"? Greek philosophy uses "being" to describe a process that occurs naturally and reveals itself naturally.

This process has a controlling and gathering power that allows all things in nature to appear and remain.

But before 'being-for-itself', I have assumed that I am 'being-for-itself' and that life is badly human, and that the choice to be human is the freedom of will.

It is I "being-for-itself" who chooses the artificiality of my birth. My different attitude towards birth in this paradox is precisely the way in which I take full responsibility for birth, and the way in which birth becomes my (self-existent) birth

Thus, action for human reality is to maintain a fundamental relationship with the world, to transcend the simple and stable determinism of the world through action in order to transform the world in its materiality.

"Being-for-itself' is a being that can detach itself from the reality of the world and from its own 'nothingness'. The eternal possibility of this detachment is the same thing as freedom.

And since what determines man's existence is his own inscrutable "nothingness", freedom of will and this inscrutable "nothingness" are also the same thing.

Man's will is free because what he manifests in his actions is not himself, but a presence of the self, which he always wants to transcend.

Man's existence should be reduced to action, and the first condition for action is the freedom of the will, then the existence of the self is the freedom of the will. Thus, "human existence precedes human nature and makes human nature possible".

Action is the action of the will to choose, and freedom is the freedom of the will to choose. The freedom of the will is absolute autonomy, so it can only exist in self-choice. Human freedom is freedom of the will only because human choice is always free.

And this freedom is not Kant's 'choice of mental characteristics', but a phenomenal, absolutely free choice.

It should be noted, however, that the freedom Scharthe discusses here is not arbitrary, lawless, or arbitrary, as some commentators have suggested.

What Schart is saying is that these reasons and grounds explode together with the actions chosen by the freedom of the will, just as the freedom of the will is expressed in the totality of motives, dynamics, and purposes.

It is true that in every activity of mine, even the smallest, the will is completely free, but this does not mean that it can be arbitrary, nor does it even mean that it is unpredictable.

The freedom of will is revealed through freely chosen actions, and is not hidden behind anything, and is revealed together with the existence of self-activity.

The characteristics of the ontology of freedom can be summarized in the following eight points.

(1) "Being-for-itself" is action.

Human existence is not to be reduced to what one does. All actions and behaviors. "To stop acting is to stop "being-for-itself".

(2) The existence of action consists in the self-discipline of "being-for-itself".

The requirement of action by "self-referential existence" is itself action, and the existence of action consists in the self-discipline of "self-referential existence". A self-determination and a freedom that is independent of all external constraint. It usually includes three meanings.

01. biologically, it means the relative autonomy (relative independence) of the infant after birth, independent of the mother and free from the complete dependence of the fetal period.

02. political and social aspects, which refers to the freedom of action without external restrictions, such as the autonomy of local educational authority from the central government.

03. in ethics or psychology, self-regulation is the main characteristic of moral action, referring to the relative autonomy of individuals in directing or regulating their own behavior. The opposite of self-regulation is "heteronomy," which has the meaning of being subject to external authority.

(3) Action is intentional 'being-for-itself'.

Action is the intentional "being-for-itself" and is prescribed by the intention of "being-for-itself". It is the ability of the mind to represent or present things, attributes, or states. Simply put, much mental activity is about the external world, and intentionality is the 'about' here.

(4) The world is revealed through the purpose of 'being-for-itself'.

The intention of "being-for-itself" is a predetermined goal and result of the behavior of the subject according to its own needs, with the help of the intermediary role of consciousness and concepts. As a conceptual form, it reflects the relationship of human practice to objective things.

Therefore, the intentional choice of purpose reveals the world, and the world is revealed as this or that according to the chosen purpose of "being-for-itself".

(5) The "being-for-itself" world has the existence of a given thing.

Action is "being-for-itself" breaking with the given, and in breaking with the given, when non-being illuminates the given, the world of "self-existence" has a pre-defined existence as a standard or goal.

(6) The negation inherent in "being-for-itself". Choice is always unconditional.

To reveal the inevitability of a given thing that is pre-determined as a standard or goal only within the scope of nihilization is the negation of the substance of "being-for-itself" that is hidden within things. Freedom is freedom only because the choice of 'being-for-itself' is always unconditional.

(7) Freedom is a choice of being-for-itself, but it is not the basis.

Freedom of choice is absurd. It is a philosophical term, derived from the Latin word absurdus, which means musically "out of tune" and is used in existentialism to describe the meaningless, paradoxical, and disordered state of life. For freedom is the choice of one's "being-for-itself" existence, but it is not the basis of this "being-for-itself".

(8) The project of freedom is the existence of the "being-for-itself" of the self.

The plan of freedom is the basic 'being-for-itself' because it is the existence of my 'being-for-itself'. That is to say, human existence can never be as full as being-in-itself (French: être-en-soi) existence; it is destined to be lacking, to use a Saudi term: being-in-itself can never fully coincide with being-in-itself, and if it does, it kills consciousness.

Freedom of will is essentially freedom of choice of

action. But sometimes people feel that their freedom is limited in different ways, and sometimes they choose the same actions and get different results.

This leads to the question of context, and the relationship between context and freedom. The freedom of choice of the will is absolute and eternal.

But the freedom of the will to choose and the freedom to obtain are different.

The fact that the will is not free to choose is actually a restriction of freedom, and the fact that a situation cannot exist without it is an accidental change in freedom.

Freedom and context cannot be separated. Without context, there will be no freedom, and without freedom, the context will not be found.

4. The basis for the limitation of freedom - the context - is in five different ways

(01) My location: This refers to the place where one normally lives, the place where one's daily activities take place, etc.

When the freedom of will presents its purpose, the freedom of will itself makes the location we are in, whether it is an insurmountable dilemma or an insurmountable resistance to our plans, the key lies in the starting point that inhibits the freedom of choice.

(02) My past: This refers to the memories that each person once had, and which are now history. The past is irremediable; it is the past that gains meaning through my own existence and comes into the world. It can be said that the past is the past that I decided to act under the guidance of the purpose of my free choice of will.

(03) My surroundings, which refers to the objects-instruments around me, including their hostile coefficients, together with their passive instrumentality. They are the things around me that are alien to me.

They are manifested by the plans that I freely choose, and because of the plans that I freely choose, they may be on the same side in favor of me, or they may be on the opposite side in opposition to me.

(04) My neighbors, I live in a world entangled with my neighbors (others), a world that has come down to the meaning of being a reference center for others before I choose to be.

But others do not limit my freedom of choice. The

freedom of others limits my freedom, which is, to put it bluntly, only a distinction between my freedom and the freedom of others.

(05) My death, death is not the expectation of life. Death is the impossibility of all possibilities, the negation of all choices.

Death is absurd, sudden and accidental, in fact a way of being born, a negation and nihilization of the existence of the self, but it comes to us from outside.

Without others, I would never have known 'death', selfhood would never have encountered it. It is the condition as a limit, but freedom never encounters this limit, it transforms life into a destiny elsewhere, "I am a free willed man who must die".

In short, the situation cannot prevent the free choice of man's will; man is absolutely free in the situation. Freedom gives meaning to the situation, but it is not the situation that determines man's free choice of will.

Therefore, man is the absolute 'author' of choice, and he must take full responsibility for the results of his actions. The freedom that Schartre describes is 'freedom to act', otherwise 'freedom of ideas' would always be a mere term.

The close combination of situation and freedom reveals the profound meaning of 'action' and gives freedom an ethical meaning.

It can be said that what Schatt describes is the ethicalization of the situation. It can also be said that the purpose of Schart's entire exposition of Being and Nothingness is to focus on the freedom of the human condition.

5. The Psychoanalytic Approach to Existence

In this section, Schart is mainly examining the purpose of the free choice of action. The purpose of defining this project of psychoanalytic freedom is to establish what type of relationship between it and existence.

Scharthe begins by criticizing empirical psychology for its error in asserting that the individual is defined by his various desires. In effect, this evades the transcendence of 'being in itself'.

It is also too arbitrary to establish a bunch of intentions through "being-for-itself" empirical observation, and to subjectively define a path for others.

That is, one should avoid listing behaviors, intentions, and preferences, and instead objectively identify them and question them. This is the method of existential psychoanalysis.

The principle of existential psychoanalysis is that man is a whole and not a collection. In his "being-for-itself" behavior, even in the most meaningless and superficial things, he has expressed his existence in its entirety. There is no human behavior that is not revealing.

The purpose of existential psychoanalysis is to identify and reveal the behavior of human experience, to understand the revealed behavior of free choice. This characterizes the phenomenological description of Schart's philosophy.

The point of departure of the existential psychoanalysis is experience, and its focus is on the fundamental understanding of the ontology of the individual's pre-possession, and the freedom of the will. It is not a rational activity, and unlike cognition and reflection, any action toward a purpose, a conception of the self itself, is understandable.

Existential psychoanalysis is still a comparative

method. Each individual action symbolizes the need to reveal the choices of the ego-obsession in the way of the freedom of one's own will, while at the same time these actions conceal the unbridled choices of the ego-obsession under the characteristics of chance, and historical contingency. The comparison of these actions through psychoanalysis allows the unique revelation of each action to burst forth.

This psychoanalytic approach actually borrows from Freud's psychoanalysis, which is mainly a "retrospective" approach to experience analysis. However, it differs significantly from Freudian psychoanalysis.

The main reason for this is that Scharthe firmly denies the existence of the subconscious, believing that the two are self-contradictory and that awareness is not the same as being known. Moreover, sexual desire is not, in Scharthe's view, a primitive scheme of 'being-for-itself'.

"Desire' is reducible, desire is to 'have', and the 'act' of desire is reducible to the means of 'having'.

Therefore, the Saudi human being is the consciousness of "being-for-itself", an intuitive center, which gives rise to a spontaneous choice, a choice that transcends towards a meaning. The ontology is to clarify in this process the relationship between 'being-for-itself'

The ontology is to clarify the relationship between 'being-for-itself' and 'being in itself' in this process.

Ontology makes it possible to explain the psychoanalytic method of 'being in itself', the human action according to choice, and to make intelligible the connection between an active and spontaneous choice, and the choice of 'being-for-itself'.

In order to know oneself, one must make oneself, and in order to make oneself, one must know oneself, and every knowledge implies action.

Only by tracing the choices made in this way with existential psychoanalysis can the free willed person be truly understood. And man can give meaning to his actions only when he freely assumes the full responsibility of this choice.

(01) The moral description of freedom

In the concluding section of Being and Nothingness, Scharthe further clarifies his basic philosophical position.

In the fourth part of the book, Scharthe defines his

ontology as "being-for-itself" and "being in itself" with respect to being as a whole.

The ontology is the various structural descriptions of being-for-itself and being in itself as a whole.

The ontology cannot express the moral law itself, so he ends up describing the ethics of the free willed person who, in the face of real situations, is actually responsible for 'being-for-itself'.

Sartre concludes by stating that the real solution to the various problems of freedom can only be found on a moral basis. Sartre foreshadowed that he would write such a work specifically, but he never completed the work that was anticipated here.

Freedom is a fundamental concept in Scharthe's existentialist philosophy. Scharthe gave a new meaning to "freedom," which is characterized by the subjectivity and transcendence of man, the pure activity of consciousness, which is identical with self.

In other words, freedom is not a certain nature of existence, but the existence itself of man's own will. Freedom is not something that one pursues or chooses, but something that one has in oneself. Freedom of will is inescapable and cannot be escaped.

The primary condition for action is freedom. For all activity should be intentional, that is, purposeful and motivated. Accordingly, Sartre makes several important propositions as follows.

First, existence precedes essence. Freedom of will has no essence, because freedom is the basis of all essences. Therefore, man is destined to be free. Second, freedom of the will is freedom of choice.

This choice has no intentional support, it is a motive that I love to impose on myself, so it can be extremely absurd. This absurdity is not because it is an irrational existence, but because it has no choice at all.

Again, freedom and responsibility. Since freedom of will is freedom to choose, some people think that freedom means freedom to do as one pleases, to do as one pleases.

He repeatedly pointed out that in making any kind of choice, one has to be responsible for oneself, for others, and for society. This responsibility is of a particularly special type, and it can be said that "we did not ask to be born", which is a childish idea that emphasizes the factuality of people.

Sartre refers to the established things that limit the human existence to its dynamism. The idea that man does not live in a situation entirely of his own choosing, that self-existence is always connected to self-existence, that 'being-for-itself' is always connected to the world and to its own past, and that this limitation is incidental to man and incomprehensible to him.

Since man's "being-for-itself" is nullified, that is, no being can determine everything about man, Scharthe makes the extreme assertion that "man is absolutely free".

However, Schart is not a crazy idealist, he does not completely deny the limitation of being in itself to human beings. In his theory, "being in itself" is the basis of man's existence.

Therefore, ontologically, man must exist in a certain situation and space-time, with its unchangeable history, and these unchangeable elements are what a man "is", or this is what the things that are man actually are.

However, because of the uniqueness of human "for oneself"-"freely" existence, human beings can always go beyond their "being". This is what makes people "absolutely free": they are absolutely free to set what they "are not but should be".

Just as the uniqueness of a time is revealed by its problems and shortcomings, so the uniqueness of a person is defined by his shortcomings, by his choice to be what he is not, not what he is.

In this way, human beings recapture the dimension of "action" and "meaning," and are not merely determined by existence - whether this existence is the so-called logic of history, or God.

(02) Absurdity

These issues have been repeatedly emphasized in the Introduction, in Vol. 1, and in Vol. 2. The "freely"-"for oneself" relationship is not a component of "freely," but a component of self-love.

The ability to make one's self nothing is in fact the recognition of a lack, of a hope, of the possibility of a better future, which is Sartre's freedom.

Therefore, Sartre calls this movement, which is constantly beyond itself by virtue of the possibility of others, and which can never get or cease to be what it is in nothingness, transcendence.

In broad or colloquial language, it means: "Man is always free to plunge ahead of himself, whether in space or time.

For example, who I am going to be or where I am going to go. Freedom takes this attribute beyond the person himself, leaving himself in another place and becoming somewhere else. Thus, my freedom provides my place and defines it as the place I am.

For without human nature, freedom would not exist-as the capacity for nihilization and choice, without which human nature would not be discovered, or even have any meaning.

Thus, Sartre analyzes the reasons why man is judged to be free:

First, God does not exist, and there is no divinely given commandment to follow. The individual has no one to turn to, and must explore and choose for himself;

Second, there is no a priori, universal human nature, and therefore no universal ethics to instruct you on what to do; your will is absolutely autonomous, and you can choose between various possibilities;

Third, the world is absurd, there is no such thing as

necessity, there is no objective necessity that can fetter the free choice of the self. If this is the case, then what else can provide a universal standard of value for the individual? Nothing.

There is no objective ethical or moral code or standard.

But Sartre believes that since man is absolutely free, each individual must take full responsibility for his own choices, actions, and values, no matter what happens.

Moreover, in the real world man's existence is not isolated and pure, but always in a certain situation.

Through the choices of one person, he or she is connected to the whole of humanity. This means that we are responsible not only for ourselves, but for all people, and this is the meaning and value of the human being. Here, Sartre turns existentialism into a kind of moral humanism.

Section 3: The Comparison of Sartre and Descartes in Freedom

Descartes begins by doubting everything and discovers that consciousness has a capacity for free choice, beyond any decision.

Whether there is a deceiver or not, it does not negate my ability to doubt freely, absolutely and instinctively.

In the case of Saudi Arabia, this absolute freedom has a profound effect. We might even say that the Saudis are making a deeper case for this absolute freedom.

However, the Saudis also reject Descartes' notion of "self" and "God". This is because, for Saud, it is a denial of absolute freedom. Therefore, the following discussion will be based on his concept of absolute freedom.

(01) Descartes' Freedom

In his Meditations, Descartes begins with a general suspicion of all things, and in particular, a rejection of all external material things.

The utility of this general skepticism, in Descartes' view, is to help us reject all unsubstantiated prejudices and to provide the mind with a convenient doorway to independence from the senses.

Descartes argues that almost all knowledge, concepts, and value judgments we learn as children are heard and uncritically analyzed and uncritical, so that their truth cannot be ascertained.

If we want to know the truth or falsity of those knowledge, we must examine them one by one. But such a task can never be completed.

Therefore, Descartes believes that it is important to do a clean sweep or a clean sweep of knowledge or opinions and start from scratch, so as to avoid the obstacles of false beliefs or prejudices. This method of clearing out is what Descartes calls "doubt".

Descartes' metaphysics begins with a thorough skepticism, the rejection of preconceived ideas that are based on unreliable sources and on premises that have not been carefully examined.

Further, the evidence provided by the senses is doubted (in accordance with the so-called dream argument), i.e. the contemplative questions the existence and nature of the world that surrounds him.

Even the fundamental truth of mathematics (how do I know that whenever I calculate two plus three or a square with several sides, the God who would deceive me does not make my calculations wrong).

Thus, Descartes temporarily brackets all past beliefs on the basis of the slightest reason, even if fictitious. Not one of my beliefs is beyond doubt.

However, what does this doubt tell us? Schart said, "This doubt explains our freedom. For, in confirming or denying, pursuing or avoiding the understanding of whatever is proposed.

Descartes says that we are completely free to act without realizing that any external pressure determines my judgment, determines my tendency to act in a particular way.

Descartes argues that even if my inclination to choose is due to my clear and obvious perception, it is still free for the choice to be made.

For, even if I always know what is true and good, the decision about what judgment I should make and what choice I should make is still a matter of will power itself.

Otherwise, the so-called choice would not be what it is. But having a clear understanding of what is true and good makes it easy for us to decide what judgment we should make and what choice we should make.

We do not take sides for lack of reason - we adopt a neutral attitude. Thus, Descartes says that free choice and judgment are therefore absolute.

The will does not need as clear and definite a conception to ratify-even to support-its inclinations. ……...he insists on the absolutely unconditional character of the will.

This so-called free will means that we have the ability to choose to do something, or to choose not to do something (the ability to affirm or deny; to pursue or avoid). In this act of choosing, affirming, or denying, I am not bound by anything.

Therefore, for Descartes, it is sufficient and vast, without any limitation. The will faculty is greater and more perfect than all the other faculties of the self-consciousness, when compared with the other characteristics of the self-consciousness.

For example, in the case of comprehension, we can find that its scope is quite narrow and limited.

Similarly, the faculties of memory or imagination, or any other faculty that I have, are found to be small and limited.

No matter how advanced technology is, we still cannot deny the finiteness of the faculties of understanding, memory, imagination, or any other faculty, and ascribe to

God an infinite and perfect character.

However, the faculty of will or choice alone is infinitely vast and rivals that of God.

For, when we are judging something to be doubtful or uncertain, when we are denying the object, we are purely free to act.

In this process of doubting and setting aside, we are judging only by ourselves, without realizing that there is any external pressure that determines our particular behavior - doubt.

In this process of doubt, we have free will, and this free will is infinitely vast and unrestricted.

This led Descartes to realize that although man is limited in his other faculties of perception, he is similar to God in this freedom of the will.

Only the faculty of the will ... is the most vast and unrestricted that I have ever experienced, so that I cannot conceive of another, more vast and infinite conception of the will.

Therefore, only my will can make me realize that I have some kind of resemblance to God. This absolute, instinctive, unconditional freedom is precisely what Schartre demands of human freedom.

The freedom of Descartes, according to Schartes, is the freedom and solitude of human thought. The reason is that there is no longer any backing (God) from which to escape.

No one can judge for me, and I must ultimately be in a solitary situation, affirming or denying according to my own freedom, and deciding the truth for the whole universe alone.

God is no longer the source of truth; only man himself can determine the laws of the world. Affirmation and negation, pursuit and avoidance, illustrate the infinite power and capacity of man in his limited capacity.

However, in Descartes, this freedom (the ability to choose, to judge) does not represent the whole nature of the mind. The whole nature of the mind, for Descartes, lies in "thought".

That is to say, man is really a thinking thing. This thinking thing doubts, understands, affirms, denies, desires, rejects, imagines, and feels (sees, hears, etc.). Not only understanding, desiring, and imagining are thoughts, but also feelings are thoughts.

Thought means all the inner activities that we are aware of, including the activities of the senses. Based on the fact that I know that I exist, and at the same time, I judge that I am a thinking thing.

There is nothing else that belongs to my nature or essence. I therefore rightly conclude that my nature is only that I am a thinking thing.

This ability to freely choose, confirm or deny something is merely a function of the will, one of the many faculties of thought. In contrast, as far as Sartre is concerned, this ability to be free is the whole of consciousness.

Consciousness is this absolute freedom. That is to say, this absolute freedom is the whole nature of man, and nothing else. Freedom is not attached to the thing that thinks, it is not a faculty.

Rather, this absolute freedom, this freedom of choice, is man himself, consciousness itself. What makes a man a man is that he is free. So, although Schartre shares Descartes' view of absolute freedom as a spontaneous (undetermined, instinctive) action, there is a difference in degree.

However, there is a clear difference in the degree of it. It can be said that Scharthe, more thoroughly than Descartes, provides a basis for this absolute freedom, not as a function of the will, but as a person himself.

(02) The Saudi Freedom

Schartre argues that the reason for man's absolute, instinctive, and unconditional freedom lies in the fact that man must be an extra-deterministic being. When Descartes doubted all beliefs.

For Schartre, this means that consciousness (I-thought) is an alienation of these perceived beings, beliefs, all knowledge, morality, values, etc. This universal doubt proves that consciousness is an alienation of these perceptions. This universal doubt proves that consciousness is another kind of existence outside all systems, and it must be another kind of existence.

Because, under the decision of a causal cycle, everything is determined in advance in a necessary certainty, all my thought activities are included in the previous state or regulation, belief, system, like a chain of chains.

It is like a chain of chains that does not allow for disorder. However, when consciousness makes a negative judgment, it is a proof that consciousness is a free existence outside of determinism, not being determined.

Because my thought activity is no longer determined by previous states or regulations, I am out of order from the chain of cause and effect.

By this doubt, by this negation, I break with all beliefs, with the state of my surroundings. The consciousness is thus detached from that sequence of results relative to causes, and is thus free from any determination of existence.

Doubt seeks to destroy all propositions that are certainly outside the scope of our thinking; in other words, anything that exists, I can put between the brackets. ……...Doubt has nothing to do with existence.

But it is because of doubt that human beings have been able to liberate themselves from the existing universe and meditate on the eternal possibility of illusion. The will, like doubt, desire, knowledge, imagination, rejection, etc., is, in Scharthe's view, merely a way of expressing this absolute freedom.

Since freedom is absolute, spontaneous, and man is this absolute freedom, it is not a special place for the second section, the ego, God, and Scharthe.

Then, for Chartres, the existence of Descartes' "ego" and the search for the nature of this ego are contradictory to absolute freedom.

Descartes concludes, after widespread doubt, that "I think, I exist". This is the point of certainty that Descartes is looking for, the first principle of philosophy.

No sceptical hypothesis, even the most absurd, can be overturned. Its veracity cannot be questioned.

For, if I doubt, it means I must exist. If I am deceived, then I must exist. If my thinking is wrong, it means that I exist.

Otherwise, there is no such thing as a mistake to be made. And if I think in a dream, I also exist, otherwise there would be no dreaming.

Therefore, the proposition "I think, I am" is absolutely indisputable. Descartes then examines what the "I" is. So, strictly speaking, I am only a thinking thing.

That is to say, a mind, or an intellect, or an

understanding, or a reason - terms whose meanings were previously unknown to me.

I am, however, a real thing, and actually exist; but what am I exactly? I have already answered: I am a thinking thing.

The "I" as a thinking thing is, for Descartes, an entity that can doubt, understand, affirm, deny, will, reject, and also imagine and feel.

In other words, to doubt, to understand, to affirm, to deny, to will, to reject, to imagine, to feel, etc., are the properties of this entity (thought-object). Descartes thinks that even if I doubt almost everything and understand something.

to affirm that only one thing is true and deny everything else; to have the will to know more of these things, but not to be deceived; to imagine many things; to feel.

To imagine many things; to feel many things, and so on, are all attributes of the "I" as a thinker. None of these attributes can be separated from my thoughts, or from the "I".

"I am as an entity attached to these attributes. For it is

I who doubts, it is I who understands, it is I who desires.

There can be no objectless doubt, objectless understanding, or objectless desire. Every attribute (including the volitional faculty) has something to which it is attached, and that is the entity on which the self is attached, the "self. I know that I am an entity whose entire nature, or essence, is simply thought.

For Schart, the existence of this entity, the ego, is another way of pushing human reality into the abyss of determinism. According to Schartre, Husserl and Descartes were wrong to seek a self through an analysis of the ego.

In Chapter 2, we have already discussed whether the ego exists in consciousness. For Schart, what makes man free is that he is in fact an extra-deterministic being, and all conscious activity is pre-reflective consciousness-directly meaning something.

The "self" emerges only in reflective consciousness. (Although, Descartes admits that the Self is known indirectly through one of its attributes.)

However, Scharthe does not think that this "ego", which appears in the second stage, exists as a subject of consciousness. On the contrary, it exists as an object, an object of conscious activity.

Like the relationship between the subject and the subject, the ego is an object of thought, a subject, not a subject.

"There is always a clear distinction between the ego and the subject of consciousness (the activity of consciousness), although the ego corresponds to the subject. Therefore, the ego and the subject of consciousness are not the same, but the subject of consciousness exists as the "not" ego.

I think, therefore I am. What am I? It is a being that is not its own foundation. What, then, is the subject of consciousness?

According to Schart, it is not some physical being, but only nothingness, a non-being, something outside of being (Descartes thinks that the "I" is at least something, not nothingness, but "something").

In other words, it is merely some spontaneous activity of consciousness that is not determined by any factor that determines its own behavior.

All behavior comes from the consciousness of

nothingness, which determines itself. Thus, contrary to Descartes, who believed that there is no action from nothing, all action must have a source.

However, in Schartre's view, if one strictly adheres to the spontaneous, unpressured, instinctive freedom of choice and judgment that Descartes emphasizes in his free will, then one should not be in a position of free will.

Then one should not add an entity before free choice - as the basis for its free choice, judgment (including all conscious faculties).

Nothing (including the ego) can be the basis of choice, affirmation, and negation (conscious activity). This foundation is established only in the choice itself, in other words, in the activity of consciousness itself.

Thus, according to Schart, we cannot find anything in consciousness, except for the manifestation of action.

So, in Scharthe, free choice is not just a capacity, an attribute. It also involves the existence of the subject of consciousness itself as freedom itself.

That is to say, the subject of consciousness must be something other than determinism, something other than being, a non-being, a nothingness.

Thus, according to Schartre, Descartes regards the entity "ego" as a subject of consciousness, which means that it destroys the instinct of choice, the instinct of affirmation and negation, the instinct of pursuit or avoidance of freedom.

Because the ego (the entity, the essence) determines its properties, choice, denial, or pursuit and avoidance are only presupposed, already defined, and there is no more freedom for man.

In conclusion, the only way to remove all nature, all nature, from consciousness is to remove all nature from consciousness.

The conclusion is that only when all nature is removed from consciousness, and the subject of consciousness is nothing, non-existent, is absolute, instinctive, spontaneous freedom possible.

After stating "I think, I am", Descartes further argues for the existence of "God". He argues that "God" as a perfect idea cannot come from an imperfect being.

The foundation of this perfection must come from a divine grace that is at least as perfect as this concept... not only does it not diminish freedom, but it increases and strengthens it.

........because if one always knows what is true and good, one can easily decide what judgment one should make and what choice one should make.

However, this human freedom becomes less absolute and less unconditional for the Saudis in the sense that "God intrinsically determines my thoughts.

For even Descartes claims that man has sufficient power to choose to do something or not to do something; to affirm or deny it; or to pursue or avoid it.

or to pursue it or to avoid it. However, in the case of "God's inner determination of my thoughts," the free choice of the will is suddenly enveloped by some inner and miraculous spiritual light, God.

Whether I affirm or deny (in terms of belief), it is God's influence on us, and it is God who affirms or denies through our will.

If I insist on affirming an idea or denying an idea, it is simply because the miraculous light of God is pressing upon me, so that I know the true and the good with clarity and understanding, and avoid making mistakes.

"God intrinsically determines my thoughts" - a question of faith. In this context, since understanding is not a sufficient reason for the act of faith, the whole will is illuminated by some inner and miraculous spiritual light, namely, "grace".

We see that this spontaneous and unlimited freedom is suddenly influenced by grace. In fact, Scharthes argues that "God" can never intrinsically determine our thoughts.

For, when Descartes argues for the existence of "God", he argues that the idea of perfection cannot find its source in me, because I am a finite, easily deceived, imperfect being, and that the same perfection can only be found in God.

Therefore, I am not the basis of the concept of perfection. This means that something that may be the foundation cannot endure any small discrepancy between what it is and what it is conceived to be.

Therefore, consciousness is "not" the idea found in itself, but consciousness is a being "different" from that perfect idea.

It does not belong to the content that expresses it, let alone to other consciousnesses (such as God). This "different from" separates the subject of consciousness from the discovered perfection, from the perfection of existence.

A gap, a fissure, a nothingness appears between the subject of consciousness and the object of thought. For the Saudis, this is proof of the absolute, spontaneous freedom of man.

For this "different from" has caused us to escape from the idea of perfection, from the existence of perfection, to free ourselves from the laws of the world (whether the physical world or the world of God), to retreat.

Therefore, all my choices and judgments are absolutely free, completely spontaneous, and unconditionally originated from the subject of consciousness.

The concept of perfection, even the perfect being (the world of God), is only the object of my thoughts. He can even retreat from everything within, from memory, imagination, and the body. ……... He acquires this unparalleled independence to resist the omnipotence of evil spirits - even to resist God.

Conclusion: A metaphysical conclusion is drawn on the relationship between "Self-being" and "Self-activity".

From the previous chapters, we can clearly see that Schartan philosophical thought combined with Heidegger's view of man's unique mode of existence, and with the majority of existentialist philosophers of his time, affirm that man's potential can transform the situation of his existence through autonomous conscious action.

Thus, human beings can only define their own existence through the existence of nothingness, unlike other species whose manifestation of existence is limited by their intrinsic framework. However, it is precisely the action of being-for-itself that defines the human being all the time.

Many people, fearful and unwilling to take responsibility in the face of this freedom of being-for-itself action, pretend to be themselves by means of mauvaise foi.

The role played by the social matrix, as if it were

someone other than oneself or the environment, determines one's life. The concept of self-deception is the key element that Scharthe uses to explain the comfort of "being-for-itself" situations for all social classes.

Thus, Scharthe constantly emphasizes the extreme, instinctive, absolute freedom derived from a transparent being-for-itself (the reality of man).

Being-for-itself, as man's real beingness, is released from the immensity and fullness of being-itself, like all objects or phenomena that exist objectively in nature, into a free, unfettered 'being-for-itself'.

This is simply because man has the ability to negate (exclude) everything that is different from his subjectivity, including all institutions, customs, concepts, mental worlds, and values.

Therefore, self existence (human subjectivity) is a lack of existence, a nothingness for the Saudis. It has been constantly throwing itself into the world, catching its past (being), and pursuing a holistic existence.

However, it is impossible for [being-for-itself] to become this integral (self-same) existence.

For "being-for-itself" always escapes from its own

past, ego, state, and motivation, and becomes a totally and absolutely free being. All actions of the being-for-itself are free actions, and no single factor can determine and influence the choices of the being.

If any chosen action of the actor "being-for-itself" seems to be influenced by any kind of "being-for-itself," says Schart, it is only because the actor "being-for-itself" has chosen that [self-existent] factor, the practice of that "being-for-itself. itself".

Conversely, without the choice of the actor's being-for-itself, external commands, values, and motives would never be possible. Therefore, the choice of an actor "being-for-itself" is always a free choice, and the human reality is completely free.

Thus, we can see from the Saudi concept of absolute freedom that even in the midst of a barren (forceful bondage), even desperate, environment, the Saudis still have an extremely positive attitude toward the human being in that environment.

The Saudi concept of absolute freedom shows that even in the midst of a barren (fettered) and even desperate environment, the Saudis still have a very positive attitude toward the "being-for-itself" of the people in that environment. Everyone is free, and by virtue of this innate freedom, determines his or her own future and keeps on starting over again.

Nietzsche's "God is dead" is not a literal reading; Nietzsche believed that a God actually existed at the beginning and then died. What he is trying to say is that God is no longer the source of the meaning of life or a moral standard.

It is precisely Scharthe who wants to affirm that the destruction of God [the source of the meaning of life or the moral standard] frees man from the limits of his "nature", his "being-for-itself", without a predefined nature.

On the contrary, Nietzsche sees the disintegration brought about by the death of God as an existing moral assumption: "When a man renounces Christianity, he takes Christian morality out from under his feet.

This morality is by no means self-evident. ...... When one breaks the Christianization of faith in God, one breaks everything: one's hands are necessarily empty." This is why in "The Madman" there is a passage addressed primarily to non-theists (and especially atheists) whose problem is to retain any non-divine value system.

The death of God is a way of saying that human beings can no longer believe in this cosmic order because they cannot recognize whether it really exists.

Nietzsche's view that "God is dead" is not only a loss of faith in the cosmic or material order, but also a denial of absolute values - a loss of belief in an objective, universal moral law that encompasses every individual.

This loss of absolute morality is the beginning of nihilism. This nihilism led Nietzsche to try to find a way to reassess basic human values, i.e., a deeper cosmology than Christian values.

Nietzsche believed that most people do not subscribe to (or refuse to recognize) the idea that "God is dead" because they have a deep-seated fear or anger in their hearts.

So when this death becomes widely known, they feel great pain, and nihilism becomes rampant, and relativism becomes the law in human society-everything is permitted.

This is part of the reason why Nietzsche thought Christianity was quite nihilistic. For Nietzsche, nihilism is a necessary consequence of all idealized philosophical systems, because all idealism has a moral code like that of Christianity.

However, there is some criticism from the Christian side: they think that since the Saudis do not recognize the eternal value of the commandments and all the rules laid down by God, all that remains is spontaneous action.

In this way, anyone can do what he likes, and according to this view, we will not be able to condemn anyone's views or actions, and all actions seem to be permissible.

Of course, Saudi Arabia does not believe that these allegations are valid. Schart believes that he has not only been working on the problem of how to make human generation possible, but has also affirmed that any truth and any action

He also affirms that any truth and any action includes objective circumstances without losing the subjectivity of man. He argues that the Christian charge of freedom to do whatever one wants is not the true meaning of the "being-for-itself" freedom he advocates.

Although Schartrecht emphasizes the absolute

freedom of choice and self-determination of man's "being-for-itself," this does not mean that the actor is not responsible for his actions.

In fact, it is precisely this absolute freedom of choice and self-determination of being-for-itself that enables everyone to understand himself as he is, and to take full responsibility for his own existence by being-for-itself.

For a free person, there is no longer any shelter that can be used as an excuse for "being-for-itself" actions, but everything is spontaneous and out of one's own choice, and one must take full responsibility for it. Moreover, this responsibility is not only for oneself, but also for all people.

For, in fact, in all the actions taken by a being-for-itself, there is not a single action that is not the creation of what the being-for-itself thinks it should be, that is, every action of the being-for-itself is the creation of its own nature or essence.

In choosing between this image and that image, he is at the same time affirming the value of the image he is choosing, which, in his view, must be "being in itself".

For, if one is free to choose, one must choose good

values, not bad ones. This good value is not only in terms of the actor's "being-for-itself".

In the actor's view, this "being in itself" is universally applicable. Therefore, when an actor chooses to be "being-for-itself," it is not only the actor alone who takes on this responsibility, but also on behalf of all humanity.

For example, when I decide to get married and have children, even though this decision is made at my own request, when I "being-for-itself" makes this choice

In fact, when I make this choice, I agree with the value of monogamy as a responsibility of being in itself, and I think that others, even all human beings, should practice monogamy and abide by the responsibility of living together and the rules and norms of their behavior.

Therefore, while one is responsible for one's own "being-for-itself", one is also responsible for the "being-for-itself" of all people.

Therefore, Saud denies God and the accompanying values (good and evil, or virtue and vice, etc.) because, in Saud's view, these values, which are superior and located in the realm of the transcendent, are not materially helpful to man in making choices.

In fact, Saud does not believe that the absence of God leads to a freedom to do whatever one wants, but rather that the absence of God makes one ultimately responsible for one's own actions and for all of humanity.

For example, a young man decides to join the army in order to avenge the death of his brother in a war, and to fight to the death against the enemy army that killed his brother.

But in making this decision, he is faced with the dilemma of being the only support for his mother. He has to choose between the two: to join the army or to be with his mother.

At this point, there was no moral value that could tell him what to choose, and he could not hope to find any ethical justification for his actions. All he can do is to choose spontaneously.

If, it is said, he can make a decision based on the depth of his feelings for his mother. But how can one gauge the depth of feeling? Only by "being-for-itself" actions! The depth of one's feelings is expressed by one's actions.

Some people may say, "Then ask someone. But is this

person any "being in itself" person? No! It is the one who is being in itself that you know what kind of answer he will give you.

Therefore, all values, and even the so-called inner feelings of people, must be expressed through "being-for-itself" choices (actions).

Therefore, when the absolute freedom of Saudi "being-for-itself" emphasizes man's self-evasion, or self-retreat from the world, it has been accused of encouraging people to adopt a desperate attitude [of inaction] toward life. For they believe that by being-for-itself, one separates oneself from the world around one, as if one has retreated into a world where one can see nothing and hear nothing.

Therefore, one's 'being-for-itself' is bound to think that any action is completely useless, and eventually one just reverts to an attitude of watchfulness.

However, this difficulty does not hold true for the Saudi "being-for-itself" theory of action. In this way, the absolute freedom of Saudi "being-for-itself" has a rather positive side. In a war-torn society, Saud does not linger in the groaning of pain.

On the contrary, he strives to break out of this suffocating reality and build a world where people feel alive and still full of hope. In this respect, Schart has indeed established a set of rules for survival in the world of experience.

He abandoned the conventional moral values and attempted to find a place for people to live in the empirical world, which was indeed his way of igniting the fire of vitality and instilling a constant stream of hope for people who were at that time desolate and isolated.

1. The main philosophical thinking of Saudi Existentialism

It is because this existentialist thought is close to the needs of each individual, and each person is his own master, without any fetters. Existence as a subject is not an unarmed subject to be slaughtered.

Rather, it is a quantitative indicator that has infinite probability of occurrence, is contained in things, and foretells the development trend of things, which is an objective argument, not a subjective verification. It is an objective argument, not a subjective test. It determines its own existence, the "what is" that it should become.

Therefore, Scharthe says that his existentialism is a humanism. Whether a person becomes a hero or a coward, it comes from the absolutely free choice and decision of "being-for-itself".

No one is born a coward or a hero, it is the coward who turns himself into a coward and the hero who turns himself into a hero, and there is no external factor that can become a pressure.

As an absolutely free being "being-for-itself", man is always open, and there is no moral prohibition, no moral value that can close off or limit the human reality.

The human reality of being-for-itself is always full of possibilities. No one is born smart or stupid. A being-for-itself concrete person is always able to change, to break away from his past and to move towards an unknown and new future.

In other words, a person "being-for-itself" has the absolute power to shape himself, and after shaping, he can always leave the past behind and move towards other possibilities.

There are always infinite possibilities in front of man, and a being-for-itself person is not limited, but is completely free. Even in a post-war, barren environment, one cannot deprive the "being-for-itself" of absolute freedom, of the right to act.

Therefore, we see that Saudi Arabia does not paint a pessimistic picture of man's being-for-itself existence; in fact, it can be said that no doctrine is more optimistic than its being-for-itself.

Nor does it pour cold water on human action, for it tells man that nothing is true except that "being-for-itself" takes action, and that the existence of man's "being-for-itself" lies in his freedom of choice of action. Thus, Sartre's existentialist philosophical thought is roughly divided into four categories:

(1) "Existence precedes essence".

This means that there is no innately determined morality or soul other than the existence of the human ego itself. Morality and soul are created in the course of one's own existence by using mental activities such as sensation, perception, thinking, and memory, and by the combined awareness and knowledge of one's own physical and mental state and the changes of people, events, and things in the environment.

Man is not obliged to follow a certain moral standard or religious beliefs, but has the freedom to choose. From the point of view of Saudi philosophy, man is like a "rapeseed", his destiny is like a rapeseed that is scattered by the wind and grows wherever it falls, accidentally drifting into the world without any essence.

To establish one's own nature, one must prove it through one's own action of "being-for-itself". Man is not something else, but simply the result of his own actions.

Schart denies the existence of God or any other pre-defined rule. He rejects any "retrograde" elements in life that limit man's freedom of choice.

If there were no such resistance, then the only question a person would have to resolve is which path he chooses for his life's journey. However, man is free to recognize his environment and his self, and to recognize his clarity; even in his self-deception, there is still potential and possibility.

Schatt also suggests that "the other is hell". This view may seem contradictory to the view that "man has the freedom to choose".

In the process of choosing, the biggest problem that

people face is the choice of others, because everyone has the freedom to choose, but each person's freedom may affect the freedom of others.

Therefore, when evaluating a person, it is his behavior, not his identity, that should be evaluated, because the essence of a person is defined through his behavior, and "a person is the sum of his behavior".

(2) "Absolute freedom" and responsibility

Freedom (Hanyu: Liberty, English: Freedom, Liberty) is a philosophical concept that means the ability to govern oneself and to act by will.

Nietzsche's "God is dead" means that man is free in this world, and that man is free to choose his actions. The death of God is a way of saying that human beings can no longer believe in this cosmic order, because they cannot recognize whether it really exists or not.

Nietzsche's view that "God is dead" is not only a loss of faith in the cosmic or material order, but also a denial of absolute values, a loss of belief in an objective and universal moral law that includes every individual.

This loss of absolute morality is the beginning of nihilism. This nihilism led Nietzsche to try his best to find a way to reassess the fundamental values of humanity, a cosmology that goes deeper than Christian values.

Nietzsche believed that most people do not subscribe to (or refuse to recognize) the notion that "God is dead" because they have abandoned the universal value of the true meaning of their own existence and have a deep-seated fear or anger in their hearts.

Therefore, when this "God is dead" death is widely known, they will feel great pain and nihilism will become rampant.

And relativism will make totalitarianism the law in human society, where everything is permitted. This is part of the reason why Nietzsche thought Christianity was quite nihilistic.

For Nietzsche, nihilism is a necessary consequence of all idealized philosophical systems, because all idealism has a moral code like that of Christianity.

However, there are some criticisms from the Christian side: they argue that since the Saudis do not recognize the commandments of God and the eternal value of all regulations, what remains is only spontaneous action.

In this way, what we think of as freedom is often what

we think of as the 'right' choice. Why do we seek freedom? Every person who feels unfree has a deep desire to

To be free from negative states (such as resentment, mistrust, oppression, depression, the expectations of others, public judgment, self-examination, and fear) and hope that freedom will help us find the ideal state.

We can do what we like, and according to this view, we will not be able to condemn anyone's views or actions, and all actions seem to be permissible.

This is because man's choices have no innate pattern, are not guided by God, nor can they be judged by others; man is the sole director of his own actions, but he is responsible for his own actions.

Although we have great autonomy in drawing our own blueprints, we are inevitably influenced by our past habits and subtleties: our upbringing, our experiences, our emotional and psychological changes, and so on, together shape us.

A large part of this is not ours to choose, but if we are free to choose our future, we must first accept this part, with its responsibilities, obligations, privileges, and oppressions.

So freedom never means freedom from constraints. When we say we want to be free to be ourselves, free to choose the life we want, most of us associate it with rights.

And we intentionally or unintentionally ignore the relative responsibilities we should take on. Therefore, when we choose our own life, it is not as simple as picking up a pen and drawing on a piece of white paper at will.

The perfect life we imagine is so simple. Freedom is not a wild horse. According to Spinoza, if one can use reason correctly, one's mind is completely under one's own right, i.e., one is completely free.

The "rightness" of freedom means the ability to recognize the objective reality, the consequences of one's actions, and to respond rationally and correctly.

We are born passive, but we can actively choose to experience and feel for ourselves, to know our lives and to know the world through our own experiences. The reason why experiencing and feeling is to know things around us through practice, rather than observation, is because we have a better understanding of the world.

It is because we are responsive to everything around us. This response determines the way we will know the things around us through practice, that is, the way we respond to life, which is a free choice for the future.

And the "right" response is called responsibility. So freedom and responsibility are not contradictory. The key is to choose the right kind of freedom and responsibility.

Therefore, true freedom is not afraid of responsibility. Responsibility helps us to realize and feel our own authenticity, our own potential, and our own freedom. Once we feel this true freedom, it is the human business of choosing ourselves.

True freedom is the ability to choose one's own life, one's own meaning and values, and the right and ability of each individual to choose the role he or she wants to play. To give full play to one's greatest passion for life and to feel the meaning of the true significance of one's own life existence.

Therefore, the reason why Schart is rejecting God and the accompanying values (good and evil, or virtues and vices, etc.) is that, in Schart's view, these values, which are superior and located in the realm of the transcendent, are not of any real help to man in making his choices.

On the contrary, the absence of God makes man

ultimately responsible for his own actions, as well as for the norms and standards by which people live together and behave.

(3) "The world is absurd"

Camus' description of absurdity in The Myth of Xerxes (Xerxes): In a universe suddenly deprived of illusion and light, man feels himself to be an outsider, a stranger, and since he is deprived of the memory of his lost home or of the hope of a promised land, his banishment is irrevocable.

This separation of human and life, of actor and scene, is the absurdity of emotion. Man comes into the world by chance, facing a rapidly changing, irrational, unordered, purely accidental, chaotic, and unreasonable objective outside world, and feels restricted and obstructed everywhere.

In this vast world, man has no control over his own destiny, and he can only feel sick and vomit. The existence of a fundamental meaning of all objects or phenomena that exist objectively in nature must be explained by a higher genre and ideological content.

But the meaning of this higher genre and content of

thought must be explained by a higher genre and content of thought than it.

This "chain of fetters and restrictions on interpretation" cannot reach a result, so that nothing can have a supreme meaning. Even if this result is discovered, it is possible that it does not satisfy us.

It's like saying that the absurdity of life probably comes from the fact that we all take life very seriously. Not just you and me, but everyone we meet takes their lives seriously.

We try to accomplish our goals, we prepare for the future we are looking forward to, and we feel extremely frustrated and lost because of the setbacks in our lives.

The reason for this attitude is that we feel that life is full of hope and meaning.

It's fun to accomplish goals because we feel it makes sense; it's frustrating to fail because we know it makes sense to accomplish what we want to do.

We often want to help others because we think they can live a meaningful life, and if that person's life (some objective law or so-called fate) is predetermined and inevitably meaningless, then we have no position to take.

Then there is no position we can take to help that person live that life. We will find that the whole world, the whole society, everyone around us, is taking life very seriously.

What everyone is doing is telling us that life should be meaningful. All we have to do is to live our lives well, to live our lives in a way that is practical and meaningful.

The irony is that this idea of "should" as a matter of course is where the absurdity of life lies. Everyone takes life seriously as meaningful, but can life really be meaningful? Whenever we do something, we can ask why we are doing it.

When I was a child, I studied hard so that I could get a good job in the future. The reason for having a good job is to make more money and live a more comfortable life. What about a more comfortable life? Is it to have a happier life? But why do we need to be happy?

From the point of view of the entire universe, after tens of millions of years, all of us may not be around anymore. So what is the point of being happy now or not? This question is not based on the meaning of life.

In the broader cultural context, it is not just a

personal problem. Before the scientific worldview fully entered our culture, religious philosophy was often the ultimate basis of meaning for that person.

Religion tells us what is meaningful, and because religion is the ultimate truth, we do not need to ask further questions.

So people who live under a religious worldview generally feel little of the absurdity of life. However, the science that takes a certain object as its scope of study and seeks to find unified and exact objective laws and truths based on experiments and logical reasoning.

The clearer the explanation of the orbit of the universe, the more it can fully replace the ability of religion to explain the world, and the more religious philosophy will slowly lose its former central position in culture.

However, the withdrawal of religious philosophy does not only occur in the interpretation of the natural world, but also in the loss of the cornerstone of the meaning of life.

Nietzsche's statement that "God is dead" is, in one sense, the same thing. And Nietzsche's philosophy can also be a response to the death of God.

If we can no longer take God as the ultimate basis for everything, how should we survive? Can we really only accept nihilism and relativism? Or can we have a different basis of value than the one we have?

So, on one level of the value of life, God is dead and the absurdity of life are two sides of the same coin. When God is dead, we naturally feel the absurdity of life.

Of course, Camus and Nietzsche's answers are not identical, but they both ask the question of how we can or should live if we cannot find the ultimate basis for the value of life.

This sense of absurdity is the starting point of their philosophical thinking. Thus, existentialism and nihilism, in this sense, consider life as meaningless and absurd.

(4) Nothingness and Self-Deception

Schart's description of "man as nothingness" is a paradoxical and even contradictory expression, because common sense suggests that "nothingness" is impractical and inscrutable.

Does Schart mean that man does not exist? No, he is simply trying to highlight the uniqueness of the human mode of existence, which is completely different from that of any other being.

In contrast to Nietzschean philosophical thinking, nihilism is the idea that everything that happens is meaningless, and therefore a kind of human nihilism.

The meaninglessness of life is usually an indicator of whether life is worth living or not, and because of the religious philosophical interpretation that life is meaningless, it is also thought that a meaningless life is not worth living.

Therefore, by listing the basic properties of an event or an object to describe or regulate the meaning of a word or nothingness, nihilism, can be divided into at least two types.

This can be divided into at least two categories.

First, nihilism, which is valid in all circumstances, i.e., life is meaningless and not worth living in any case.

The second is nihilism, which occurs only under certain circumstances, that is, when the meaning of our life is lost under certain circumstances or reasons, making life not worth living.

Nietzsche is talking about the latter kind of nihilism,

the nihilism that occurs at the time of the death of God. It must be made clear here that nihilism is not a direct result of the death of God.

After all, the phenomenon of God's death can be said to be a certain kind of human knowledge of natural or social things all over the world, a meaning given to the object by human beings, a common experience of the spiritual content transmitted and communicated by human beings in the form of symbols, but not everywhere, as in Europe, nihilism appears.

Therefore, if we say that God is dead, it will lead to nihilism, which is too much of a selfish view of things or issues from a certain position or perspective.

It is the tendency to look at things from the perspective, experience, consciousness, spirit, feelings, desires, or beliefs that an individual can have as the basis of an arbitrary attribute.

The emergence of nihilism requires, in addition to the death of God, another condition, that we believe in God (and the values He promotes), i.e., that man believes in God, but God is dead, in order to cause the loss of values and the derivation of nihilism.

According to Bernard Reginster's interpretation, in the view of Nietzsche's philosophical thinking

Nihilism can be divided into two aspects.

One is the nihilism that teaches people to lose "the positive meaning and usefulness of the object to the subject".

The second is the nihilism that makes people despair of the positive meaning and usefulness of the object to the subject.

The first is the nihilism that teaches people to lose "the positive meaning and usefulness of the object to the subject" (disorientation) because people no longer believe in God.

Therefore, when [God's values and His worldview] values, that is, all objects or phenomena that exist objectively in nature, have positive meaning for people or groups, are valued by people, or can make people feel satisfied, and become something that people respect or are interested in pursuing or thinking about, the things that are the target also fall apart at the same time.

If we use the position on which Nietzsche based his research and analysis or criticism of problems and issues, it is that value as a philosophical concept is a being, and this being is a number, a number, a numerical value; value is the appearance, the depreciation of the presence, or the loss of the value (devalue) of the number, number, or numerical value of the being in the world.

The result of this situation is that the value that man used to believe is universal and general, not only in the subject (man) as it is now misinterpreted, but it is a kind of being that exists in parallel with other beings, and the value, the quantity of value, is the loss of the name of these two parallel lines.

From now on, we do not know what to do. If the realization of the values believed in the past is the meaning of living, then the loss of values means that people have lost the meaning of living.

Secondly, it is the nihilism that makes people despair of the positive meaning and usefulness of the object to the subject.

Not only as a human linguistic term, but also as a term expressing the presence of a certain purpose or situation that is expected to be achieved, it appears in people's thinking and speech as a loss, and the death of

God means at the same time that values cannot be realized.

With the death of God, man's belief in the other side of the world also collapses. In the past, life under the Christian worldview was painful, but at least we believed that there was a being as a philosophical concept.

This kind of being is the presence of a certain purpose that is desired or a certain situation that arises; value is the appearance and presence of the number of beings, numbers, and numerical values in the world, which can be realized elsewhere.

But because we no longer believe in the other side, as a human linguistic term of belief, it is also a term expressing a certain purpose desired to be achieved, or a certain situation present, which appears in people's thinking and speech.

It is now impossible to realize. The value of not making a situation that did not exist a reality, where only the present world and its various sufferings remain, causes despair. 2.

2. A metaphysical conclusion to the whole book.

As described earlier in Schart's Being and Nothingness, "man is nothingness," and at the same time a combination of two kinds of nothingness.

On the one hand, man denies his own reality, for example, his past; on the other hand, he is at the same time a lack of imagination about the future. In short, what man lacks is his own unrealized purpose and plan.

Therefore, Schartre describes man's existence as "is what it is not, is not what it is" (is what it is not, is not what it is). Man is what he is not, because what he is is not determined by his past, but by the plans and purposes that he is not, but should be.

At the same time, he "is not what he is" because man is always beyond what he is, denying his reality, denying his actuality. In this sense, man can act to transcend his reality.

(1) "Self-existence" and "Self-action

Scharthe believes that the fundamental problem of the discipline that uses analytical thinking to investigate and reflect on fundamental questions of life, knowledge, and value is the problem of being, and distinguishes between [being] and 'being-for-itself'. To understand Schart's existentialism, we must first clarify the meaning of these two concepts.

Being in itself" refers to the objective world, the objective existence, which is not transferred by human consciousness. It exists independently of man, and although it can be manifested by man's consciousness, it cannot be created by man's consciousness, "being in itself" is not yet manifested by man's consciousness, and is not given meaning by man's consciousness.

"Being in itself" means that it is completely within itself, without any relationship. It is neither created nor created; it is neither active nor passive; it is neither affirmative nor negative; it is neither internal nor external. It is an undifferentiated, amorphous, fully realized being.

Therefore, 'being in itself' is 'being as it is', without change or development, without past, present or future.

It is something beyond consciousness, equivalent to the unknown root and essence of things and people, which is "beyond the present" and hidden in the depths.

It is always and absolutely present, but it is contingent, without necessity, and cannot be truly known.

It is therefore an alien, absurd, and repulsive world.

Being-for-itself" means human consciousness, human ego. Contrary to being in itself, it is non-existent, it is self-existent, it is what it is not or what it is not.

First of all, 'being-for-itself' is 'non-being', 'nothingness'. Consciousness is the absence of being, the negation of being, the emptiness. But consciousness, as the absence of being, tends towards being.

As a "hole" in existence, it wants to fill this hole. It is "nothingness", but it wants to be something; it is never something, but it tends to be something.

This tendency is an activity of nihilization or negation. Human consciousness is able to ask questions, to negate, to make existence nothing.

The subjective consciousness of man. It is real existence and at the same time a kind of nothingness. Because it is free and indeterminate. It is always something that it is not, and it is always going to be something.

So it makes the "now" always a negation and dissatisfaction of the self, and thus it becomes "nothingness".

"Being in itself' and 'being-for-itself' are opposed in nature, but they are unified. "Being-in-itself" and "being-for-itself": "being-in-itself" is full and without cracks, while the cognitive ability of human beings to recognize the environment and the self, as well as the clarity of cognition, make "being in itself" split.

As our mind reflects the objective material world to an object, it separates it from something else, reveals the object, takes it out of its context, and removes everything else, so that it does not exist.

This "removing and making non-existent" is the process of becoming nothingness, and the process of becoming nothingness is also the process of negation. It is through one's ability to recognize the environment and the self, and the clarity of that recognition, that nothingness comes into the world.

Self-existence is only meaningful when associated with "being in itself". "Being in itself' is something that is all-in-one, lifeless, unconnected, meaningless, and worthless, and it is only through the emergence of 'being-for-itself', that is, consciousness, that 'being in itself' acquires meaning and value.

On the one hand, 'being-for-itself' cannot be left alone, self-existence exists independently, is an absolutely free subjectivity or set of subjectivities, they make an entity or a substance its fundamental place, and it necessarily exists, without which it loses its identity.

"Being-for-itself' is contrasted with contingency: contingency is the nature of an entity or substance to exist contingently, and without the nature of 'being in itself', the substance can still retain its identity, but without being in itself, self-being would be abstract, like a sound that cannot exist without pitch and tone.

Self-existence, on the other hand, is only something that is distinguishable and exists independently within itself. But it need not be a physical being. In particular, abstraction is usually regarded as a physical entity.

As to what it is, it is given by self or consciousness. Without the presence of Self, the Self would be a meaningless, valueless, monolithic thing that contains no distinction in itself. Self-existence can only be explained by human consciousness, and can only become a meaningful and real existence.

(2) Existence precedes essence

Schart is to take pure subjectivity, i.e., self-being, as the basic existence of man; to take it as the starting point of philosophy, and to deduce from it the existence of the external objective world. Thus "human existence" or "pure subjectivity" is taken as the basis and origin of all things in the universe.

Human existence is considered to be the first. Based on this, Sartre, contrary to the traditional concept of "essence before existence", puts forward the view that "existence is prior to essence" on the question of human existence and essence.

The so-called "existence" refers to "human reality" or "self". This means the subjectivity of an individual. The so-called "essence" refers to man's own will in action, creating his own regularity. The basic meaning of the proposition that existence precedes essence is:

The essence of man, the various characteristics of man, are all chosen and created by subjectivity later. Therefore, there is no human nature in the world, because there is no God who set human nature in the world.

Man is nothing but something created by himself. Unlike things, they cannot choose and make their own nature.

The nature of things is given by the will of man, and is the result of his own needs, purposes, and mental activities such as sensation, perception, thinking, and memory.

The role of comprehensive awareness and recognition of changes in people, things, and objects to make it have some kind of meaning and value, that is, to make it have the essence, and then appear in the world as something.

Obviously before something appears, its essence has existed in the personal use of sensory, perceptual, thinking, memory and other mental activities, their physical and mental state and the environment in the comprehensive awareness and knowledge of people, things, and changes in things.

Therefore, the nature of things is prior to existence. In this way, human beings and things are clearly distinguished, thus affirming the value and dignity of human beings, emphasizing their subjectivity, their status and role, and inspiring people to take positive actions to create their own essence. Therefore, Scharthe's "existentialism" is in fact a philosophy of action.

(3) Theory of Freedom

Scharthe believed that "being-for-itself" is absolutely free, not bound by anything, including oneself, but constantly denying, creating and developing oneself.

At the same time, man's freedom is prior to his nature. Man does not first exist in order to become free, but there is no difference between man's existence and his "being free. But by freedom, Schartrecht does not mean the achievement of ends and success, but only the autonomy of choice.

Freedom is freedom as long as one can choose, and even not choosing is a choice, that is, choosing not to choose is also freedom.

From Schart's philosophical point of view, the absolute freedom of man to be "being-for-itself" only means that man is thrown into the dusty world in isolation. He is not determined by anything. He has to decide for himself, to choose for himself, to make himself.

Man's life is a project of constant choice, constant free choice, creation of his own nature, and constant self-making towards the future.

(4) Responsibility and Humanism

Jean-Paul Sartre's motto: "Existence precedes nature". By this he means that there is no innate morality or soul other than the human being himself. Morality and soul are created by man in the course of his existence.

Man is not obligated to conform to a moral standard or religious belief, but has the freedom to choose. When evaluating a person, it is his behavior, not his identity, that is to be evaluated.

The nature of man is defined by his actions, and "man is the sum of his actions".

Therefore, Schart denies the existence of God or any other pre-defined rule. He rejects any "perverse" elements in life because they reduce man's freedom of choice.

If there were no such differences between others and oneself to identify other human beings, it would be a cumulative component of the self-image.

And as a resistance to the recognition of reality, then the only problem a person has to solve is which path he chooses to take. Yet man is free; even in his self-deception, he still has potential and possibilities.

Schartre also suggests that "the other is hell". This view may seem to contradict the view that "man is free to choose", but in fact everyone is free to choose.

In fact, everyone is free to choose, but everyone has an inescapable responsibility for the outcome of his or her choice.

Because everyone has the freedom to choose, but each person's freedom may affect the freedom of others, so called "others is hell". Therefore, Schart analyzes the reasons why man is judged to be free:

First, God does not exist: God is a labelled value and is defined as a characteristic possessed by a person who acts in a chemical way, different from one's social identity and different from the state of self-identity.

In the philosophical "inquiry," morality is defined and defined as the standard of behavior that is based on the values of a certain society or class (form of life), social opinion, traditions and customs, and the power of one's inner beliefs to adjust the evaluation and determination of the relationship between the behavior of others and oneself, good and bad, honor and disgrace, right and wrong, etc.

By establishing certain standards of good and evil and standards of conduct, to restrain people's mutual relations and personal behavior, regulate social relations, and with the law on the normal order of social life, produce a protective role.

It is a symbolic order that God does not exist, and there is no God-given commandment that can be followed. Individuals have no one to turn to, and must explore and choose for themselves;

Second, there is no a priori, universal human nature. Schart believes that the focus on moral sentiment exists in the a priori, universal human nature of all human societies, and that moral sentiment is part of the general principles of universal culture.

Some studies even consider that moral-related behaviors such as honesty, helpfulness, tolerance, loyalty, responsibility, social justice, equality, family and national security, stability of social order, and repayment of kindness are part of universal values.

In other words, these a priori, universal human behaviors may be the universally recognized virtues of all societies.

The ethical position that advocates such a priori, universal human principles is called moral universalism.

The philosophy of moral relativism, however, is opposed to moral universalism, which holds that there is no universal moral principle, and thus no a priori moral principle.

Thus, there is no a priori, universal ethics of human nature that instructs you what to do, and your will is absolutely autonomous to choose among various possibilities;

Third, the world is absurd, and there is no such thing as necessity, which refers to a state of things that "must not be", derived from the Latin words "ne" (not) and "cedere" (to go away), meaning "cannot leave".

The Greek word for inevitability is anagke, which originally meant fate and destiny. Basically, necessity includes the necessity of propositions, things or gods.

In the case of propositional necessity, when a proposition is said to be necessary, it means that the proposition cannot be denied.

At this point, in the language of Aristotle (384-322 B.C.), the opposite of necessity is not "contigency" or "appropriateness," but "impossibility" (the absence of which is impossible).

For the opposite of a contingency is another

contingency. G.W. Leibniz (1646~1716) also distinguished between "absolute necessity" and "hypothe-tical necessity".

The former necessarily leads to rational truth, the opposite of which necessarily leads to contradiction, while the opposite of the latter does not necessarily lead to contradiction.

Imm. Kant (1724~1804) thought that so-called necessity does not exist only in "analyticaljudgements", but also in "synthetical a priori judgements", so that human knowledge has the possibility of being established.

In terms of the necessity of things, it means that other states of things are impossible. The atomism and Stoicism of Greek philosophy advocate a physical causality.

They believe that the creation and operation of the world are governed by the law of necessity. B. Spinoza (1632-1677), on the other hand, believed that the existing world was derived from the nature of God, just as the geometry of the mathematical system was deduced to be necessary.

D. Hume (1711~1776) and P. Holbach (1723~1789) also elaborated on necessity from the viewpoint of empirical connection and materiality. J.G. Fichte (1762-1814) emphasized the need for "necessity" between the Ego and the non-Ego.

As for the necessity of God, Aristotle's analysis leads to the notion of a necessary being, which is the "first cause" of all other beings.

Avicenna (980~1037), St. Thomas Aquinas (1225~1274) and Leibniz also had similar views. existence.

Philosophically, it refers to the decisive connection or trend that is prescribed by the intrinsic nature of things. In the development of things, necessity and contingency are linked to each other, interact, and transform each other under certain conditions.

As opposed to "chance", there is no objective necessity that can fetter the free choice of the self. If this is the case, what else can provide a universal standard of value for the individual? Nothing. There is no objective ethical or moral code or standard, and all standards of human behavior, all distinctions between right and wrong, are determined and freely chosen by the individual.

But Sartre argues that since man's "self-existence" is

absolutely free, each person must take full responsibility for his own choices, actions, and values, no matter what happens.

Moreover, in the real world, the existence of human "self-existence" is not isolated and pure, but is always in some kind of suspended situation.

Through a person's choice, he or she is involved in the whole of humanity. In other words, we are not only responsible for ourselves, but also for all people, and this is the meaning and value of human beings.

It is because the focus on human beings is basically a study of the nature, composition, criteria, and evaluation of values, and a study of the meaning of various material and spiritual phenomena, as well as people's behavior, to individuals and groups from the perspective of the subject's needs and whether and how the object can satisfy the subject's needs.

In particular, we are concerned with the basic human life and the basic living conditions. The concern is for human happiness, emphasizing mutual assistance and care among human beings, and valuing human values. Here, therefore, Sartre turns existentialism into a kind of humanism.

The idea that values do not depend on the existence or nature of the logic of knowledge, but in fact values, too, do not have a physical state of existence and role

Thus, the framework of Sartre's existential theory is based on a conceptual division of "essence before existence. For Scharthe, at the most basic theoretical level, there are only two kinds of being, namely, being in itself and being for itself.

Being in itself, that is, "being what it is", either equates "its being with its essence" or "essence precedes being".

For example, before a glass is made, its essence already exists in the mind of the designer; otherwise, it is a material without any meaningful "existence", a material that cannot be easily combined with other elements or compounds.

However, self-existence, or consciousness, is characterized by the fact that "existence precedes essence". Thus "existence" does not have any given essence and creates itself through action. In Saudi Arabia there are systematic and organized laws or arguments.

In the doctrine that is derived from actual verification or deduced from concepts, human existence, consciousness, and self-existence emphasize that they are one and the same thing, not determined by any existence.

"Self-existence" is like a cloth without flaws, full and without defects or holes, so that one can conform to one's own mind; man, as an existence of nothingness, is a hole in the cloth.

This hole can only exist in the cloth and does not in itself determine how the hole exists. This is exactly the same as the Shah's concept of freedom, which corresponds to each other.

"Self-existence" can only provide the basis for man's existence, but it cannot determine how man exists. In short, one does not decide what one "is" when one is born into the world.

For example, the background of birth and the family situation, because they are the actual state of existence (the likeness), are objective descriptions of natural and social phenomena.

However, the objectivity of the actual state of existence does not determine what one is not, what one lacks, or what one should be, which is the uniqueness of being human, that is, what one "is not but should be".

Therefore, self-existence is not simply "non-existence", but, to use Schart's philosophical point of view, it is the temporal space that temporarily comes to man, the existence that moves from elsewhere to here, dependent on self-existence and not affected by the latter.

To regard man as a being determined solely by external objects is to regard consciousness as some kind of being determined by other beings. But consciousness is not this being, but nothingness or existence.

It is a void in existence, and cannot be understood in terms of the law of cause and effect at all. Under the law of cause and effect, everything is a "full" existence, and there is no place for man as a "real being" of nothingness.

Man exists first in the world, facing the absurdity that existentialism uses to describe the meaningless, paradoxical, disordered state of life.

Without any objective value, he calls it being-in-itself and then chooses what is his own nature and gives meaning and value to the world. In addition to abstract ontological research, Schart's theory contains a great deal of theoretical analysis of real human emotions, such as anxiety, fear, pride, submission, love, and sexual desire.

This is why intellectuals find in Scharthe's theory not only a purely intellectual satisfaction, but also a way to reacquaint themselves with the actual state of existence.

Although human beings have a facticity that they cannot control, for example, where they are born, whether they are rich or not, or what their country is like, they say that this is how things actually are.

But unlike reality (which means that it has rationality and objectivity), this actuality is usually used in philosophical logical analysis and is not the nature of human beings, who always have the qualities of transcendence and negation.

This is why the French always say that "there are no ugly women, only lazy women". The nature of I depends on how I plan to move towards my future, not on how my past exists.

Nietzsche believed that human beings could still find positive possibilities without God. The abandonment of belief in God opens the first doorway for humans to develop their own creative capacities.

The Christian God often has arbitrary commands and prohibitions, but He is no longer able to sway man, so man can give up seeking help from supernatural forces and instead come to know a new set of values in the world.

Therefore, the actual life and actions of human beings need to be explained by another dimension. According to Schart's ontology, this dimension necessarily includes "freedom".

Once we start from the viewpoint that "man is nothingness" and abandon the mechanical law of causality to explain the state of man's existence, we have to ask what determines nothingness, i.e., in what way man himself exists.

Thus, Schartre argues that it is ontologically wrong to consider human beings as links in a purely causal chain, e.g., to consider all human actions as determined by unconsciousness or past experiences.

On the ontological level, the "freely"-"for oneself" relationship is not independent of each other, but is self-referentially related to the self through internal relations, and strives to form an ideal whole.

If there is no freely, there will be no concrete

self-realization. In other words, there is no common measure between the reality of man and the self-caused existence he wants to become.

Each person has a different perspective of the God he wants to incarnate into, and each self-caused being has a different process of becoming nothing.

Man himself is a moral subject, the existence on which all kinds of values depend. Therefore, the psychoanalysis of existence reveals that man is to pursue to become "freely"-"for oneself" as an integrated being.

The process of human being is the pursuit and transcendence of the ideal goal, that is, the metamorphosis of "being-for-itself" into the holistic process of sublimation and self-realization.

The whole human being. This means that we are responsible not only for ourselves, but for all people, and this is the meaning and value of the human being. Here, Sartre turns existentialism into a kind of humanism.

References

1. ^ E Keen. Suicide and Self-Deception. psychoanalytic Review. 1973 [2017-08-19]. (Original content archived 2018-09-18).
2. ^ E Keen. Suicide and Self-Deception. psychoanalytic Review. 1973 [2017-08-19]. (Archived from the original on 2018-09-18).
1. Copleston, F.C. Existentialism. philosophy. 2009, 23 (84): 19-37. jstor 4544850. doi:10.1017/S0031819100065955.
2. ^ See James Wood's introduction to Sartre, Jean-Paul. nausea. london: Penguin Classics. 2000. isbn 978-0-141-18549-1.
3. ^ Tidsskrift for Norsk Psykologforening, Vol 45, nummer 10, 2008, side 1298-1304, Welhaven og psykologien: Del 2. (Page archived for backup in the Internet Archive)
4. ^ Lundestad, 1998, p. 169
5. ^ Slagstad, 2001, p. 89
6. ^ Seip, 2007, p. 352
7. ^ Stanford Encyclopedia of Philosophy, Existentialism, 3.1 Anxiety, Nothingness, the Absurd (page archived in the Internet Archive)
8. ^ Bassnett, Susan; Lorch, Jennifer. Luigi Pirandello in the Theatre. Routledge. March 18, 2014 [26 March 2015]. (Archived from the original on 2021-02-05).
9. ^ Thompson, Mel; Rodgers, Nigel. Understanding

Existentialism: Teach Yourself. Hodder & Stoughton. 2010 [2017-08-19]. (Original content archived 2021-02-05).

10. ^ Caputi, Anthony Francis. Pirandello and the Crisis of Modern Consciousness. university of Illinois Press. 1988 [2017-08-19]. (Original content archived 2021-02-20).

11. ^ Mariani, Umberto. living Masks: The Achievement of Pirandello. university of Toronto Press. 2010 [26 March 2015]. (Archived from the original on 2021-02-05).

12. ^ Jean-Paul Sartre. Existentialism is a Humanism, Jean-Paul Sartre 1946. marxists.org. [2010-03-08]. (Archived from the original on 2011-06-14).

15. Revised translation of Being and Nothingness, Sartre, Chen, Xuan-Liang, Du, Xiao-Zhen, and Sanlian Publishing Co.

The Revised Translation of Existence and Time by Martin Heidegger, translated by Chen Jiaying, Wang Qingjie, Sanlian Publishing House

Aversion, Sartre

18. The Transcendence of the Self, Sartre

19. Du Xiaozhen: The Burden of Existence and Freedom by Du Xiaozhen

The Theory of Existentialism, edited by Li Jun, Wang Yuechuan, Shan Dong Publishing House

21.... Zhang Zhiwei's Interpretation of Existence and Time

METAMORPHOSIS:
The Reality of Existence and Sublimation of Life
(Volume 5)

蛻變：生命存在與昇華的實相（國際英文版：卷五）

出版者/美商 EHGBooks 微出版公司

發行者/美商漢世紀數位文化公司

臺灣學人出版網：http://www. TaiwanFellowship.Org

地　　址/106 臺北市大安區敦化南路 2 段 1 號 4 樓

電　　話/02-2701-6088 轉 616-617

印　　刷/漢世紀古騰堡®數位出版 POD 雲端科技

出版日期/2023 年 2 月

總經銷/Amazon.com

臺灣銷售網/三民網路書店：http：//www. Sanmin.com. Tw

　　　　三民書局復北店

　　　　地址/104 臺北市復興北路 386 號

　　　　電話/02-2500-6600

　　　　三民書局重南店

　　　　地址/100 臺北市重慶南路一段 61 號

　　　　電話/02-2361-7511

全省金石網路書店：http://www.kingstone.com. Tw

定　　價/新臺幣 3000 元（美金 100 元/人民幣 600 元）

2023 年版權美國登記，未經授權不許翻印全文或部分及翻譯為其他語言或文字。
2023© United States，Permission required for reproduction，or translation in whole or part

CPSIA information can be obtained
at www.ICGtesting.com
Printed in the USA
BVHW021938290123
657071BV00033B/324